Managing Employee Relations in the Hotel and Catering Industry

A selection of related titles published by Cassell

Marketing in Hospitality and Tourism *Richard Teare, Josef A. Mazanec, Simon Crawford-Welch and Stephen Calver*

Strategic Hospitality Management *Richard Teare and Andrew Boer*

Managing Projects in Hospitality Organizations *Richard Teare, with Debra Adams and Sally Messenger (eds)*

Hotel and Food Service Marketing *Francis Buttle*

Management of Foodservice Operations *Peter Jones, with Paul Merricks*

Principles of Hotel Front Office Operations *Sue Baker, Pam Bradley and Jeremy Huyton*

Achieving Quality Performance *Richard Teare, Cyril Atkinson and Clive Westwood*

The Dynamics of Workplace Unionism *Ralph Darlington*

Keyguide to Information Sources in Labour Studies *John Bennett and Tony Elgar*

Understanding Industrial Relations (fifth edition) *David Farnham and John Pimlott*

MANAGING EMPLOYEE RELATIONS IN THE HOTEL AND CATERING INDUSTRY

Rosemary E. Lucas

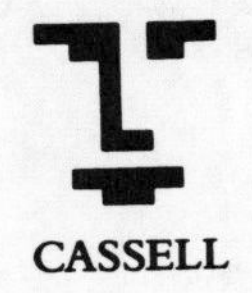

CASSELL

Cassell
Wellington House
125 Strand
London WC2R 0BB

PO Box 605
Herndon
VA 20172

First published 1995
Reprinted 1997

British Library Cataloguing-in-Publication Data
A catalogue record for this book is available from the British Library.

Library of Congress Cataloging-in-Publication Data
Lucas, Rosemary E.
Managing employee relations in the hotel and catering industry / Rosemary E. Lucas.
p. cm.
Includes bibliographical references and index.
ISBN 0–304–32910–X : $60.00. — ISBN 0–304–32897–9 (pbk.) : $30.00
1. Hospitality industry — Personnel management. I. Title.
TX911.3.P4L83 1995 95–2056
647.94'068'3—dc20 CIP

ISBN 0-304-32910-X (hardback)
0-304-32897-9 (paperback)

Typeset by York House Typographic Ltd, London
Printed and bound in Great Britain by Redwood Books, Trowbridge, Wiltshire

CONTENTS

ACKNOWLEDGEMENTS

This book emanates from over twenty years' experience of industrial/employee relations and personnel/human resource management, time served as both practitioner and academic. During that journey a number of people have contributed directly and indirectly to this book, of whom a few deserve special mention. First is the late Bill Davidson, Industrial Relations Manager of the Cooperative Wholesale Society, who had the courage to appoint me to the post of Industrial Relations Officer in 1972. It is only now, with the benefit of hindsight, that I can really appreciate how much I have learnt from him. Second is Eric Armstrong, Emeritus Robens Professor of Industrial Relations, Manchester Business School, who opened my window on to academia in the 1980s and, I suspect, must have spotted some potential in me that I did not know existed. I hope I have proved to be a worthy pupil.

Roy Wood of the Scottish Hotel School, University of Strathclyde helped sow the seeds for the book, though not in this form. Once the proposal had been shaped up and accepted, Mick Marchington of the Manchester School of Management, UMIST, provided constructive advice on how to proceed and, most importantly, pointed me in the direction of the Workplace Industrial Relations Survey 1990 (WIRS3) as a means of giving the book a really distinctive edge. The WIRS3 data could not have been produced without the help of Neil Millward of the Policy Studies Institute and funds from the Economic and Social Research Council's Research Resources Board; this help is gratefully acknowledged. Neil Millward, Philip Wilkinson, Roy Wood and Jenny Wynne have read and offered valuable comments on earlier drafts of different parts of the text, but particular thanks must extend to Mick Marchington for reading the whole manuscript and providing incisive comments throughout. Colleagues at work, particularly Howard Hughes, Sue Gilpin and Jenny Wynne, have also contributed invaluable moral support.

This book would not have been possible without the love, support and encouragement of all my family. I know my late father, Bob Lucas, would have been very proud; I am sure Elsie and Nick Lucas are too. Most of all, my husband Chris Adams tolerated my need for time and space and, as he never ceases to remind me (thanks initially to Uncle Lenny), provided the word-processing facilities to enable me to undertake most of this task. Wasdale has been a constant and faithful companion throughout, and it is a pity that he cannot speak, because he could justifiably lay claim to being the most informed member of the canine species on industrial relations. His contribution may be measured from the number of paw marks left on the many earlier drafts of the manuscript that littered the floor of my study.

Last, but not least, my thanks go to David Royle, formerly of Cassell, for commissioning the book, to Steve Cook for getting the go-ahead to start writing, and to Judith Entwisle-Baker and her enthusiasm for aiding the book through to completion.

Any mistakes or errors are my own. I hope these are few and far between.

PREFACE

This book seeks to fill a gap in the literature of 'people' management in the hotel and catering industry – managing employee relations. The book is intended for lecturers and students (mainly at final-year undergraduate and postgraduate level) who are involved with hotel and catering degrees or diplomas within specialist departments or business schools in UK universities. The text will also be of interest to lecturers and students in other subject areas, such as human resource management, personnel management, industrial relations, business studies, humanities and social science, particularly those who wish to undertake an hotel and catering industry-focused study.

The book has both primary and secondary aims:

1 To examine key issues in the management of employee relations, and thereby to offer a new and discursive insight into issues in employee relations management that reflect the distinctive characteristics of a significant employment sector.
2 To provide substantial referencing to assist the reader to explore chosen topics in greater depth, perhaps for an extended essay, a project or a dissertation.

There are four underlying themes to the book:

1 To discuss employee relations issues which are of significance to managing in the hotel and catering industry in the 1990s and beyond.
2 To place such employee relations issues within a more clearly defined and identifiable industry and national context than hitherto.
3 To identify particular problem areas and make summary propositions throughout.
4 To provide subject matter at a relatively advanced level for discussion and debate, and for the further investigation of topics in greater depth.

Thus, the book is intended neither to constitute a textbook (in the prescriptive sense of following a set course or syllabus) on employee relations in the hotel and catering industry, nor to be a 'how-to-do-it' management manual. But it should serve as an important sourcebook. Material has been drawn from a wide variety of sources, including government and official surveys, previously unpublished 1990 Workplace Industrial Relations Survey (WIRS3) data, authoritative textbooks, journal articles and empirical and theoretical research findings.

The book is divided into three main parts.

Part 1, 'The employee relations, managerial and employment contexts', comprises three chapters. It offers a context against which all the issues discussed in Parts 2 and 3 can be better understood and includes some important material that does not 'fit' easily into the more 'conventional' analysis of employee relations issues covered in Chapter 4 and beyond. This context should benefit both those with relatively little knowledge about the industry and others who are more familiar with the industry, because the material in these chapters has not previously been assimilated in this form.

Chapter 1 sets out the basic philosophy of approach, defines appropriate terms and briefly outlines the development of employee relations in the hotel and catering industry. Chapter 2 examines the essential characteristics and features of managing in the hotel and catering industry, and the management organization for employee relations. Chapter 3 outlines employment trends, key employment issues and the main forms of employment and work in the industry.

Part 2, 'The employment relationship', comprises five chapters. It breaks new ground and represents a detailed and in-depth analysis of approaches to employee relations in the hotel and catering industry, which are compared with those in other industries.

Chapter 4 identifies the nature of the employment relationship, considers the perspectives from which it may be managed and discusses the role and attitudes of the institutions and parties involved in it. Chapter 5 identifies and discusses the main elements of substance in the employment relationship, which centre on pay and conditions of employment. Chapter 6 discusses the main forms of decision-making that can be used to determine remuneration and other terms of employment, the complex issue of power, interpersonal skills and decision-making processes. Chapter 7 considers the main types of procedure and process that are used to make continual adjustments in, and to, the employment relationship. Chapter 8 outlines the variety of methods that are available for managing employee involvement and maximizing commitment, and considers the role of communications.

Part 3, 'Legal and social issues', comprises three chapters. This part examines the subject of law in the employment relationship, which is becoming increasingly complex, largely as a result of European developments. Employment law has been broken into three discrete, but interrelated, areas to aid clarity of presentation. The legal aspects reported are current to February 1995.

Chapter 9 outlines the role of individual and collective law in the employment relationship, reviews the most significant aspects of workplace employment law and assesses the extent of influence of developments in EC law and social policy on British law and employment practice. Chapter 10 outlines the issue of gender in the employment relationship, considers the British and EC legal framework for equalizing the employment relationship through measures relating to anti-discrimination, equal pay and pregnancy and maternity rights, and discusses issues in managing equality and the workplace diversity more effectively. Chapter 11 reviews the importance of health, safety and welfare legislation, considers the effect of European social policy on British health, safety and welfare law and practice, and discusses contemporary issues in occupational health and welfare.

Each substantive chapter covers the main subject matter, contains an introduction, argument development, summary propositions, discussion points, extensive referencing and suggested further reading.

Chapter 12 is a brief concluding chapter that draws together some of the main themes and key issues from the main body of the text which merit further analysis, evaluation and empirical study.

There are six appendices. Appendix 1 summarizes the main workplace and workforce characteristics of the WIRS3. Appendix 2 outlines the author's perspective on employee relations. Appendix 3 summarizes the effect of EC law in the UK. Appendix 4 summarizes the main aspects of workplace employment law discussed in Chapter 9. Appendix 5 summarizes the main aspects of equality legislation discussed in Chapter 10. Appendix 6 summarizes the main aspects of health and safety legislation discussed in Chapter 11.

A list of abbreviations, lists of legal cases and UK statutes and regulations, and bibliography are included at the end of the book.

LIST OF ABBREVIATIONS

ACAS	Advisory, Conciliation and Arbitration Service
AIDS	acquired immune deficiency syndrome
AIS	all industries and services
APL	Accreditation of Prior Learning
CA	Court of Appeal
CBI	Confederation of British Industry
CHME	Council for Hospitality Management Education
CIR	Commission on Industrial Relations
CMLR	Common Market Law Reports
COHSE	Confederation of Health Service Employees
COSHH	Control of Substances Hazardous to Health Regulations
CRE	Commission for Racial Equality
EAT	Employment Appeal Tribunal
EC	European Community
ECJ	European Court of Justice
ECR	European Court Reports
EEC	European Economic Community
EOC	Equal Opportunities Commission
EPA	Employment Protection Act
EPCA	Employment Protection (Consolidation) Act
EPPTER	Employment Protection (Part-time Employees) Regulations
EU	European Union
GCSE	General Certificate of Secondary Education
GMWU	General and Municipal Workers' Union
GNVQ	General National Vocational Qualifications
HCEDC	Hotel and Catering Economic Development Committee
HCI	hotel and catering industry
HCIMA	Hotel, Catering and Institutional Management Association
HCITB	Hotel and Catering Industry Training Board
HCTB	Hotel and Catering Training Board

HCTC	Hotel and Catering Training Company
HCU	Hotel and Catering Union
HCWU	Hotel and Catering Workers' Union
HIV	human immunodeficiency virus
HL	House of Lords
HMSO	Her Majesty's Stationery Office
HRM	human resource management
HSC	Health and Safety Commission
HSE	Health and Safety Executive
ICR	Industrial Court Reports
IDS	Incomes Data Services
IPD	Institute of Personnel and Development
IPM	Institute of Personnel Management
IRLB	Industrial Relations Law Bulletin
IRLR	Industrial Relations Law Reports
IT	industrial tribunal
ITR	Industrial Tribunal Reports
JP	justice of the peace
LEC	Local Enterprise Company
NALGO	National Association of Local Government Officers
NALHM	National Association of Licensed House Managers
NEDC	National Economic Development Council
NEDO	National Economic Development Office
NES	New Earnings Survey
NIHC	Northern Ireland High Court
NUPE	National Union of Public Employees
NVQ	National Vocational Qualification
PAYE	pay-as-you-earn
PHR	Pre-hearing Review
PLC	public limited company
PSS	private sector services
RIDDOR	Reporting of Injuries, Diseases and Dangerous Occurrences Regulations
SI	service industries
SIC	Standard Industrial Classification
SMP	statutory maternity pay
SNMW	Statutory National Minimum Wage
SSCBA	Social Security Contributions and Benefits Act
SSP	statutory sick pay
SVQ	Scottish Vocational Qualification
TEC	Training and Enterprise Council
TGWU	Transport and General Workers' Union
TQM	total quality management
TUC	Trades Union Congress
TULRCA	Trade Union and Labour Relations (Consolidation) Act
TUPE	Transfer of Undertakings (Protection of Employment) Regulations
TURERA	Trade Union Reform and Employment Rights Act
UK	United Kingdom
UMIST	University of Manchester Institute of Science and Technology
USA	United States of America
VAT	Value Added Tax

VDU	visual display unit
WIRS	Workplace Industrial Relations Survey

This book is dedicated to Kate and Caroline Aubrook: women for the twenty-first century

PART ONE

THE EMPLOYEE RELATIONS, MANAGERIAL AND EMPLOYMENT CONTEXTS

ONE

The Employee Relations Context

The purposes of this chapter are to:

- explain the rationale for the book;
- provide outline definitions of the terms used in the title;
- present a background overview of developments in employee relations from literature between the late 1960s and the early 1990s.

1.1 EMPLOYEE RELATIONS – A NEGLECTED AREA

A working definition

Defining employee relations is problematic. The problem with any definition is that it implies certain things should be in and others out, but it is not always clear-cut what could or should be included. While a definition does serve to focus attention on the broad principles of any topic, it can also mislead, an issue explored more fully below. However, as a starting point, it is proposed that employee relations in hotels and catering is about the management of employment and work relationships between managers and workers and, sometimes, customers; it also covers contemporary employment and work practices. The pattern of such relationships is multidimensional and occasionally triadic, particularly at the point of service and/or production.

These relationships can be individually or collectively based, involving informal or formal arrangements, and will be shaped by a variety of internal and external factors, including political, economic and social influences. The detailed issues that permeate and pervade such relationships centre on the effort–reward bargain and include remuneration, grievances, disputes and discipline, securing commitment and involvement, equality and the quality of working life. The ways in which work can be arranged, constituted, organized, controlled and managed are also multivariable and complex. Contemporary employment and work practices are a reflection of what exists at any point in time, for which there are no absolute given circumstances or conditions.

A literature gap

Managing employee relations is not a subject that has received much explicit consideration in hotel and catering 'textbook' literature, particularly recently, although some of the issues addressed in this book are mentioned, in one way or another, in a number of the current texts. These include those that address the subject of human resources management (including Riley, 1991; Boella, 1992; Mullins, 1992), and those that are sociologically based and concerned with work and the employment experience (such as Gabriel, 1988; Wood, 1992a) and organizational behaviour (Wood, 1994).

The only coherent attempt to consider employee relations as a whole has been in the form of a user friendly handbook for managers and students, now entitled *Employee Relations in the Hotel and Catering Industry*. Published by the Hotel and Catering Training Company (HCTC) – the former Hotel and Catering Industry Training Board (HCITB) and Hotel Catering Training Board (HCTB) – this is now in its seventh edition (HCTC, 1990), having first appeared in 1976 under the title *Industrial Relations – a Guide for Managers* (HCITB, 1976). It gives outline, but basic, guidance on good employee relations policies and practices, both of a voluntary nature and conforming to employment law, and addresses both individual and collective issues. The HCTC defines employee relations as 'all the factors in the relationships between employers, staff, and other individuals and institutions involved in the employment process' (HCTC, 1990, p. 9), but this definition is considered to be rather too woolly to be of much use in this analysis.

Elsewhere, recent more generally based texts still adopt a predominantly pluralist/collectivist framework for analysis and prescription of issues in the employment relationship (for example, Kessler and Bayliss, 1992; Gospel and Palmer, 1993). Even though these and other texts report a changing national scene in industrial and employee relations towards practices and behaviours that embrace a greater element of unitarism and individualism, making them increasingly appropriate (for example, Storey and Sisson, 1993), much of the material is still presented in a form that holds limited relevance for those interested in examining the employment relationship in hotels and catering. Thus trade unions, collective bargaining and industrial disputes are not issues in the employment relationship in the hotel and catering industry in the sense that they are largely absent, although the reasons for their absence are, of course, an interesting subject for analysis and debate.

It is the absence of detailed analysis, discussion and evaluation of issues in the employment relationship in an industry such as hotels and catering that have provided the motivation for writing this book. The industry has been, and continues to be, isolated from the mainstream of British employee relations and industrial relations practices. Indeed, hotels and catering may be said to represent industrial relations practices that exemplify one extreme of British industrial relations (Lucas, 1994c). As a general rule, there is more emphasis on unitarist/individualist employment practices. Few methods of formal regulation have evolved. Wages councils were arguably the most important formal method of regulation to have existed, but they were abolished in 1993.

Special problems

The attempt to write this book was not without problems. First, there is a dearth of in-depth research, particularly empirical research, on key subject areas in the employment relationship, and many employee relations references have a tendency to be relatively 'hidden' in the literature. Furthermore, the research that is relevant has been approached from a wide range of perspectives, theoretical constructs and methodological bases that are not always easy to reconcile coherently. The research outcomes are of variable quality. Such drawbacks could invite criticism that a text like this contains rather too many mixed metaphors and metonymies, but no author is ever in possession of perfect information. Theoretical perspectives on the management of the employment relationship in the industry are poorly, if at all, developed (Lucas, 1994c).

Second, the research that has been carried out has tended to concentrate on the hotel industry, particularly the large national companies. This cannot be said to be representative of the hotel sector, let alone the industry as a whole. Small, independently owned and/or managed businesses still predominate in the hotel industry (HCTC, 1994a) and the other sectors, but *an employment relationship still exists even if an organization employs only a handful of employees*.

Third, and related to this second point, employment research among small firms has been relatively limited in nature and in scope. For example, the research efforts of Curran and Stanworth (1979); Goss (1988) and Rainnie (1989) concentrated on manufacturing firms employing relatively large proportions of skilled workers that were unionized to a greater or lesser extent. So, although these studies offer an interesting insight into employment relations in some small firms, the nature of employment relations in these workplaces is subject to a number of influences that differ from those that impact on the employment relationship in much of hotels and catering.

Recent small-firm studies have included samples from hotels and catering (for example, Abbott, 1993; Kitching, 1994), but there has been a tendency for this research to take a labour market perspective concerned to examine the complex interaction between small firms and the labour market (for a recent, more general example, see Atkinson and Storey, 1994), rather than to examine the nature of the employment relationship within the workplace. This is not to suggest that 'smallness' *per se* is the only factor that influences employment and the nature of employment relations, but it is undoubtedly a significant one.

Finally, factors connected to this 'smallness' structural characteristic of the hotel and catering industry pose significant methodological problems for employment researchers. Such difficulties have undoubtedly contributed to the lack of research across this sizeable industry (Lucas, 1992b, 1993a). In other words, there is a possibility that what we do know about the industry may be telling only part of the story.

A breakthrough: the Workplace Industrial Relations Survey 1990 (WIRS3)

Prima facie the existence of relatively little 'representative' and, arguably, coherent knowledge upon which to base understanding was a significant inhibiting factor in the writing of this book. However, the extraction of hotel and catering industry data from the Workplace Industrial Relations Survey 1990 (WIRS3) datasets provided a significant breakthrough. The WIRS3 data undoubtedly constitute the most comprehensive account of employee relations management practices and procedures in the industry to date and enable these to be compared with the national scenario and, perhaps more appropriately, with all service industries and private sector services. Thus, for the first time, it is possible to give answers to the question, 'Are employee relations in hotels and catering really different?', with considerably more authority than has ever been possible before.

Most of the data drawn from the WIRS3 are discussed throughout the text in the context of particular issues. All the WIRS3 data could have been presented separately but, on balance, incorporating them at appropriate points in the text makes for a better structured book. However, the WIRS3 data discussed in this book will be analysed and evaluated separately in a forthcoming journal article. The more general characteristics of the hotels and catering sample, including the size and ownership patterns of establishments and workforce characteristics, are, however, shown in Appendix 1.

The purpose of this book is to provide an introductory forum for the main features of the WIRS3 data on hotels and catering on a comparative basis with broader sectors in the economy as a whole. Many of the WIRS3 data are presented in tabular form, although some pertinent comments are made in the text about the significant issues that arise from these tables. The reason why such detailed tables have been presented is to enable the reader to undertake further analysis from them. Deeper analysis and evaluation could be undertaken from a more comprehensive analysis of the WIRS3 datasets but that does not fall within the central focus of this text. The small size of the hotel and catering sample must inevitably place limitations on the potential gains that can be achieved from a more searching analysis of a number of the issues addressed in this text.

Thus the WIRS3 survey data enable valid comparisons to be made between the hotel and catering industry and other sectors of the economy for the first time. However, the WIRS3 sample is limited to workplaces with 25 or more employees, thus excluding the vast majority of workplaces in the industry which employ fewer than 25 employees (97 per cent – see Table 1.1). The effect of this is that in the WIRS3 sample nearly three-quarters of the responding workplaces are hotels and restaurants, whereas these sectors actually account for around 58 per cent of all workplaces in the industry as a whole. Therefore the WIRS3 data cannot be said to be fully representative of the industry by size of establishment or by industry sector.

Structural issues

Because of the special problems identified, it has been necessary to draw on examples from outside the hotel and catering industry to illustrate a number of the topics covered in the book. A number of non-UK studies (mainly from the USA) are also mentioned. Additionally, the changing contours of the subject of industrial/employee relations (and personnel management/human resource management (HRM)) make for difficulties in determining what topics should be included, how they should be labelled, the amount of depth required for discussion and where particular issues should be placed in the text.

The book comprises three parts. The approach taken for the main body of the text on employee relations (Part 2) has been to work from a broad analytical framework that has its roots in more conventional industrial relations texts and to modify this, as appropriate, to reflect the needs of hotels and catering and to incorporate relevant contemporary developments in human resourcing issues (see below). The increasing role and relevance of the law in employee relations can be discerned from the inclusion of an increasingly sizeable, and complex, amount of material on employment legislation (Part 3).

However, it is considered that the analysis and discussion contained in Parts 2 and 3 can be better understood, and will be more informed, if it is set in some kind of context. This should be of benefit not only to those who know relatively little about hotels and catering but also to those who know more, because this material has not been assimilated in this way before. Thus Part 1 offers analysis and evaluation of contextual issues, and of contemporary practices that do not necessarily 'fit' easily into the main body of the text thereafter.

Summary propositions

- A thorough analysis and evaluation of employee relations in the hotel and catering industry is a long overdue topic.
- Although there are knowledge gaps about employee relations in the industry, the extraction of WIRS3 (1990) data relating to hotels and catering makes for a considerably more well-informed analysis and evaluation than has been previously possible.

Discussion points

- Define employee relations from your own experiences of hotels and catering.
- Suggest different types of research methods that could be used to find out more about employee relations and employment issues in hotels and catering. Evaluate the benefits and drawbacks of these methods.

1.2 MANAGING EMPLOYEE RELATIONS IN THE HOTEL AND CATERING INDUSTRY

The managerial context

The title 'Managing employee relations in the hotel and catering industry' requires a little amplification. The book focuses on the subject of employee relations within the context of a specific industry sector. 'Managing' is not used in the sense of 'how-to-do-it', but rather to signify that employee relations is something that can be, and should be, managed. Implicit in much of the commentary on the hotel and catering industry is the suggestion that employee relations are not (and have not been) managed well, with the result that organizations have not functioned well. Thus 'managing' is viewed from the dual perspective of what may be appropriate practice and the reasons that may militate against 'good' practice. However, what may be regarded as 'good' and 'bad' management and employment practice is highly subjective, a point on which the book ends.

The issue of the organizational framework and management is developed separately and more fully in Chapter 2 because it is considered to form an important prerequisite to understanding the nature of employee relations in hotels and catering. Employee relations and its contextual setting are now explained.

Why employee relations?

Part of personnel management

Industrial relations is the more traditional, and still most widely used, term for discussing issues in the employment relationship, although employee relations is becoming more widely used. Employee relations has become increasingly used by personnel practitioners to signify the aspect of their work that is concerned with the regulation of individual and collective relations. This is reinforced within the Institute of Personnel and Development's (IPD) (formerly the Institute of Personnel Management (IPM)) Professional Education Scheme, where one of the Stage Two examinations is called Employee Relations, as opposed to Industrial Relations. Employee relations is therefore part of personnel management and is also a fundamental issue in the more topical and controversial HRM.

Not necessarily industrial relations

The terms 'employee relations' and 'industrial relations', though broadly similar, are not necessarily synonymous. Perhaps most importantly, there are more fundamental academic differences between them. In employee relations, the focus falls 'on management alone (rather than on all parties to the employment relationship) and on contemporary practices ... also greater emphasis on informal and individual aspects of management–employee relations rather than on collective bargaining and representative institutions' (Marchington and Parker, 1990, p. 8). Employee relations is also 'seen as more appropriate in the context of small and medium-sized enterprises' (Roberts *et al.*, 1992, p. 240).

This academic terminological difference is reflected in the contents of the leading academic journals in this field. The *British Journal of Industrial Relations* is still very much concerned with 'traditional' issues centring on formal institutions such as trade unions and collective bargaining, largely in manufacturing and predominantly, but not exclusively, in Britain. The *Industrial Relations Journal* has developed a broader perspective within a more international framework; the employment issues that are addressed are less 'traditional' and include many service sector examples. *Employee Relations* is the most diverse in terms of coverage of issues in the employment relationship, covering topics as multifarious as childcare provisions, smoking at work, quality circles, flexible teamwork, stress at work, industrial conflict and negotiations, and has also recently become more 'international' in scope.

The nature of employee relations as presented in this book has, perhaps, been best encapsulated in a special edition of *Employee Relations* devoted to 'Employment in hotel and catering services' (Lucas and Wood, 1993). This included contributions on labour economics in the hotel and catering industry (Riley, 1993), the influence of law on employment practice (Price, 1993), HIV/AIDS (Adam-Smith and Goss, 1993), ageism (Lucas, 1993b), power relationships and empowerment (Wynne, 1993) and the views of hotel managers towards trade unions (Aslan and Wood, 1993). Employee relations thus embodies a broad-ranging agenda and is a more appropriate term than industrial relations for discussing contemporary issues in the employment relationship in the hotel and catering industry.

What is the hotel and catering industry?

Emphasis on the commercial sector

On its broadest definition, the hotel and catering industry employs an estimated 2.4 million workers (10 per cent of the UK workforce) and can be divided into two sectors (HCITB, 1985a; HCTC, 1993, 1994a). The *commercial sector*, employing an estimated 1.3 million workers, includes hotels, guesthouses, restaurants, cafes, snack bars, public houses, clubs and contract catering (canteens). This sector mirrors the Employment Department's Standard Industrial Classification (SIC) Class 66 – hotels and catering, which is also the basis for the classification of the main WIRS3 data used throughout this book. A more detailed analysis of the nature and structure of employment in this activity sector features in Chapter 3, so that the reader will be more fully informed of the complexities that impinge on the employment relationship. The employment characteristics of the WIRS3 data are contained in Appendix 1.

The *catering services sector*, employing an estimated 1.1 million workers, includes tourism and travel catering, catering in education, medical and other health services, public administration and national defence, retail distribution, personal and domestic service, and industrial and office catering. This is not identified separately in Employment Department statistics, or in the Workplace Industrial Relations Survey series (1980, 1984, 1990) (Daniel and Millward, 1983; Millward and Stevens, 1986; Millward *et al.*, 1992; Millward, 1994) because catering employees working in these sectors are included under the SIC for the main business.

Therefore this book deals predominantly with catering businesses in the commercial sector. The industry on its broadest definition is simply too large and diverse for the contents of one book. However, this does not preclude the use of examples of practice from the catering services sector by way of comparison.

Main structural characteristics

As there may be an element of misconception about what the hotel and catering industry is and does, particularly among those with little direct knowledge of the industry, it is instructive to outline the main characteristics of the industry. Although these features may alter over time, it is unlikely that overall shifts will occur quickly.

Identifying the structural characteristics of the commercial sector with any degree of accuracy is difficult, because different sources show significant variations. The most recent figures produced by the HCTC (1994a) that give as comprehensive a view as possible are drawn from the Census of Employment (1991) and the Labour Force Survey (1992). The Census of Employment is based on a large sample of establishments with 25 or more employees on the Inland Revenue's PAYE register plus a sample of small establishments on the register. The Labour Force Survey gives information on 'establishments owned or managed by self-employed proprietors or partners – which means there is an element of "double-counting" as establishments in this group with employees may also be included in the Census of Employment sample' (HCTC, 1994a, p. 22). This combined analysis has been used to compile Table 1.1.

Table 1.1 *Numbers of establishments by size (1991)*

Total number of establishments		Size of establishments (by number of employees)					
		1–10	(%)	11–24	(%)	25+	(%)
Hotels	52,200	45,400	(87)	3,600	(7)	3,200	(6)
Restaurants	99,900	92,300	(92)	5,600	(6)	2,000	(2)
Public houses and bars	77,100	65,500	(85)	9,600	(12)	2,000	(3)
Clubs	16,100	11,400	(71)	3,700	(23)	1,000	(6)
Contract catering	19,200	16,800	(88)	1,600	(8)	800	(4)
Total	264,500	231,400	(88)	24,100	(9)	9,000	(3)

The figures are based on the Census of Employment and Labour Force Survey, which means there is an element of double counting (see text above). Census of Employment figures alone show that 64 per cent of employees work in establishments employing fewer than 25 employees (HCTC, 1994a, p. 44).

Source: Adapted from Table 3.1, p. 22 and Figure 3.4, p. 24 (HCTC, 1994a).

Regardless of which analysis is used, the vast majority of businesses are still very small, highly fragmented and independently owned and/or managed. The average number of employees per establishment is 4.5 (HCTC, 1994a, p. 44). Key Note has identified that the industry comprised 126,000 businesses subject to VAT, 77 per cent of which were operated by sole proprietors and partnerships (Key Note, 1992). If smaller businesses that are not registered for VAT are included, the total rises to around 200,000 establishments. The HCTC's (1994a) estimates, shown in Table 1.1, are higher – the industry comprises 264,500 establishments, which represents an increase of 5 per cent since 1984. 'The highest proportion of establishments were located in the South East (31% in 1991), where around one-third of the total population of Great Britain resided' (HCTC, 1994a, p. 23). The main characteristics of each sector are now summarized; the figures given in Table 1.1 inform much of this analysis.

The hotels and other accommodation sector is extremely fragmented; it comprises an estimated 52,200 establishments of considerable diversity, including hotels, motels, holiday camps, guesthouses, boarding houses, hostels and bed and breakfast establishments. There are around 35,000 hotels and guesthouses (Key Note, 1992). Although 87 per cent of hotels employed between 1 and 10 employees, a relatively high proportion of hotels (6 per cent) employed more than 25 employees (see Table 1.1). The vast majority of hotels are independently owned (average size around 20 bedrooms). This sector contains the highest proportion of self-employed owners/managers (70 per cent) (HCTC, 1994a, p. 23). Despite the growth of hotel chains, fewer than 10 per cent of hotels are owned by companies (average size around 90 bedrooms), although they have over one-quarter of the market share (Roper, 1988). The top 25 hotel groups, dominated by Forte, controlled 1,130 hotels, a small proportion of the total (see HCTC, 1994a, p. 26). The highest concentration of hotels, which are proportionately larger than elsewhere, is found in London and the South East of England.

Restaurants, fast-food outlets, roadside catering, cafes and snack bars comprise 99,900 establishments. A vast majority (90 per cent) of such establishments are independent, privately owned small businesses (Key Note, 1989). There are a number of restaurant chains, many of which are still classified as small to medium-sized businesses, with the exception of chains operating within a larger overall group of companies and, of course, the ubiquitous McDonald's. An overwhelming majority (92 per cent) employed one to ten employees, although changes in the sector during the 1980s may have reduced this proportion slightly. The fast-food sector – hamburgers, chicken and pizza – has enjoyed rapid growth: franchising is a particularly popular means of 'managing' such units. Cafes and take-aways have declined. Most ethnic restaurants staff their businesses with family members.

There are some 71,100 *public houses* and bars. The five big multiples together and the regional breweries each owned around one-quarter of them while the remaining half were owned by independent brewers (Key Note, 1993). The ownership structure changed radically following the relinquishing of over one-third of their tied estate by the national brewers, resulting from the Monopolies and Mergers Commission report. Between 1989 and 1994 the number of independent outlets increased by 69 per cent (see HCTC, 1994a, p. 29), although a number of public houses have also closed. Declining beer sales have forced many public houses to diversify by increasing their provision of catering facilities. Although a large majority of establishments are 'small', managed houses (around 40 per cent of the national and regional brewers' outlets) they are subject to brewery policy influence over staffing issues (Key Note, 1993).

There are approximately 16,100 *clubs* (the total rises to 32,000 if sporting clubs are included), many of which have a turnover below the VAT threshold and are staffed by members. Most employ one to ten employees.

There are an estimated 19,200 *contract catering* units in this growing catering sector (non-permanent contract catering units account for an additional 2,500 units). Four groups hold around 80 per cent of market share (one group holds nearly 40 per cent). Corporate image is an important factor to these organizations, whose staff are generally contracted out in small numbers.

Summary proposition

Defining the terms employee relations, managing and the hotel and catering industry is problematic, but some attempt is necessary in order to give the reader a clear focus for understanding the subject matter of this text.

Discussion points

- In what way does the size and structure of the industry make it difficult for researchers to gather reliable and valid employment data about hotels and catering?
- Suggest how you might approach the design of a research sample to undertake a piece of employment research in any sector of hotels and catering.

1.3 THE DEVELOPMENT OF EMPLOYEE RELATIONS

Although there are examples in the literature about employment issues in the hotel and catering industry that predate relatively recent times (Orwell, 1933; Whyte, 1948, 1949), the most significant body of literature has developed over the past 25 years or so. The most important sources, mainly published, are reviewed briefly below.

The 1960s and 1970s

Employee relations (under the more traditional label used then of 'industrial relations') in the hotel and catering industry assumed prominence in the wake of the Donovan Report (Royal Commission on Trade Unions and Employers' Associations, 1968), at a time when the industry was beginning to emerge as a significant employment sector in its own right. Much of this early examination came from studies or reports produced by state-created institutions, reflecting the increasing involvement of governments (mainly Labour) in the regulation of the employment relationship throughout the 1960s and 1970s.

Reflecting Donovan and the mood of the times, the emphasis was on procedural and institutional reform and policy development. However, formal methods of regulation were largely absent from the industry. Some specific employment practices that were thought to be problematic, such as tipping and staff turnover, were examined by the Hotel and Catering Economic Development Committee (HCEDC) (HCEDC, 1968a, 1969). More general manpower issues, including the need to develop comprehensive policies, were also addressed (HCEDC, 1967, 1968b, 1975, 1977; Department of Employment, 1971, 1976). The pioneering Commission on Industrial Relations (CIR) examined the potential for replacing three of the four wages councils with voluntary collective bargaining arrangements (CIR, 1971, 1972, 1973) with the view to the 'securing of any improvements in industrial relations that appear necessary and desirable' (CIR, 1971, p. iv). Latterly, the Advisory, Conciliation and Arbitration Service (ACAS) also considered the possibilities for extending voluntary collective bargaining in hotels and restaurants (ACAS, 1980). Apart from the abolition of one of the wages councils in the mid-1970s, there is scant evidence to indicate that reforms of any real substance actually occurred.

Probably the most distinctive early industrial relations contributions were made by Mars and others. First, *Room for Reform?* sought 'to explore and explain the paradoxes within this industry by examining them in the context of changes that are altering its industrial relations system' (Mars and Mitchell, 1976, p. 3). Although influenced by Dunlop's (1958) systems approach, Donovan (1968) and the conventional wisdom of collectivism, the case study identified many of the special features of the employment relationship in hotels, including *ad hoc* management, the triadic nature of the employment relationship and individual contract making, which underpin much of the analysis throughout this book. This study was later combined with other material to form a problem solving guide for managers on managing human relations at unit level (Mars *et al.*, 1979).

Other academic studies with a collective focus also contributed to the developing body of knowledge about the industry, particularly trade unions (or rather the lack of them) (Chopping, 1977) and collective disputes (Palmer, 1968; Wood and Pedler, 1978). Labour turnover, which featured in a number of the HCEDC reports, began to be examined in more detail by others (Mars *et al.*, 1979). Other sociological studies also provided insights into the nature of working in the industry, including illicit employment practices (Mars, 1973; Shamir, 1975; Bowey, 1976).

The newly commissioned New Earnings Survey (NES) (1968) began to supply authoritative data on pay and earnings on an annual basis. In the mid-1970s, the Low Pay Unit began to champion the cause of the low paid (for example, Brown and Winyard, 1975; Mars and Mitchell, 1977; Jordan, 1978; Thomas and Erlam, 1978); and the HCITB produced *Industrial Relations – a Guide for Managers* (1976).

Nevertheless, by the end of this period, although there was extensive literature centring on problems in the industry, it was not thought to be 'particularly relevant to industrial relations or manpower problems' (Mars *et al.*, 1979).

The 1980s

The employee relations agenda in hotels and catering, as reflected in the literature, began to broaden during the 1980s at a time when a radical programme of national industrial relations reforms was being pursued by a new Conservative administration. The government's espoused policy of deregulation sought to effect a gradual removal of state involvement from issues in the employment relationship and aimed to shift the balance of power from collectively determined practices towards those that were more individually based.

Newly instigated national surveys on workplace industrial relations practices did not include specific hotel and catering industry examples (Daniel and Millward, 1983; Millward and Stevens, 1986). However the analysis of macro-data on employment and pay (Robinson and Wallace, 1983, 1984) and various HCITB reports on manpower in the industry (including HCITB, 1983, 1984a, 1984b, 1985a, 1985b; HCTB, 1987) established a base for developing further understanding of employment relationship issues, including female and part-time working. It is perhaps significant that the HCITB changed the title of its managerial guide from *Industrial Relations* to *Employee Relations* in 1981.

Many industry studies continued to be dominated by the 'problem' issues of the 1970s: high labour turnover (Riley, 1980, 1981a, b; Johnson, 1980, 1981, 1985, 1986) and the limited presence of trade unions (Johnson and Mignot, 1982; Macfarlane, 1982a, b; Jameson and Johnson, 1985, 1989; Riley, 1985). Other individual and collective issues that were addressed over the decade included payment and reward systems (Butler and Skipper, 1981; Johnson, 1982, 1983a, b), low pay, discrimination and poor working conditions (Dronfield and Soto, 1980; Byrne, 1986), industrial relations in hotels (Snow, 1981) and wages councils (Craig *et al.*, 1982; Johnson and Whatton, 1984; Lucas, 1989). The work of Analoui and Kakabadse (1989) on defiance in the workplace developed some of the ideas about unconventional workplace behaviours first mooted by Mars and others (Mars and Mitchell, 1976; Mars *et al.*, 1979).

In the later part of the 1980s, in the wake of Atkinson (1984), the subject of employment flexibility assumed prominence (Bagguley, 1987; Kelliher and McKenna, 1987, 1988; Lockwood and Guerrier, 1988, 1989; Guerrier and Lockwood, 1989a, b; Kelliher, 1989b). Other topics of interest to researchers, a number of which began to break new ground, included personnel management (Kelliher and Johnson, 1987; Ralston, 1989), the management of labour and human resource management (Croney, 1988a, b; Kelliher, 1989a), internal labour markets (Simms *et al.*, 1988) and the quality and experiences of working life (Hales, 1987; Gabriel, 1988).

What was beginning to emerge from some of these studies was that the reform of industrial/employee relations along the lines of traditional collectivism might not be appropriate. In relation to wages councils, while the decision to abolish the staff canteen wages council was found wanting (Craig *et al.*, 1982), desirable though minimum wage protection might be, perhaps 'Britain's wages council system had provided an inappropriate model for wage-fixing in the catering industry' (Lucas, 1989, p. 199). Furthermore, although high labour turnover was perceived as problematic to the industry by some (Mars *et al.*, 1979; Johnson, 1980, 1981, 1985), others viewed the phenomenon as unavoidable, necessary, desirable and a deliberate part of management strategy (Bowey, 1976; Riley, 1980; Simms *et al.*, 1988). Most studies on employment flexibility concluded that Atkinson's model of the flexible firm (Atkinson, 1984) was found wanting in terms of the hotel and catering experience. In other words, a growing body of literature was beginning to suggest that perhaps the hotel and catering industry was 'different', and that those seeking to 'improve employee relations' might need to review the basis on which this might be achieved.

These 1980s examples are not exclusive, but they indicate a relatively restricted field of studies connected with the employment relationship in the hotel and catering industry. Although the agenda had broadened over the decade, major topics notable by their absence or limited coverage included: power issues in the employment relationship; managerial and workforce strategies; conflict and cooperation; securing commitment through participation and employee involvement; grievance, discipline, dismissal and redundancy; and the role and implications of employment legislation. The late arrival and relative dearth of information on personnel, human resource or employee relations management is even more surprising given the oft repeated generic statement that 'people are our most important resource'. Finally, most studies concentrated on the hotel industry, leaving vast areas of the industry still 'undiscovered'.

The 1990s

The early indications are that the employee relations agenda in the hotel and catering industry has broadened even further in the 1990s, but it still remains limited in terms of traditional British industrial relations issues. A shifting agenda is likely to be a reflection of changing patterns of employee relations that have been developing elsewhere since the 1980s (Millward *et al.*, 1992), the emergence of human resource management and the 'new industrial relations', a growing interest in the labour process debate (see Thompson, 1987) and growing internationalization, particularly the creation of the single European market.

At national level, the WIRS3, although the third of its kind, was the first to provide direct published evidence of employee relations practices by SIC Division, although in the overall analysis the hotel and catering industry is grouped with repairs (Millward *et al.*, 1992). Fortunately access to the WIRS3 datasets has enabled a separate analysis for hotels and catering to be produced, and this is presented and discussed throughout this text.

Conversely, analysis of the industry at a general level is likely to have been reduced by the demise of the National Economic Development Office (NEDO), which had contributed many valuable reports on manpower issues since the 1960s. A similar break with the past has occurred as a result of the abolition of the wages councils, which means that wage data by collective agreement (the councils regulated the minimum pay of nearly one million catering workers) will no longer be reported in the NES. Although the HCTC has a changed, more commercialized, role, statistical reports on employment and other related topics are still produced (HCTC, 1992, 1993, 1994a, b), but are less comprehensive than a number of those produced in the 1980s.

Personnel practice has been examined on a case study basis in a budget hotel company (Lucas and Laycock, 1991) and across a larger sample of high-quality hotels and restaurants (Price, 1993, 1994). Flexibility remains highly topical but has been further developed in regard to gender (Bagguley, 1990; Crompton and Sanderson, 1990; Walsh, 1990). The issues of wages councils, collective bargaining and low pay have assumed a higher profile in the wake of government policy and the abolition of the wages council system (Lucas, 1990, 1991a, b, 1992a, b, 1994a; Macaulay and Wood, 1992b; Lucas and Bailey, 1993). Discrimination from a wider base has been addressed, embracing race, age and HIV/AIDS (Commission for Racial Equality (CRE), 1991; Adam-Smith and Goss, 1993; Lucas, 1993a, b, 1995). Employment and demographic trends have been widely discussed, some of which link closely to issues of employment policy, gender and age discrimination (Johnson and Thomas, 1990; Lucas and Jeffries, 1991; Mogendorff *et al.*, 1991; NEDC, 1991; Lucas, 1992c, 1993d, 1994b; Wood, 1992b; Dieke, 1993). The influence of the social dimension of the single European market has also been considered (Kelliher and Blackman, 1991; Lucas, 1993c). Quality (Lucas and Laycock, 1992; Price and Tempest, 1992), employment law and employment practice (Price, 1993, 1994), power relationships and empowerment (Wynne, 1993, Lashley, 1994), communications and social skills in managers (Clark, 1993) and trade unions (Macaulay and Wood, 1992a; Aslan and Wood, 1993) are among other employee relations topics that have received attention.

Academic research still remains relatively selective in nature and scope, and is still largely the preserve of researchers within hotel and catering departments, mostly within the 'new' universities. The prominence of discussions about labour turnover and trade unions has been superseded by discussions on other topics, although no one subject area appears to predominate. While many of these topics fall within the broad field of labour market studies, reflecting both individual and collective issues, the subject areas have become less clearly delineated and frequently overlap, making for difficulties of classification. This point is not exclusively an hotel and catering issue, but is a reflection of the changing boundaries of the 'people' subjects: industrial relations, employee relations, personnel management and human resource management.

The future

As the summary of material in this chapter shows, employee relations in the hotel and catering industry is a significant subject, but it has not received sufficient attention holistically in terms of analysis, discussion and evaluation (see Lucas, 1994c). This book aims to begin to remedy this deficiency. As will become evident, many of the issues considered in this book pose more questions than answers. Much of this is deliberate because this book is intended to provoke thought, provide a basis for discussion and point to areas for further study. Thus, it is hoped that the subject of employee relations in this important industry sector will be given more attention in academic institutions (in teaching and through research) and within the industry itself.

Although the main audience for the book must remain students and academics who have a particular interest in the industry, it is largely the case that the broader academic world has failed to recognize the potential offered by the industry for employee relations research. It is conceivable that the industry does present something of a 'back to the future' scenario (Riley, 1993) but, arguably, these issues justify more detailed analysis.

This scenario offers a challenge to others to direct more of their attention to an industry which, despite some clear problems of accessing data, may prove to be something of a Pandora's Box, and from which the rewards may prove to be rich.

Finally, the author's perspective on employee relations is contained in Appendix 2. It can be best understood after reading the section 'Perspectives on the employment relationship' in Chapter 4, but has been presented separately so as not to break up the flow of that chapter.

Summary propositions

- Employee relations in the hotel and catering industry has not received sufficient attention holistically in terms of analysis, discussion and evaluation.
- There is considerable opportunity to carry out more employee relations research, although methodological problems have to be faced.

Discussion points

- Have employee relations issues in hotels and catering changed since the 1960s?
- What do you consider to be the key employee relations issues of the 1990s?

REFERENCES

Abbott, B. (1993) Small firms and trades unions in services in the 1990s. *Industrial Relations Journal*, **24** (2), 308–17.

Adam-Smith, D. and Goss, D. (1993) HIV/AIDS and hotel and catering employment: some implications of perceived risk. *Employee Relations*, **15** (2), 25–32.

Advisory, Conciliation and Arbitration Service (1980) *Licensed Residential Establishment and Licensed Restaurant Wages Council Report No. 18*. London: HMSO.

Analoui, F. and Kakabadse, A. (1989) Defiance at work. *Employee Relations*, **11** (3), 1–62.

Aslan, A.H. and Wood, R.C. (1993) Trade unions in the hotel and catering industry: the views of hotel managers. *Employee Relations*, **15** (2), 61–70.

Atkinson, J. (1984) Manpower strategies for flexible organizations. *Personnel Management*, August, 28–31.

Atkinson, J. and Storey, D. (eds) (1994) *Employment in the Small Firm and the Labour Market*. London: Routledge.

Bagguley, P. (1987) *Flexibility, Restructuring and Gender: Changing Employment in Britain's Hotels*. Lancaster Regionalism Group, University of Lancaster.

Bagguley, P. (1990) Gender and labour flexibility in hotel and catering. *The Service Industries Journal*, **10** (4), 737–47.

Boella, M.J. (1992) *Human Resource Management in the Hospitality Industry*, 5th edition. Cheltenham: Stanley Thornes.

Bowey, A. (1976) *The Sociology of Organizations*. London: Hodder and Stoughton.

Brown, M. and Winyard, S. (1975) *Low Pay in Hotels and Catering*. London: Low Pay Unit.

Butler, S.R. and Skipper, J. (1981) Working for tips. *The Sociological Quarterly*, **22**, 15–27.

Byrne, D. (ed.) (1986) *Waiting for Change*. London: Low Pay Unit.

Chopping, B.C. (1977) Unionisation in London hotels and restaurants. BPhil thesis, University of Oxford.

Clark, M. (1993) Communication and social skills: perceptions of hospitality managers. *Employee Relations*, **15** (2), 51–60.

Commission for Racial Equality (1991) *Working in Hotels*. London: Commission for Racial Equality.

Commission on Industrial Relations (1971) *The Hotel and Catering Industry, Part I. Hotels and Restaurants, Report No. 23*. London: HMSO.

Commission on Industrial Relations (1972) *The Hotel and Catering Industry, Part II. Industrial Catering, Report No. 27*. London: HMSO.

Commission on Industrial Relations (1973) *The Hotel and Catering Industry, Part III. Public Houses, Clubs and Other Sectors, Report No. 36*. London: HMSO.

Craig, C., Rubery, J., Tarling, R, and Wilkinson, F. (1982) *Labour Market Structure, Industrial Organization and Low Pay*. Cambridge: Cambridge University Press.

Crompton, R. and Sanderson, K. (1990) *Gendered Jobs and Social Change*. London: Unwin Hyman.

Croney, P. (1988a) An investigation into the management of labour in the hotel industry. MA thesis, University of Warwick.

Croney, P. (1988b) An analysis of human resource management in the UK hotel industry. Paper presented to the International Association of Hotel Management Schools Autumn Symposium, Leeds.

Curran, J. and Stanworth, J. (1979) Work relations and social involvement in the small firm. *Sociological Review*, **27** (4), 317–42.

Daniel, W.W. and Millward, N. (1983) *Workplace Industrial Relations in Britain*. London: Heinemann.

Department of Employment (1971) *Manpower Study No. 10: Hotels*. London: HMSO.

Department of Employment (1976) *Manpower Study No. 11: Catering*. London: HMSO.

Dieke, P.U.C. (1993) Tourism policy and employment in The Gambia. *Employee Relations*, **15** (2), 71–80.

Dronfield, L. and Soto, P. (1980) *Hardship Hotel*. London: Counter Information Services.

Dunlop, J. (1958) *Industrial Relations Systems*. New York: Henry Holt and Company.

Gabriel, Y. (1988) *Working Lives in Catering*. London: Routledge & Kegan Paul.

Gospel, H.F. and Palmer, G. (1993) *British Industrial Relations*, 2nd edition. London: Routledge.

Goss, D. (1988) Social harmony and the small firm: a reappraisal. *Sociological Review*, **36**, 114–32.

Guerrier, Y. and Lockwood, A. (1989a) Managing flexible working. *The Service Industries Journal*, **7** (3), 406–19.

Guerrier, Y. and Lockwood, A. (1989b) Core and peripheral employees in hotel operations. *Personnel Review*, **18** (1), 9–15.

Hales, C. (1987) Quality of Working Life: job redesign and participation in a service industry: a rose by any other name? *The Service Industries Journal*, **7** (3), 253–73.

Hotel and Catering Economic Development Committee (1967) *Your Manpower: a Practical Guide to the Manpower Statistics of the Hotel and Catering Industry*. London: HMSO.

Hotel and Catering Economic Development Committee (1968a) *Why Tipping?* London: HMSO.

Hotel and Catering Economic Development Committee (1968b) *Service in Hotels*. London: HMSO.

Hotel and Catering Economic Development Committee (1969) *Staff Turnover*. London: HMSO.

Hotel and Catering Economic Development Committee (1975) *Manpower Policy in the Hotel and Catering Industry – Research Findings*. London: HMSO.

Hotel and Catering Economic Development Committee (1977) *Employment Policy and Industrial Relations in the Hotels and Catering Industry – Research Findings*. London: HMSO.

Hotel and Catering Industry Training Board (1976) *Industrial Relations – a Guide for Managers*. Wembley: HCITB.

Hotel and Catering Industry Training Board (1983) *Manpower Changes in the Hotel and Catering Industry*. Wembley: HCITB.

Hotel and Catering Industry Training Board (1984a) *Manpower Flows in the Hotel and Catering Industry*. Wembley: HCITB.

Hotel and Catering Industry Training Board (1984b) *Manpower Forecasts for the Hotel and Catering Industry: Supplementary Report to Hotel and Catering Skills – Now and in the Future. Report Prepared for the Education and Training Advisory Council*. Wembley: HCITB.

Hotel and Catering Industry Training Board (1985a) *Hotel and Catering Manpower in Britain*. Wembley: HCITB.

Hotel and Catering Industry Training Board (1985b) *Hotel and Catering Establishments in Great Britain, Part 1: Hotels and Guesthouses, Restaurants and Cafes, Pubs and Clubs*. Wembley: HCITB.

Hotel and Catering Training Board (1987) *Women in the Hotel and Catering Industry*. Wembley: HCTB.

Hotel and Catering Training Company (1990) *Employee Relations in the Hotel and Catering Industry*, 7th edition. London: HCTC.

Hotel and Catering Training Company (1992) *Meeting Competence Needs in the Hotel and Catering Industry – Now and in the Future*. London: HCTC.

Hotel and Catering Training Company (1993) *Employment Forecasts Update 1992–2000 – Meeting Competence Needs in the Catering and Hospitality Industry and Licensed Trade*. London: HCTC.

Hotel and Catering Training Company (1994a) *Catering and Hospitality Industry – Key Facts and Figures*. London: HCTC.

Hotel and Catering Training Company (1994b) *Employment Flows in the Catering and Hospitality Industry*. London: HCTC.

Jameson, S.M. and Johnson, K. (1985) The hotel shop steward – an emerging role in British industrial relations. *International Journal of Hospitality Management*, **4** (3), 131–2.

Jameson, S.M. and Johnson, K. (1989) Hotel shop stewards – a critical factor in the development of industrial relations in hotels? *International Journal of Hospitality Management*, **8** (2), 167–77.

Johnson, K. (1980) Staff turnover in hotels. *Hospitality*, February, 28–36.

Johnson, K. (1981) Towards an understanding of labour turnover? *Service Industries Review*, **1** (1), 4–17.

Johnson, K. (1982) Fringe benefits: the views of individual hotel workers. *Hospitality*, June, 2–6.

Johnson, K. (1983a) Payment in hotels: the role of fringe benefits. *The Service Industries Journal*, **3** (2), 191–213.

Johnson, K. (1983b) Trade unions and total rewards. *International Journal of Hospitality Management*, **2** (1), 31–5.

Johnson, K. (1985) Labour turnover in hotels – revisited. *The Service Industries Journal*, **5** (2), 135–52.

Johnson, K. (1986) Labour turnover in hotels – an update. *The Service Industries Journal*, **6** (3), 362–80.

Johnson K. and Mignot, K. (1982) Marketing trade unionism to service industries: an historical analysis of the hotel industry. *Service Industries Review*, **2** (3), 5–23.

Johnson, K. and Whatton, T. (1984) A future for wages councils in the hospitality industry in the UK. *International Journal of Hospitality Management*, **3** (2), 71–9.

Johnson, P. and Thomas, B. (1990) Employment in tourism: a review. *Industrial Relations Journal*, **21** (1), 36–48.

Jordan, D. (1978) *Low Pay on a Plate*. London: Low Pay Unit.

Kelliher, C. (1989a) Management strategy in employee relations: some changes in the catering industry. *Contemporary Hospitality Management*, **1** (2), 7–11.

Kelliher, C. (1989b) Flexibility in employment: developments in the hospitality industry. *International Journal of Hospitality Management*, **8** (2), 157–66.

Kelliher, C. and Blackman, D. (1991) Human resource challenges for the 1990s. *International Journal of Contemporary Hospitality Management*, **3** (2), 4–9.

Kelliher, C. and Johnson, K. (1987) Personnel management in hotels – some empirical observations. *International Journal of Hospitality Management*, **6** (2), 103–8.

Kelliher, C. and McKenna, S. (1987) Contract caterers and public sector catering. *Employee Relations*, **9** (1), 8–13.

Kelliher, C. and McKenna, S. (1988) The employment implications of government policy: a case study in public sector catering. *Employee Relations*, **10** (1), 8–13.

Kessler, S. and Bayliss, F. (1992) *Contemporary British Industrial Relations*. London: Macmillan.

Key Note (1989) *Key Note Report – an Industry Sector Overview: Restaurants*, 5th edition. Hampton: Key Note Publications Ltd.

Key Note (1992) *Key Note Report – UK Catering Market*. Hampton: Key Note Publications Ltd.

Key Note (1993) *Key Note Report – A Market Sector Overview: Public Houses*, 9th edition. Hampton: Key Note Publications Ltd.

Kitching, J. (1994) Employers' work-force construction policies in the small service sector enterprise. In J. Atkinson and D. Storey (eds), *Employment, the Small Firm and the Labour Market*, pp. 103–46. London: Routledge.

Lashley, C. (1994) Is there any power in empowerment? Paper presented to the Third Annual CHME Research Conference, Napier University, April.

Lockwood, A. and Guerrier, Y. (1988) Underlying personnel strategies in flexible working. Paper presented to the International Association of Hotel Management Schools Autumn Symposium, Leeds.

Lockwood, A. and Guerrier, Y. (1989) Flexible working in the hospitality industry: current strategies and future potential. *Contemporary Hospitality Management*, **1** (1), 11–16.

Lucas, R.E. (1989) Minimum wages – straitjacket or framework for the hospitality industry into the 1990s? *International Journal of Hospitality Management*, **8** (3), 197–214.

Lucas, R.E. (1990) The Wages Act 1986: some reflections with particular reference to the Licensed Residential Establishment and Licensed Restaurant Wages Council. *The Service Industries Journal*, **10** (2), 320–35.

Lucas, R.E. (1991a) Promoting collective bargaining: wages councils and the hotel industry. *Employee Relations*, **13** (5), 3–11.

Lucas, R.E. (1991b) Remuneration practice in a wages council sector: some empirical observations in hotels. *Industrial Relations Journal*, **22** (4), 273–85.

Lucas, R.E. (1992a) Minimum wages and the labour market – recent and contemporary issues in the British hotel industry. *Employee Relations*, **14** (1), 33–47.

Lucas, R.E. (1992b) Minimum wages, the labour market and the hotel industry; some research issues. Paper presented to the Council for Hospitality Management Education (CHME), Hospitality and Tourism Industry Research Conference, Birmingham Polytechnic, April.

Lucas, R.E. (1992c) Employment trends in the hotel and catering industry in the 1980s. Paper presented to the International Association of Hotel Management Schools Conference, Manchester, May.

Lucas, R.E. (1993a) Some age-related issues in restaurant employment in Greater Manchester. Paper presented to the CHME Second Research Conference, the Manchester Metropolitan University, April.

Lucas, R.E. (1993b) Ageism and the UK hospitality industry. *Employee Relations*, **15** (2), 33–41.

Lucas, R.E. (1993c) The Social Charter – opportunity or threat to employment practice in the UK hospitality industry? *International Journal of Hospitality Management*, **12** (1), 89–100.

Lucas, R.E. (1993d) Hospitality industry employment – emerging trends. *International Journal of Contemporary Hospitality Management*, **5** (5), 23–6.

Lucas, R.E. (1994a) Part-time youth pay in catering: issues for the 1990s. Paper presented to the 3rd CHME Research Conference, Napier University, April.

Lucas, R.E. (1994b) Trends in British hotel and catering employment in the 1980s. *Tourism Management*, **15** (2), 145–50.

Lucas, R.E. (1994c) Industrial relations theory, discourse and practice – are hotels and catering merely a case of oversight? Paper presented to the British Universities Industrial Relations Association Conference, Worcester College, Oxford, July.

Lucas, R.E. (1995) Some age-related issues in hotel and catering employment. *The Service Industries Journal*, **5**(2), 234–50.

Lucas, R.E. and Bailey, G. (1993) Youth pay in catering and retailing. *Personnel Review*, **22** (7), 15–29.

Lucas, R.E. and Jeffries, L.P. (1991) The 'demographic timebomb' and how some hospitality employers are responding to the challenge. *International Journal of Hospitality Management*, **10** (4), 323–37.

Lucas, R.E. and Laycock, J. (1991) An interactive personnel function for managing budget hotels. *International Journal of Contemporary Hospitality Management*, **3** (3), 33–6.

Lucas, R.E. and Laycock, J. (1992) Quality and the human resourcing function in Campanile

UK. Paper presented to the Second Annual Conference on Human Resource Management in the Hospitality Industry, 'Quality and Human Resources', London, December.

Lucas, R.E. and Wood, R.C. (eds) (1993) Employment in hotel and catering services. *Employee Relations*, **15** (2), 1–80.

Macaulay, I.R. and Wood, R.C. (1992a) Hotel and catering industry employees' attitudes towards trade unions. *Employee Relations*, **14** (3), 20–8.

Macaulay, I.R. and Wood, R.C. (1992b) *Hard Cheese: a Study of Hotel and Catering Industry Employment in Scotland*. Glasgow: Scottish Low Pay Unit.

Macfarlane, A. (1982a) Trade union growth, the employer and the hotel and restaurant industry: a case study. *Industrial Relations Journal*, **13** (1), 29–43.

Macfarlane, A. (1982b) Trade unionism and the employer in hotels and restaurants. *International Journal of Hospitality Management*, **1** (1), 35–43.

Marchington, M. and Parker, P. (1990) *Changing Patterns of Employee Relations*. Hemel Hempstead: Harvester Wheatsheaf.

Mars, G. (1973) Hotel pilferage: a case study in occupational theft. In M. Warner (ed.), *The Sociology of the Workplace*, pp. 200–10. London: Allen & Unwin.

Mars, G. and Mitchell, P. (1976) *Room for Reform?* Milton Keynes: Open University Press.

Mars, G. and Mitchell, P. (1977) *Catering for the Low Paid: Invisible Earnings. Low Pay Bulletin No. 15*. London: Low Pay Unit.

Mars, G., Bryant, D. and Mitchell, P. (1979) *Manpower Problems in the Hotel and Catering Industry*. Farnborough: Gower.

Millward, N. (1994) *The New Industrial Relations?* London: Policy Studies Institute.

Millward, N. and Stevens, M. (1986) *British Workplace Industrial Relations*. Aldershot: Gower.

Millward, N., Stevens, M., Smart, D. and Hawes, W. R. (1992) *Workplace Industrial Relations in Transition*. Aldershot: Dartmouth Publishing Company.

Mogendorff, D., Lyon, P. and Cowls, C. (1991) Older workers and fast food: human resources practices during a period of demographic change. Paper presented to the Third International Journal of Contemporary Management Conference, October, Bournemouth.

Mullins, L. (1992) *Hospitality Management: a Human Resources Approach*. London: Pitman.

National Economic Development Council (1991) *Developing Managers for Tourism*. London: NEDC Tourism and Leisure Industries Sector Group, NEDO.

Orwell, G. (1933) *Down and Out in Paris and London*. Harmondsworth: Penguin.

Palmer, G. (1968) Inter-union dispute in the Torquay hotel industry. *British Journal of Industrial Relations*, **6** (2), 250.

Price, L. (1993) The limitations of the law in influencing employment practices in UK hotels and restaurants. *Employee Relations*, **15** (2), 16–24.

Price, L. (1994) Poor personnel practice in the hotel and catering industry: does it matter? *Human Resource Management Journal*, **4** (4), 44–62.

Price, L. and Tempest, I. (1992) Achieving quality through good employment practices: a snapshot of current practices and a model for generating change. Paper presented to the Second Annual Conference on Human Resource Management in the Hospitality Industry, 'Quality and Human Resources', London, December.

Rainnie, A. (1989) *Industrial Relations in the Small Firm*. London: Routledge and Kegan Paul.

Ralston, R. (1989) The changing nature of personnel management in the hotel and catering industry. MSc thesis, University of Manchester.

Riley, M. (1980) The role of mobility in the development of skills for the hotel and catering industry. *Hospitality*, March, 52–3.

Riley, M. (1981a) Recruitment, labour turnover, and occupational rigidity: an essential relationship. *Hospitality*, March, 22–5.

Riley, M. (1981b) Labour turnover and recruitment costs. *Hospitality*, September, 27–9.

Riley, M. (1985) Some social and historical perspectives on unionisation in the UK hotel industry. *International Journal of Hospitality Management*, **4** (3), 99–104.

Riley, M. (1991) *Human Resource Management: a Guide to Personnel Practice in the Hotel and Catering Industries*. Oxford: Butterworth Heinemann.

Riley, M. (1993) Back to the future: lessons from free market experience. *Employee Relations*, **15** (2), 8–15.

Roberts, I., Sawbridge, D. and Bamber, G. (1992) Employee relations in small and medium-sized enterprises. In B. Towers (ed.), *A Handbook of Industrial Relations Practice*, 3rd edition, pp. 240–57. London: Kogan Page.

Robinson, O. and Wallace, J. (1983) Employment trends in the hotel and catering industry in Great Britain. *The Service Industries Journal*, **3** (3), 260–78.

Robinson, O. and Wallace, J. (1984) Earnings in the hotel and catering industry in Great Britain. *The Service Industries Journal*, **4** (2), 143–60.

Roper, A. (1988) *The British Hotel Industry 1988*. Bristol: Jordan and Sons Ltd.

Royal Commission on Trade Unions and Employers' Associations 1965–1968, Chairman Lord Donovan (1968) *Report*. London: HMSO.

Shamir, B. (1975) A study of working environment and attitudes to work of employees in a number of British hotels. PhD thesis, University of London.

Simms, J., Hales, C., and Riley, M. (1988) Examination of the concept of internal labour markets in UK hotels. *Tourism Management*, **9** (1), 3–12.

Snow, G. (1981) Industrial relations in hotels. MSc thesis, University of Bath.

Storey, J. and Sisson, K. (1993) *Managing Human Resources and Industrial Relations*. Buckingham: Open University Press.

Thomas, C. and Erlam, A. (1978) *Unequal Portions: a Survey of Pay in the Hotel and Catering Industry*. London: Low Pay Unit.

Thompson, P. (1987) *The Nature of Work: an Introduction to Debates on the Labour Process*, 2nd edition. London: Macmillan.

Walsh, T. (1990) Flexible labour utilization in the private service sector. *Work, Employment and Society*, **4** (4), 517–30.

Whyte, W.F. (1948) *Human Relations in the Restaurant Industry*. New York: McGraw-Hill.

Whyte, W.F. (1949) The social structure of the restaurant. *American Journal of Sociology*, **54**, 302–10.

Wood, R.C. (1992a) *Working in Hotels and Catering*. London: Routledge.

Wood, R.C. (1992b) Hospitality industry labour trends: British and international experience. *Tourism Management*, **13** (3), 297–301.

Wood, R.C. (1994) *Organizational Behaviour for Hospitality Management*. Oxford: Butterworth Heinemann.

Wood, S. and Pedler, M. (1978) On losing their virginity: the story of a strike at the Grosvenor Hotel, Sheffield. *Industrial Relations Journal*, **9** (2), 15–37.

Wynne, J. (1993) Power relationships and empowerment in hotels. *Employee Relations*, **15** (2), 42–50.

TWO

Managing in the Hotel and Catering Industry

> **In a service industry the most important ingredient in the product is people. The quality of our people determines the quality of the service we give to customers and thus our success in the market place ... Doing that means having a caring and efficient personnel function to assist line managers in what is one of their primary responsibilities. (Forte, 1982, p. 32).**

The purposes of this chapter are to:

- identify some of the essential characteristics and features of managing in the hotel and catering industry;
- consider the management organization for employee relations;
- point to the potential significance of these two points for the management of the employment relationship.

2.1 INTRODUCTION

Central to this chapter is the hypothesis that any analysis, evaluation or discussion of managing employee relations in the hotel and catering industry cannot be undertaken without some appreciation of what managing in the industry entails and how it is organized. This begs the question, 'Is managing in the hotel and catering industry different or special in any way?', to which the answer is probably both 'yes' and 'no'. Similarities and differences in 'management' can be discerned among particular industries, as well as in different parts of the hotel and catering industry, and these are shaped by a complex mix of external and internal factors. The aim here is to see if such similarities and differences can be combined to create something that approaches a management typology that will fit the theoretical perspectives on managing employee relations discussed in Chapter 4.

Thus this chapter first considers the broad organizational framework within which managers operate and the nature of managerial work in the hotel and catering industry. Second, from the Workplace Industrial Relations Survey 1990 (WIRS3) and other sources, the nature of generalist and functional managerial work related to employee relations and how that work is organized are discussed. Although key issues of significance for managing employee relations will be drawn from this analysis, an evaluation of how these relate to theoretical perspectives on the management of the employment relationship will be presented in Chapter 4.

Since employee relations is taken to be part of personnel management and HRM, all these terms are used. The inclusion of material under appropriate headings largely reflects the way in which particular commentators have chosen to label their material.

2.2 THE ORGANIZATIONAL FRAMEWORK

It is instructive to preface the broader discussion of managing in the hotel and catering industry with an outline summary of what can be termed the organizational framework. Factors external to the organization also influence how the organization is constituted and managed, some of which are noted here and others of which are considered more specifically in Chapter 4.

Determining what the organizational framework is about is a problem in itself, but from the literature, organization structure, culture, strategy and policy would seem to be the main constituent parts that influence the job of managing generally, and managing employee relations specifically. They are subjects of an ever-growing body of literature that seems to produce more dissent than consent about their meaning, relationship and importance, made more confusing by the frequency with which some commentators interchange terms; for example, strategy and policy.

Organization structure

Although an organization's culture may 'have a more profound effect on people's behaviour than the formal structure of reporting relationships' (Torrington *et al.*, 1989, p. 120), organization structure is seen to be an important framework which defines and guides the ways in which organizations operate. Since 'it is through an organization's structure that a framework for integrating the organization's strategic plans for the allocation of its resources is achieved' (Schaffer, 1984, p. 159), 'success in strategy implementation depends partly on whether a firm's strategy is congruent and complementary with its structure' (Tse and Olsen, 1990, p. 17). Organizational effectiveness results where there is a good fit between strategy and structure (Mintzberg, 1979).

'Structures are differentiated according to the ways in which a division of labour is established and then controlled and coordinated' (Thompson, 1989, p. 51). Torrington *et al.* (1989) identify four main types of organization structure: entrepreneurial, bureaucratic, matrix and differentiated, the main characteristics of which are now summarized.

Most organizations begin with an entrepreneurial structure revolving round a 'powerful' individual, and are associated with risk and initiative. Such a structure is generally most viable in smaller businesses, although there are many examples of entrepreneurial structures at a micro-level in larger organizations. There are few rules and procedures and there is an emphasis on individual power and *ad hoc* decision making. The restaurant chain My Kinda Town, inspired and run by the late Bob Payton, has operated on these principles, as do innumerable other restaurants and hotels. Lennon and Vannocci (1989, p. 25) suggest that a branding strategy among larger organizations 'develops a non-corporate image which has been reinforced by encouraging unit managers to manage with creativity and flair (i.e. to act as entrepreneurs)' and suggest how management flair and innovation might be encouraged in smaller firms.

Bureaucracy is thought to be the most common organizational form. Jobs are grouped according to some common feature, usually by function (personnel, finance, marketing, etc.) and ranked in a hierarchy of responsibility to distribute power between individuals. In theory, it is best suited to organizations where the working routine is predictable, jobs are closely defined and efficiency derives from rational allocation of work; in other words, standardized performance is required. McDonald's is a good example of a commercial catering organization that is highly bureaucratic.

The matrix structure is found where different methods of organizing activities overlay one another. This results in a joint decision making process which cuts across conventional lines of authority by reporting to a project team as well as a functional department. Such organizations face 'problems in balancing workloads between individuals and groups and in managing the conflict that such a diverse group of people generate' (Torrington *et al.*, 1989, p. 129).

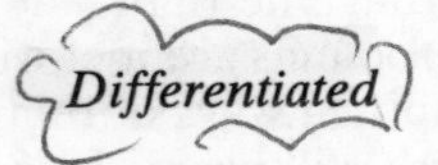

Differentiated structures, which may be any combination of the above-mentioned structures, are related to environmental requirements (developed as *contingency* theory) and the needs of organizational members. Burns and Stalker (1961) found that organizations were more likely to have a bureaucratic or *mechanistic* structure if they operated in relatively stable and predictable markets, whereas uncertain markets characterized by rapid change generated *organic* structures to maximize flexibility and orientation. Lawrence and Lorsch (1967) identified that companies operating in unstable, unpredictable environments had the greatest variations in organizational structure and style. Mintzberg's variant of the matrix form – *adhocracy* – is characterized by an organization which comprises many highly trained specialists with a flexible organic structure.

Studies of organization structure in the industry have tended to focus on hotels within the broad framework of contingency theory, and have generally rejected the bureaucratic model. Shamir (1978) found, contrary to expectation, that although the 'traditional hotel organization displays an essentially bureaucratic "mechanistic" structure, albeit with some deviations (like the high degree of autonomy in some departments) ... it also displays some operational practices of an organic "flexible" nature' (p. 297). The more 'flexible' an hotel was required to become in some respects, the more 'mechanistic' structures were in others. Thus hotels did not fit the contingency models proposed by Burns and Stalker (1961) and Lawrence and Lorsch (1967). Newer hotels facing different situations resolved their conflicting constraints and demands with a different structural solution. Riley and Jauncey (1990) have rejected the notion of bureaucracy in hotels and place them at the adhocracy end of the spectrum. Analoui and Kakabadse (1993) imply that adhocracy is likely to be associated with workforce conflict that is more diversified in nature, a theme that will be considered in later chapters.

Schaffer's (1984) examination of the relationship between strategy, structure and performance in three hotel chains has illustrated 'the role that structural analysis can play in organizational effectiveness' (p. 164). There is some evidence to suggest that organizations are developing less hierarchical structures (Fender and Litteljohn, 1992). Ferguson and Berger (1984) maintain that the organization structure of restaurants contributes to a style of management known as 'being there'; this allows managers to substitute their time and skills for craft and semi-skilled tasks when required (Baum, 1989). For further discussion on organization structure and design see Wood (1994, pp. 32–40).

The significant aspects of structure in terms of the industry would appear to centre on the terms 'flexibility' and '*ad hoc*'. What will also become clear later in the chapter is that one outcome of the way most hotel and catering organizations are structured results in relatively little functional specialism in personnel management or employee relations.

Culture

Most people would concur that 'culture', whether it is corporate or organizational, is about a sense of shared values, beliefs and assumptions. It embodies a sense of common purpose, often expressed as 'the right way to do things'. 'A particular strategy demands that characteristic problems are met with an appropriate reaction, which in turn may be dependent on "habitual ways of thinking and feeling" amongst organisational members' (Ogbonna and Wilkinson, 1988, p. 10). Culture is manifested in the organization through stories, routines and rituals, symbols, organizational structure, control systems and power structures (Johnson and Scholes, 1993).

The creation of a strong culture, with an emphasis on leadership, was found to be an essential quality of 'excellent' companies which overrode the need for detailed policy manuals, detailed procedures or rules (Peters and Waterman, 1982, p. 75). The creation of a strong organizational culture is also perceived as a means to secure the commitment of staff to remain with that organization (Brody, 1993, p. 22). On the other hand, culture can be weak and dysfunctional. Here the focus would be on internal politics at the expense of customers, and on numbers to the detriment of the service and the people involved in service provision, resulting in poor company performance (Peters and Waterman, 1982, p. 76).

Much of the recent emphasis has centred on the need to change organizational cultures to secure the successful implementation of corporate strategies (for example, Schein, 1985). Purcell (1986) maintains that one of the major roles of a centralized personnel department is to represent the organization's value systems and culture and to develop organizational culture. Ogbonna (1992/3) believes that confused meanings of culture – it may about behaviour or values – inhibit the development of a convincing conceptual model that renders managing culture more of a fantasy than a reality.

There has been little systematic analysis of culture in relation to managing hotel and catering organizations from a unit perspective (see, for example, Lundberg and Woods, 1990), although it is possible to identify the hints of dysfunctional cultural behaviour at unit level elsewhere (for example, Dann and Hornsey, 1986; Analoui and Kakabadse, 1989). Marshall's (1986) ethnographic study of workplace culture in a restaurant offers an insight into an alternative model of workplace culture. The highly entrepreneurial manager, Dixie, managed through a combination of authoritarianism, paternalism, erratic beneficence, autocracy and insulting behaviour. Yet this idiosyncratic management style was associated with high levels of job satisfaction among staff, in spite of their subjection to long hours, hard work and low pay (Marshall, 1986, pp. 38-9). Although some of these 'negative' rewards were compensated for by an informal workplace economy (fiddles, etc.; see Chapter 5), Marshall believed that the key to job satisfaction lay in the fact that staff were engaged in activities that they would have readily accepted as leisure, such as eating the same food as customers and playing dominoes during quiet periods.

Dann (1991) sees corporate culture as a manifestation of organizational and operational strategies which is ultimately reflected in patterns of managerial work. An important part of cultural norms and philosophy is learnt through working experience or 'socialization'. Put another way, managers learn what the culture is in order to work with it (Handy, 1985).

This 'culture of management' can also be referred to as management style. Thus the culture of hotel management emphasizes 'being there' as the most important aspect, while the achievement of results is secondary (Guerrier and Lockwood, 1989b, p. 86). The ubiquitous multi-skilled manager is thus able to substitute for operatives who may be missing or who are not fully able to perform appropriate tasks (Guerrier and Lockwood, 1989a). Drawing on the work of Ouchi (1981), Guerrier (1986, pp. 232-3) notes that hotels operating with an informal and open style of management conform to an organization culture described as type Z. Since the way people work together is essential to the success of the company, there is a tendency for management to be homogeneous. The dominant culture is likely to be Anglo-Saxon male, with a racist and sexist bias (see also Hicks, 1990; CRE, 1991; Purcell, 1993).

There is, however, a growing body of literature that addresses the management of culture from an international perspective. This is most evident in relation to fast-food or hotel operations that are expanding worldwide (see, for example, Love, 1987; Pizam, 1993; Jones *et al.*, 1994). Jones *et al* (1994) foresee that large international hotel organizations will resemble 'cultural mosaics' rather than 'cultural melting pots' and that trends to increasing standardization and 'developments in international human resource management may in the longer term lead to the emergence of a common philosophy and management style in this sector' (p. x).

The significance of culture in the hotel and catering industry, manifested in managerial work, particularly 'being there' and 'conservatism', will become clearer later on. The area would certainly seem to be a topic justifying much more study (see also Wood, 1994, pp. 147-58).

Strategy

Corporate strategy and strategic management have become increasingly used 'buzz-words' in managerial literature since the 1970s. Since the 1980s, businesses have been exhorted to develop a strategic approach to meet the changes and challenges of the times. Corporate strategy is essentially about an organization's response to its environment over time, its main objectives and its means of accomplishing them that stem from the organization's long-term goals. Strategy is a stratified concept and may operate at different levels in the organization (Miller, 1991), and is not the exclusive preserve of the large organizations.

To some, strategy is a deliberate, formal process that is concerned with the rational exercise of managerial choice (Porter, 1980). To others, strategy is an abstract, emergent and evolutionary process whereby managers seek to manage a changing complexity of considerations. Here strategy is detectable as a 'pattern' in behaviour and activity (Mintzberg and Waters, 1985). However conceptualized, strategic management is a complex subject that may be more rhetorical than real; suffice it to say that the reader must look elsewhere for more detailed analysis and discussion (see, for example, Johnson and Scholes, 1993).

There is some doubt about the extent to which hotel and catering organizations 'view their business in explicitly strategic ways, although the terminology has become increasingly part of management vocabulary' (Webster and Hudson, 1991). Nevertheless, there are undoubtedly examples of organizations that utilize some form of strategic approach (Forte and Teare, 1990, 1991; Fender and Litteljohn, 1992), although Tse and Olsen (1990) question the applicability of Porter's generic strategies to service industries. Fast-food operators such as McDonald's could not have achieved worldwide expansion and maintained consistent levels of 'quality, service, cleanliness and value' without a concerted approach (Peters and Waterman, 1982). Dann (1991, p. 24) believes that 'strategy in the hotel and catering industry is frequently centred on the relative degree of centralisation or decentralisation which in turn influences managerial divisions of labour and methods by which the company chooses to measure the effectiveness of its managers.'

In other words, although strategy is generally deemed to be a 'good thing', there is little evidence to suggest that an explicitly strategic approach is used to any extent in the industry. If this is the case, then the industry may be more *non-strategic* than strategic and it would seem that a strategic approach to managing employee relations, expressed through an HRM strategy, is unlikely to be a prominent feature. This dimension will be developed in Chapter 4.

Focusing on a 'rational' strategic approach in the hotel and catering industry may well be inappropriate for a variety of reasons, and it could be the case that Mintzberg's '*emergent*' strategic model has more to offer in terms of developing an explanation for, and an understanding of, what really goes on in hotel and catering organizations. Certainly Mintzberg's work has been influential in contributing to an understanding of organizational behaviour and what managers actually do in the hotel and catering industry, areas which are discussed later in the chapter.

What has also clearly emerged from Mintzberg's studies of managerial work, power in organizations and strategy, is that there are gaps between the theory and practice of 'management'. Thus strategic management cannot be a rational process if its successful implementation is frustrated by the contextual factors which shape the way in which managers 'manage'. These themes are now developed further.

Policy

Although the business world does not always clearly distinguish policy from strategy, policy can be said to provide the general guidelines for managerial decision making that moves the organization in the direction of its objectives. Thus organizations will formulate an overall business policy from which other functional policies will be developed; for example, financial policy, marketing policy and personnel or human resources policy. Emphasis on the need to develop effective personnel and employment policies to improve labour problems in the hotel and catering industry has been a recurrent theme for three decades (for example, CIR, 1971; HCEDC, 1975, 1977; HCTC, 1990; CRE, 1991). But even where policies exist, the practical means to secure successful implementation of such policies have been largely overlooked (the same is also true of strategy: Miller, 1991).

The problem of matching the 'theory' of policy intentions to the 'practice' of management actions has been examined by others, of whom Brewster and Richbell (1982) offer a persuasive and interesting analysis which focuses on personnel policy; this helps to aid understanding of the status of people issues in the hotel and catering industry. Their research led them to define policy as 'A set of proposals and ideas that act as a reference point for managers in their dealings with employees' (Brewster and Richbell, 1982, p. 35). They then differentiate between two forms of policy to explain how and why sound corporate policy implementation is frustrated.

'Espoused' policy is the corporate approach towards employees, manifested in formal written mediums, and the verbally stated intentions of top managers. 'Operational' policy is about how managers deal with 'espoused' policy and helps to explain why 'espoused' policies are not fully implemented. First, working under pressure, managers have to deal with a plethora of policy demands, some of which may conflict or simply be too time consuming to deal with. Their intuitive response to what to do is based on an interpretation of how they perceive top management's values, or put simply, what top managers would most want to happen.

In reality top managements afford low priority to personnel matters. For example, the absence of a personnel specialist at board level suggests little or no input of 'people' policies into the corporate plan, which serves to reinforce the impression in managers' minds that personnel policies are less important than those of other functions. Additionally, the more 'intangible' nature of personnel problems like improving communications leaves them prone to be ignored when the more visibly tangible problem of machine breakdown is perceived as requiring immediate attention. The upshot of this cautionary tale is that personnel practitioners 'must devote more attention to strategies which will secure implementation of their policies' (Brewster and Richbell, 1982, p. 37) and ensure there is clear commitment from top management to personnel issues.

The notion of a *policy gap* between the organization's stated human resource policy intentions and how these are implemented at an operational level is a recurring theme in hotel and catering literature. This has been most sharply observed by Croney (1988b). Although human resource policy requirements were specified by corporate management, there was little head office guidance, formality or control of operational managers, who were left with considerable freedom to do as they pleased in a largely *ad hoc* way. Ralston (1989) considered that such line management behaviour was tantamount to that of Legge's (1978) 'conformist innovator'. A similar mismatch of intentions and deeds has been noted elsewhere; thus 'the hotel and catering industry has succeeded in talking about training and the need for training whilst pursuing, at many levels, development and employment policies designed to eliminate the need for motivated and accomplished employees' (Wood, 1992, pp. 161-2).

Brewster and Richbell's analysis implies that policy makers need to develop a greater understanding of what managers actually do and, given that an overriding concern of operational managers relates to 'keeping the boss' happy, how their effectiveness and performance are measured. These themes are developed shortly in relation to the work of line managers (see Hales and Nightingale, 1986; Hales, 1987).

Summary propositions

- The organizational framework within which most hotel and catering managers operate is most likely to be flexible, *ad hoc*, conservative, non-strategic and reactive.
- There is likely to be a policy gap that inhibits the successful implementation of espoused policies, if they exist, at operational level.

Discussion points

- Identify and describe the different types of organization structure you have come across, and evaluate their effectiveness.
- Do managers think and act strategically? If they do, how is this manifested? If they do not, what factors inhibit such behaviours and actions?
- Identify the positive and negative characteristics and features of organizational and workplace culture.
- Have you come across instances where organizational personnel policy is not being put into practice by line managers at workplace level? Can you explain why this is happening and suggest steps to remedy the situation?

2.3 MANAGERIAL WORK

A typology of fragmentation?

Although the notion of a theory of hospitality management (Nailon, 1982) remains poorly developed (Wood, 1983), managing from a line management perspective in the industry has been a topic of considerable interest to a number of researchers (Nailon, 1968; Ley, 1978, 1980; Arnaldo, 1981; Ferguson and Berger, 1984; Koureas, 1985; Hales and Nightingale, 1986; Hales, 1987; Baum, 1989; Shortt, 1989; Worsfold, 1989; Dann, 1990a, b, 1991). Unfortunately the balance of the evidence is not weighted evenly by sector or job type, and most of the studies make little or no reference to employee relations. More studies are needed that include other occupational groups in a wider variety of sectors and contexts (Dann, 1991; Wood, 1992) and, particularly, how employee relations is managed by line managers. Nevertheless, this section considers some of the main characteristics that have been established that may be of potential significance in understanding the management of employee relations at line management level.

Managerial work exists within an organizational context of cultural norms and philosophy (Dann, 1991). In a useful review of managerial work up to 1987, Dann (1990b) notes the influence of Mintzberg (1973), Machin *et al.*, (1981), Machin (1982) and Stewart (1970, 1976, 1982) on studies of managerial work in the hotel and catering industry. Mintzberg, concerned primarily with similarities in managerial work, was among the first to assemble data to form a picture of the functions of management. He classified ten managerial roles under three headings: *interpersonal roles*, figurehead, liaison, leader; *informational roles*, monitor, disseminator, spokesman; and *decisional roles*, entrepreneur, disturbance handler, resource allocator and negotiator. By contrast, although sharing Mintzberg's aim of determining what managers actually do, Stewart concentrated on differences in managerial work, initially in terms of demands, constraints and choices. She concluded latterly that choice was the most important factor, and developed the notion that managers are given a substantial degree of choice in how they operate, the functions on which they concentrate and how they spend their time. In exercising choice, they respond intuitively rather than consciously. Machin's expectations approach is more concerned with demands placed upon the manager, using the managers' role set to assess their expectations of the manager and vice versa. This approach was the first to place the work of the manager in some sort of context in an attempt to explain what managers are expected to do.

Dann has concluded that hotel and catering studies present a fragmented picture, but there is an indication that there are sectoral differences in terms of both the job demands made upon the job holder and the way in which the job is actually carried out. Thus, in terms of the overall characteristics of managerial work in the hotel and catering industry:

> The job is likely to be reactive, fragmented, subject to many interruptions, involve large numbers of contacts, be highly concerned with information gathering and dissemination, and be variable in terms of content and activities. What is important about these characteristics is that they have traditionally been connected with the 'bad' manager; the 'good' manager being the one who sat in his office and planned, organized, commanded, controlled, delegated, and made decisions in a calm proactive manner. (Dann, 1990b, pp. 328–9).

Although Dann does not use the term adhocism (a special type of crisis management or a 'for this purpose' response to each new situation encountered) popularized by Mars and Mitchell (1976), it is clearly implied. Adhocism is justified because managers 'must be flexible enough to coordinate the available resources so that they can cope with the varying demands of their customers' (Mars *et al.*, 1979, p. 3). Ralston (1989) argues that the 'main form of this adaptation appears to be in the human resource' (p. 171). (Atkinson and Meager (1994) also note that *ad hoc* or informal methods are used to determine labour requirements and their implementation in small firms.) We can now begin to discern the influence of customers on the management of the employment relationship, a point discussed more fully in Chapter 4.

Amateur management

More critical observations of management and managing have been noted by others. Noting that fewer than 10 per cent of managers held relevant qualifications, NEDC concluded that 'a spirit of amateur management characterises the small businesses which dominate much of the industry' (NEDC, 1991, p. 2). Even where formal course training exists (generally separated from general business studies), it encompasses two distinctive elements – an industrial placement which provides a period of pre-entry socialization, and a high degree of technical competence in order to exercise control over other powerful work groups such as chefs – which create 'notoriously insular' managers (Wood, 1992, pp. 79–83). These aspects of managerial training have been called into question by others (Baum, 1989).

Macfarlane (1982) suggests that because the industry insists on managers having practical experience before entering a managerial position, managers are likely to be socialized into acceptance of the subculture of illicit practices (see also Analoui and Kakabadse, 1989). As former operatives, managers will have been involved in fiddles, and accept this as the natural order of things, a theme that is developed more fully in Chapters 4, 5 and 6. Analoui and Kakabadse (1989) point to the need for managers to have more in-depth and relevant interpersonal skills and knowledge to manage people and the related aspects of their work relationships.

Others have identified other characteristics or idiosyncrasies of management and managing in the hotel and catering industry. Hotel managers come to their careers via a range of routes, reach senior positions at a relatively young age as a result of substantial mobility and are given a significant degree of latitude in running their units (Wood, 1992, pp. 79–83). They engage in a much larger number of activities than their counterparts in other industries (Nailon, 1968) and are dominated by operational demands (Baum, 1989). Using Mintzberg's ten managerial role categories, Ley (1980) found more emphasis on entrepreneurship than leadership, while Arnaldo (1981) found that managers set great store by leadership that is more people oriented. Such managers suggested that people-related criteria were considerably more important than operational-related criteria for measuring successful job performance.

Active and authoritarian management

Managers personify active management by 'being there' (Guerrier and Lockwood, 1989b, p. 85), and the ability of managers to get involved in basic operative work is likely to invite the respect of colleagues and subordinates (Guerrier, 1987). Lee-Ross (1993a, b) challenges the notion that a 'hands on, being there' style of hotel management is ineffective, because it may create a culture which engenders job satisfaction among hotel workers. Conversely, Baum (1989) and Guerrier and Lockwood (1989a) have found evidence to suggest that 'being there' styles of management are not cost effective.

'"Being there" styles of management give rise to procedures characterised by informality of communication between management and operatives and a paternalistic and authoritarian (or at least directive) approach to staff, although hotel managements understandably tend to dissent from these views' (Wood, 1992, p. 84). Gabriel (1988) has identified examples of 'machomanagement', *laissez-faire* management and the deployment of favouritism and arbitrariness as a method of control. Snow (1981) observed that the management style of hotel managers had the effect of 'divide and rule'. Analoui and Kakabadse (1993, p. 57) note that the presence of strong, authoritarian management – a 'telling' style of leadership – is more likely to lead to workers taking covert forms of industrial action, such as pilfering, rather than more overt forms of industrial action (see Chapter 9). Yet Marshall (1986) has shown that in spite of the active presence of authoritarian and active management, staff may enjoy high levels of job satisfaction if they do not perceive their jobs as 'real' work.

Guerrier and Lockwood (1989b) found that staff saw management as being rather critical, autocratic and controlling, and although managers saw the development and care of staff as a central part of their role, little was done to convert this into direct action. On the other hand, the 'hands on' hotel managers in Campanile spent between 40 and 50 per cent of their time on personnel matters, thought to be an important part of the job, although the definitions of what constituted 'personnel work' varied markedly (Lucas and Laycock, 1991a, b).

A case study in role demands and role performance

The problem with many of the management studies is that their outcomes are a function of their methodologies, which have been directed primarily at management training, development, performance and appraisal. Nevertheless, the methodology initiated by Hales and Nightingale (1986), and subsequently refined by Hales (1987), has produced probably the most instructive data on employee relations as part of managerial work in terms of managerial role demands and role performance. This also provides a useful developmental point to Brewster and Richbell's (1982) analysis of how managerial work is shaped by perceptions about superiors' demands, because it shows how managerial work is also shaped by managers responding to pressures from subordinate staff.

Hales and Nightingale (1986) identified five common core elements in unit managers' jobs, one of which was the field of 'training, motivating, recruiting and disciplining staff'. Superiors' expectations focused on tasks and achievements, rather than on activities or what is actually being done. Thus as jobs were constituted and practised, managers were subjected to 'a mass of competing, often contradictory or conflicting, demands and expectations from a multiplicity of sources, both inside and outside the managers' organization' (Hales and Nightingale, 1986, p. 10).

In one example, a superior manager expected a unit manager to do 25 separate things, only four of which were directly related to operational 'people' activities and two of which were administrative (for example, to pay wages). Of the operational tasks, only the personnel oriented tasks of selection and training were 'musts', whereas the more employee relations oriented activities of producing staff rosters, ensuring adequate staffing levels, communicating with staff and holding staff meetings were 'shoulds'.

In a more closely focused study of two unit management jobs, a family restaurant chain manager and a hospital domestic services manager, on the face of it similar in level and experience required, Hales (1987) identified that staff administration was a key area in the role demands of both managers, although the pattern of demands varied. He hypothesized that if both worked in a way consistent with satisfying those demands, each would spend a considerable amount of time on staffing matters, although they would differ in the emphasis given to routine and *ad hoc* matters. This was borne out. The restaurant manager devoted more time to routine 'clerical' administrative matters, such as recruitment, to maintain a stable operating system. By contrast, the domestic services manager managed within an inherently unstable system, and spent more time dealing with long-term employee relations problems, particularly individual and collective grievances and disputes over redundancy from subordinates. 'Of the real omissions in the two managers' performance, an almost total unconcern with staff training was the most notable' (Hales, 1987, p. 32), yet this was a clear role demand from superiors, seemingly neglected by the managers to deal with day-to-day requests for help from subordinates. Thus the way in which an organization constitutes a job will be a prime factor in how it is performed.

Summary propositions

- Sectoral differences will influence the nature of management and the way it is performed.
- Managing in the hotel and catering industry is generally a fragmented process.
- Management can be described as amateur, unsophisticated and *ad hoc*.
- A management style of 'being there' contributes to a degree of informality that is associated with authoritarianism and autocracy.
- How and what managers 'manage' are shaped by a variety of complex factors that are both external and internal to the organization.

Discussion points

- Identify the most and least effective characteristics of management style that you have experienced.
- Is managerial work in hotels and catering inevitably fragmented or can it be made more cohesive?
- Prepare a managerial job specification for a hotel and catering unit of your choice.

2.4 MANAGEMENT ORGANIZATION FOR EMPLOYEE RELATIONS

Issues of generalism and specialization

The management of employee relations, part of personnel management and HRM, is both a specialist and generalist managerial function. It is generally agreed that line managers at all levels in the organization assume prime responsibility for managing employee relations as part of managerial work (Forte, 1982; Guest, 1989), although we have noted that the industry studies of managerial work have not been hugely informative on this point.

Specialist 'people' managers, whose job title includes one or more terms such as human resources, personnel or employee relations, are generally found in medium- to large-sized organizations that are larger in terms of workforce size. Since the majority of hotel and catering organizations are small (by size of unit (bedroom size or numbers of restaurant covers) and number of persons employed) and independently owned, most probably function without professional human resource support facilities at unit, divisional and corporate level, although a line manager may be designated as responsible for personnel matters.

This section begins with a summary of the management organization for employee relations found in the WIRS3, and continues with a discussion of the state and effectiveness of the personnel function from other hotel and catering industry studies.

How typical is the WIRS3?

The WIRS3 not only provides the most comprehensive and recent analysis of employee relations covering the hotel and catering industry, but also enables specific industry practice to be compared with national and service sector practice for the first time. Although, as has already been noted, the exclusion of workplaces employing fewer than 25 employees means that the data cannot be said to be wholly representative of the industry, in a number of respects the distribution of workplace size among the responding population is broadly similar to that of its three comparators: all industries and services, all service industries and private sector services. The main differences are that hotels and catering has slightly more workplaces of 25 to 49 employees and considerably fewer workplaces employing 200 or more employees (see Table A1.1, Appendix 1).

A more significant difference between hotels and catering and its comparators may lie in the formal status of the workplaces. In hotels and catering a large majority (over three-quarters) belonged to limited companies or public limited companies (PLCs) and a significant minority (all employing 25 to 49 employees) belonged to partnerships or sole proprietorships. Although workplace status was more diverse among all industries and services and all service industries, the pattern in private sector services was broadly similar to that of hotels and catering (see Table A1.2, Appendix 1). The high proportion of companies with more than one establishment would seem to suggest that the responding workplaces might be more 'formal' than is typical of the whole hotel and catering industry. Workplaces employing 25 to 49 employees were more likely to be single independent establishments (see Table A1.3, Appendix 1).

Hotels and catering also employs more women, part-timers and manual workers, many of whom are unskilled, than its comparators (see Tables A1.4 to A1.9, Appendix 1). These issues are developed in Chapter 3.

Specialists and non-specialists

The WIRS3 data reported here are based on interviews with the most senior person at the workplace with responsibility for industrial relations, employee relations or personnel matters, and are based on establishment level responses only. This is because the data about the respondent are being used to characterize the personnel function at the establishment and the head office interview respondents are irrelevant to this. Over three-quarters of the hotel and catering respondents were titled either unit manager (48 per cent) or general manager (29 per cent). Job titles were more varied in all industries and services, all service industries and private sector services. Fewer than one-fifth of hotel and catering respondents (18 per cent) were titled personnel manager (no one had the job title of industrial/labour/employee relations manager), which was marginally higher than among all comparative sectors. Such personnel managers were most likely to be found in hotels.

Just above one-fifth of hotel and catering respondents (21 per cent) were 'strong' specialists: their job title reflected responsibility for personnel, human or manpower resources, or industrial, employee or staff relations. This level of strong specialism was slightly greater in hotels and catering than among all industries and services, all service industries and private sector services. Strong specialists were most likely to be found in workplaces employing 100 employees and above. Additionally there were considerably more 'weak' specialists (those who did not have personnel, etc. in their job titles but who spent 25 per cent or more of their time on personnel: 49 per cent) in hotels and catering than elsewhere, and these were also most likely to be found in workplaces employing fewer than 100 employees. When added together, 70 per cent of establishment level respondents in hotels and catering were strong or weak specialists, which is significantly higher than across all other sectors (at or around 50 per cent). Higher proportions of hotel and catering managers also had one or more staff to assist them in personnel matters than was the case among all industries and services, all service industries and private sector services.

In terms of qualifications and experience, significantly more hotel and catering managers (38 per cent) claimed to hold formal qualifications in personnel management than elsewhere (around 20 per cent in all industries and services, all service industries and private sector services). Given the similar proportions of personnel managers in all sectors noted above, it seems improbable that such a high proportion is IPD qualified and therefore the qualification is more likely to refer to membership of the HCIMA, the professional body of hospitality managers, whose syllabus includes a sizeable input of human resource management. (Price (1994, p. 523) also found a higher than might be expected level of personnel-related qualifications among hotel and catering respondents, although this was low among proprietors/partnerships.) The average number of years of experience in personnel management was lower among hotel and catering respondents (6.9 years) than among those in all other sectors (typically 10 years).

Personnel responsibilities: theory and practice

As a general rule, a greater proportion of hotel and catering respondents stated that they had responsibilities across the full range of personnel work than was the case in all industries and services, all service industries and private sector services (see Table 2.1), which is not surprising given the level of specialism identified above. In terms of work responsibilities, the most frequently mentioned were recruitment and selection (95 per cent), pay or conditions of employment (92 per cent) and procedures: grievance, discipline or disputes (89 per cent). The least mentioned activity for which managers had responsibility was systems of payment (68 per cent). Different patterns of responsibility are discernable in different sizes of establishment.

However, the level of responsibility for particular activities and the amount of time spent on them do not necessarily correlate. When hotel and catering managers were

Table 2.1 *Respondents' work responsibilities.*

			PSS (by no. of employees)				HCI (by no. of employees)			
	AIS	SI	All	25–49	50–99	100+	All	25–49	50–99	100+
Unweighted base[a]	1697	1013	535	139	120	276	52	19	16	17
Weighted base[a]	1644	1147	715	424	176	115	91	55	24	12
Pay or conditions of employment (%)	72	65	83	84	78	90	92	93	94	85
Procedures: grievance, discipline or disputes (%)	88	87	89	89	87	92	89	94	74	95
Recruitment and selection (%)	91	94	94	95	94	91	95	96	100	82
Training (%)	80	83	78	77	80	77	81	75	100	73
Systems of payment (%)	60	53	71	73	64	74	68	75	51	68
Job evaluation or grading (%)	68	68	67	68	65	69	70	65	88	58
Disciplinary cases (%)	87	88	86	85	83	92	85	87	84	77
Staffing or manpower planning (%)	84	84	85	85	85	85	88	88	94	78

[a]Establishment-level respondents only.

AIS, all industries and services; SI, service industries; PSS, private sector services; HCI, hotel and catering industry.

Table does not include percentage figures for missing cases.

Source: WIRS3 (1990).

asked to specify the two most time-consuming activities, their responses showed that training, recruitment and selection, and staffing and manpower planning predominated, a pattern that was also observed elsewhere, although hotel and catering managers spent more time on training (see Table 2.2). The main difference between hotel and catering managers and their comparators is that they spent more time on procedures: grievance, discipline or disputes, which points to less satisfaction in the employment relationship, and less time on pay and conditions of employment, which may indicate a greater element of *ad hoc* management.

When added together, managers in the hotel and catering industry spent rather more time on personnel or employee relations matters than their counterparts in all industries and services, all service industries and private sector services, shown in Table 2.3. It is thus interesting to note the while managers in all industries and services, all service industries and private sector services claimed a lower level of responsibility for personnel work as part of formal work activities than hotel and catering managers, they also spent less time practising it than hotel and catering managers.

Table 2.2 *Two most time-consuming activities.*

			PSS (by no. of employees)				HCI (by no. of employees)			
	AIS	SI	All	25–49	50–99	100+	All	25–49	50–99	100+
Unweighted base[a]	1690	1011	534	139	120	275	52	19	16	17
Weighted base[a]	1639	1144	714	424	176	114	91	55	24	12
Pay or conditions of employment (%)	27	22	29	29	25	36	10	15	–	9
Procedures: grievance, discipline or disputes (%)	8	8	8	9	5	10	16	22	–	21
Recruitment and selection (%)	39	42	43	41	43	50	44	44	39	51
Training (%)	38	42	39	42	40	26	53	56	51	40
Systems of payment (%)	10	7	10	11	10	10	9	13	–	5
Job evaluation or grading (%)	12	12	8	7	13	5	11	–	33	15
Disciplinary cases (%)	4	4	4	4	2	7	4	6	–	5
Staffing or manpower planning (%)	41	43	39	42	36	30	40	37	54	24

[a]Establishment-level respondents only.

Table does not include percentage figures for missing cases.

Source: WIRS3 (1990).

The WIRS series (1980, 1984, 1990) have shown that professional specialist personnel expertise is far from being the norm at workplace level anywhere in the economy and that, in 1990, hotels and catering is little different from elsewhere in terms of 'strong' specialism in workplaces employing 25 employees or more. In terms of organization, employee relations is primarily a line management activity which is central to managing in hotels and catering. The manager may or may not be supported by the presence of others, particularly functional specialists at varying levels of seniority in the organization. Here the existence of more partnerships, self-proprietorships and single independent establishments may mean that more hotel and catering units lack the back-up facility of a head or regional office specialist. Yet the fact that more hotel and catering managers claimed to hold formal qualifications in personnel management ought to compensate in some way for the absence of personnel support from elsewhere.

As the WIRS3 does not attempt directly to evaluate the effectiveness of the management organization for employee relations, or consider some of the wider managerial issues beyond the workplace, these points are developed below.

Table 2.3 *Proportion of time spent on personnel and employee relations matters (percentages).*

	AIS	SI	PSS (by no. of employees) All	25–49	50–99	100+	HCI (by no. of employees) All	25–49	50–99	100+
Unweighted base[a]	1697	1013	535	139	120	276	52	19	16	17
Weighted base[a]	1644	1147	715	424	176	115	91	55	24	12
25% or more[b]	46	53	48	39	53	73	66	65	65	71
Over 50%[b]	24	26	24	18	21	50	27	21	28	51
Over 90%[b]	10	10	11	7	8	30	8	6	6	23

[a]Establishment-level respondents only.
[b]These categories are not mutually exclusive.
Source: WIRS3 (1990).

Summary propositions

- Employee relations is primarily a line management function that is the responsibility of unit managers and general managers in the hotel and catering industry. The industry contains a higher proportion of 'strong' and 'weak' specialist managers than elsewhere.
- 'Strong' specialist professional personnel expertise is a minority practice at workplace level in the hotel and catering industry.
- Hotel and catering managers have more formal responsibilities for employee relations than managers across the economy, and spend more time managing employee relations than other managers.

Discussion points

- How can the personnel function be organized in hotels and catering in order to be most effective?
- How far do the WIRS3 findings on the management of personnel/employee relations match your own experiences?

2.5 THE PERSONNEL FUNCTION: STATE AND EFFECTIVENESS

Although functional specialization in employee relations in the hotel and catering industry is by no means the norm, it cannot be ignored. Where personnel specialists exist, they are deemed to contribute to improving employee relations by 'assisting' line managers in managing issues in the employment relationship through the provision of specialist expertise. In practice, the roles of line and functional managers, and the relationship between them, are rather more complex than has just been implied (see, for example, Legge, 1978; Farnham and Pimlott, 1990; Torrington and Hall, 1991).

A lack of specialist personnel expertise, particularly at very senior levels in the organization, has been thought to inhibit policy development (for example, CIR, 1971). By all accounts, this situation still persists (Kelliher and Johnson, 1987; Ralston, 1989; Price, 1994). More particularly, there is considerable doubt about the level and effectiveness of specialist personnel expertise at other levels in the hotel and catering industry. In short, it may be presumed that the presence of specialists who are ineffectual, or the absence of persons with specialist knowledge (for example, employment law and human resource planning techniques) and skills (including interviewing, training and negotiating) will undermine managerial effectiveness and disadvantage many businesses in the industry.

Working models

By identifying generic models of specialist expertise, we can attempt to evaluate the state and effectiveness of specialist personnel management in the hotel and catering industry. Here, Tyson and Fell's (1986) three 'models' of personnel management, which have been extended to four by Monks (1992/3), offer a useful framework.

Tyson and Fell have proposed three forms or levels of personnel management that are both discrete and contiguous to the extent that all three could be found in the same organization. In its most basic form the 'clerk of the works' or *administrative/support model* deals with basic administration and record keeping (for example, personnel records, absence and sickness statistics) and routine matters of recruitment and training. 'All authority for action is vested in line managers' (Tyson and Fell, 1986, p. 24). The 'contracts manager' or *systems reactive model*, associated with highly unionized organizations, has a heavy industrial relations emphasis on resolving disputes through formal procedural means. Authority is vested in senior line managers, for whom the personnel specialist is effectively the agent to keep the existing system operational. The 'architect' or *business manager model* involves a personnel presence at the highest level which inputs strategic human resource policies into the corporate plan. There is an emphasis on rational decision-making and personnel managers act as partners with senior line managers. The organization may or may not be unionized but managers hold strong values about managing people.

Monks's (1992/3) two 'traditional' models (*traditional/administrative* and traditional/industrial relations) are broadly synonymous with Tyson and Fell's *administrative/support and systems/reactive models. However, she identifies two types of 'innovative' model. Although the innovative/professional* model utilizes complex, sophisticated personnel systems, it does not operate the strategic, integrated approach of the *innovative/sophisticated* or business manager models. A similar differentiation between HRM that is strategic and HRM that is non-strategic has been identified by Miller (1989, 1991).

Over-optimism?

In order to evaluate the current state and effectiveness of the personnel function in the hotel and catering industry, a brief background discussion is instructive. Undoubtedly some organizations in the industry did begin to develop more professional personnel functions in the 1970s, although we know little about their organization or effectiveness; this also remains largely true of the 1980s and the early 1990s. Boella (1986) saw the 1980s as

> a period of consolidation and possibly maturation. There are now hundreds of managers with major or total responsibility for personnel matters. Their responsibilities go well beyond those of the few personnel managers of twenty years ago. Many more are now recruited from outside the industry, as senior managers recognise the value of employing professional personnel managers rather than making one of the new junior managers responsible for personnel. (Boella, 1986, pp. 34–5).

Boella's suggestions are not based on empirical findings but on a literature review, and represent, at best, a highly over-optimistic and simplistic view (Kelliher and Johnson, 1987).

More importantly, Boella's claims do not square with the more authoritative findings of the earlier WIRS (1980, 1984), which noted 'the persisting rarity of professional personnel management in Britain' (Millward and Stevens, 1986). A major research project on the changing nature of personnel management found that 'much of the distinctive contribution of the personnel function has been lost ... which may have undermined the value and potential contribution that personnel specialists offer organizations' (Mackay, 1987, p. 3).

Simplistic, reactive and poor

Specific empirical studies of the personnel function in the hotel and catering industry are few in number, but are nevertheless revealing, even though mention of employee relations is either missing or limited. This is highly significant given that Mackay and Torrington's study (1986) of the changing nature of personnel management in 350 organizations found that employee relations was the most important activity in terms of managerial discretion, time spent and importance to organizational objectives.

In the Ealing and Leeds surveys (Kelliher and Johnson, 1987), the personnel function at unit level relied very heavily on personnel manuals issued by head office; these were adhered to vigorously, leading to a form of personnel management that was highly simplistic and reactive rather than systematic and strategic. The limited impact of personnel management was attributed to the narrow way in which the function was defined: the main activities were recruitment and training. This, and the way labour was managed in four of the largest hotel groups (Croney, 1988a) would seem to match most closely with Tyson and Fell's (1986) 'clerk of the works' model of personnel management.

A similarly simplistic personnel function 'of almost unbelievable unsophistication', with a high degree of emphasis on recruitment and selection, has been observed by Ralston (1989). Employee relations was not considered to be particularly central to the personnel function and a disproportionate amount of time was spent on employee relations to what was seemingly justified by its perceived importance to organizational objectives. This dissonance would appear to be explained by significant differences between the amount of time that respondents spent on employee relations. Three National Health Service personnel spent up to 50 per cent of their time on employee relations, whereas 25 per cent of respondents spent no time on it at all.

Although 75 per cent of the organizations responding to Ralston's survey had a director or someone at the highest level with special responsibility for personnel and/or employee relations, fewer than half these 'designated' specialists had sole responsibility for personnel matters. Nearly half the respondents to Ralston's survey were professional personnel specialists, and the vast majority of the remainder were managers (typically general managers) with designated responsibility for personnel. She argues that this is a poor state that has resulted from the way in which the personnel function has been defined and operationalized; it has had little focus or direction. Ambiguity and ambivalence can only be avoided if the personnel function exists within a coherent framework of objectives, and 'current personnel practice at unit level is a recipe for disaster' (Ralston, 1989, p. ii) for the 1990s.

In a case study of the personnel function at corporate and unit level in a foreign owned budget hotel company – Campanile – a high degree of emphasis on recruitment and training-related activities was observed in the corporate function, which was a clear outcome of the company's strategy and culture, based on developing all unit managers from within the organization (Lucas and Laycock 1991a, b). Employee relations was not an activity of much importance: it took up only 5 per cent of the head office specialist's time and this activity was given a low ranking in terms of being critical to business success (Lucas and Laycock, 1991b, p. 34).

Price (1994) provides the most recent review of personnel practice from literature and a sizeable sample of hotels and restaurants (241 establishments) operating at the 'quality' end of the market. From this it might be posited that a relatively good state of personnel practice could be expected to be found. Yet Price found that 'the broad picture of personnel practice in the commercial sector of the UK hotel and catering industry and the more focused picture within hotels and restaurants are poor' (Price, 1994, p. 56). Interestingly, although present only in small numbers, the foreign-owned establishments had considerably better developed personnel management than the non-foreign-owned establishments (pp. 58-9).

Although her survey did not cover as wide a range of personnel activities as Ralston's (1989), it aimed to explore good practice 'defined as that which is required or encouraged (in terms of policies, procedures and other arrangements) by legal provisions' (Price, 1994, p. 48). Written contracts, equal opportunities policies and disciplinary procedures were more likely to be found in larger organizations but often did not meet the full requirements or spirit of the legislation (p. 50). Drawing on the National Economic Development Office's (NEDO) (1992) recommendations that quality and the offering of best practice are essential to improve the competitiveness of the UK tourist industry, Price concludes that poor practice does matter because ultimately it threatens the UK's success in hospitality and tourism markets.

A sorry state?

Arguments will continue to rage about what personnel management, employee relations and HRM actually are, whether they are strategic, what their organizational role and purpose should be, how their responsibilities should be structured and how their effectiveness can be assessed. Wood (1992, pp. 94-5) remains sceptical that better personnel management or HRM (Wood, 1994, pp. 170-2) would reduce the labour problems in the industry, and he is right to the extent that the 'professionals' cannot agree whether personnel managers are effective, and if they are not, how they can be made more effective (see Guest, 1990).

Although Forte (1982) remains unequivocal that the personnel function should be decentralized and non-specialist, Ralston (1989) believes that the power of the personnel function has been reduced as a result of personnel management being performed by the non-specialists, often as an 'add-on' to other managerial responsibilities. Constituted thus, personnel work is seen as an easy task with a minimum skill requirement that can be done by anyone.

Finally, in relation to small firms generally, Atkinson (1994) has noted that in spite of clear labour market difficulties and managerial shortcomings in the personnel function, 'most businesses do not make use of external formal sources of advice and support with their employment problems' (p. 169). Price (1994, p. 54) thought it was most worrying that 22 per cent of limited companies' units and 39 per cent of proprietor-run units did not indicate that they were using any source of advice about legal changes that affect personnel and employee relations matters. Although the main source of advice in the limited companies was head office (34 per cent of cases), in both samples the general and catering press were used as the main source of advice in around 30 per cent of cases. There was considerable ignorance about proposed changes under the Social Charter (Price, 1994, p. 55). Although external agencies such as the HCTC, ACAS and Job Centres are available to assist managers with employment problems, their use may remain relatively restricted in much of the hotel and catering industry.

Summary propositions

- The personnel function in the hotel and catering industry is poorly developed, unsophisticated, simplistic and reactive.
- Much personnel work is devoted to routine matters of an administrative nature.
- The state, role and effectiveness of the personnel function in the industry needs to be put under scrutiny.

Discussion points

- Does specialist personnel/employee relations inevitably have a limited function in hotels and catering?
- How can the poor state of personnel practice be improved?

REFERENCES

Analoui, F. and Kakabadse, A. (1989) Defiance at work. *Employee Relations*, **11** (3), 1–62.

Analoui, F. and Kakabadse, A. (1993) Industrial conflict and its expressions. *Employee Relations*, **15** (1), 46–62.

Arnaldo, M.J. (1981) Hotel general managers: a profile. *Cornell Hotel and Restaurant Administration Quarterly*, November, 53–6.

Atkinson, J. (1994) Labour market support and guidance for the small firm. In J. Atkinson and D. Storey (eds), *Employment in the Small Firm and the Labour Market*, pp. 147–71. London: Routledge.

Atkinson, J. and Meager, N. (1994) Running to stand still: the small firm in the labour market. In J. Atkinson and D. Storey (eds), *Employment in the Small Firm and the Labour Market*, pp. 28–102. London: Routledge.

Baum, T. (1989) Managing hotels in Ireland: research and development for change. *International Journal of Hospitality Management*, **8** (2), 131–44.

Boella, M.J. (1986) A review of personnel management in the private sector of the British hospitality industry. *International Journal of Hospitality Management*, **5** (1), 29–36.

Boella, M.J. (1992) *Human Resource Management in the Hospitality Industry*, 5th edition. Cheltenham: Stanley Thornes.

Brewster, C. and Richbell, S. (1982) Getting managers to implement personnel policies. *Personnel Management*, December, 34–7.

Brody, R. (1993) *Effectively Managing Human Service Organizations*. Newbury Park, California: Sage Publications.

Burns, T. and Stalker, G.M. (1961) *The Management of Innovation*. London: Tavistock.

Commission for Racial Equality (1991) *Working in Hotels*. London: Commission for Racial Equality.

Commission on Industrial Relations (1971) *The Hotel and Catering Industry, Part 1. Hotels and Restaurants, Report No. 23*. London: HMSO.

Croney, P. (1988a) An investigation into the management of labour in the hotel industry. MA thesis, University of Warwick.

Croney, P. (1988b) An analysis of human resource management in the UK hotel industry. Paper presented to the International Association of Hotel Management Schools Autumn Symposium, Leeds.

Dann, D.T. (1990a) The process of managerial work in the hospitality industry. PhD thesis, University of Surrey.

Dann, D. (1990b) The nature of managerial work in the hospitality industry. *International Journal of Hospitality Management*, **9** (4), 319–33.

Dann. D. (1991) Strategy and managerial work in hotels. *International Journal of Contemporary Management*, **3** (2), 23–5.

Dann, D. and Hornsey, T. (1986) Towards a theory of interdepartmental conflict. *International Journal of Hospitality Management*, **5** (1), 23–8.

Farnham, D. and Pimlott, J. (1990) *Understanding Industrial Relations*, 4th edition. London: Cassell.

Fender, D. and Litteljohn, D. (1992) Forward planning in uncertain times. *International Journal of Contemporary Hospitality Management*, **4** (3), i–iv.

Ferguson, D.H. and Berger, F. (1984) Restaurant managers: what do they really do? *Cornell Hotel and Restaurant Administration Quarterly*, May, 27–38.

Forte, R. (1982) How I see the personnel function. *Personnel Management*, August, 32–5.

Forte, R. and Teare, R. (1990) Responding to the competitive challenge of the 1990s. *International Journal of Contemporary Hospitality Management*, **2** (3), i–ii.

Forte, R. and Teare, R. (1991) Strategic planning in action: the Trusthouse Forte approach. In R. Teare and A. Boer (eds), *Strategic Hospitality Management – Theory and Practice for the 1990s*, pp. 3–8. London: Cassell.

Gabriel, Y. (1988) *Working Lives in Catering*. London: Routledge & Kegan Paul.

Guerrier, Y. (1986) Hotel manager: an unsuitable job for a woman. *The Service Industries Journal*, **6** (2), 227–40.

Guerrier, Y. (1987) Hotel managers' careers and their impact on hotels in Britain. *International Journal of Hospitality Management*, **6** (3), 121–30.

Guerrier, Y. and Lockwood, A. (1989a) Managing flexible working. *The Service Industries Journal*, **7** (3), 406–19.

Guerrier, Y. and Lockwood, A. (1989b) Developing hotel managers – a reappraisal. *International Journal of Hospitality Management*, **8** (2), 82–9.

Guest, D. (1989) Personnel and HRM: can you tell the difference? *Personnel Management*, January, 48–51.

Guest, D. (1990) Human resource management and the American dream. *Journal of Management Studies*, **27** (4), 378–97.

Hales, C. (1987) The manager's work in context: a pilot investigation of the relationship between managerial role demands and role performance. *Personnel Review*, **16** (5), 26–33.

Hales, C. and Nightingale, M. (1986) What are unit managers supposed to do? A contingent methodology for investigating managerial role requirements. *International Journal of Hospitality Management*, **5** (1), 3–11.

Handy, C.B. (1985) *Understanding Organizations*, 2nd edition. Harmondsworth: Penguin Books.

Hicks, L. (1990) Excluded women: how can this happen in the hotel world? *The Service Industries Journal*, **10** (2), 349–63.

Hotel and Catering Economic Development Committee (1975) *Manpower Policy in the Hotel and Catering Industry – Research Findings*. London: HMSO.

Hotel and Catering Economic Development Committee (1977) *Employment Policy and Industrial Relations in the Hotels and Catering Industry – Research Findings*. London: HMSO.

Hotel and Catering Training Company (1990) *Employee Relations in the Hotel and Catering Industry*, 7th edition. London: HCTC.

Johnson, G. and Scholes, K. (1993) *Exploring Corporate Strategy*, 3rd edition. London: Prentice Hall.

Jones, C., Nickson, D. and Taylor, G. (1994) 'Ways' of the world: managing culture in international hotel chains. In A.V. Seaton, C.L. Jenkins, R.C. Wood, P.U.C. Dieke, M.M. Bennett, L. R. MacLellan and R.Smith (eds), *Tourism: the State of the Art*, pp. 626–34. Chichester: John Wiley & Sons.

Kelliher, C. and Johnson, K. (1987) Personnel management in hotels – some empirical observations. *International Journal of Hospitality Management*, **6** (2), 103–8.

Koureas, G. (1985) The effect of type, category, and geographical location on the work activities of hotel managers. MSc thesis, University of Surrey.

Lawrence, P.R. and Lorsch, J.W. (1967) *Organization and the Environment*. Cambridge, Massachussetts: Harvard University Press.

Lee-Ross, D. (1993a) An investigation of 'core job dimensions' amongst seaside hotel workers. *International Journal of Hospitality Management*, **12** (2), 121–6.

Lee-Ross, D. (1993b) Two styles of hotel managers, two styles of worker. *International Journal of Contemporary Hospitality Management*, **5** (4), 20–4.

Legge, K. (1978) *Power, Innovation and Problem-solving in Personnel Management*. London: Heinemann.

Lennon, J. and Vannocci, I. (1989) Entrepreneurs – reality and rhetoric. *Contemporary Hospitality Management*, **1** (2), 25–7.

Ley, D.A. (1978) An empirical examination of selected work activity correlates of managerial effectiveness in the hotel industry using a structured observation approach. PhD thesis, University of Michigan.

Ley, D.A. (1980) The effective GM: leader or entrepreneur? *Cornell Hotel and Restaurant Administration Quarterly*, November, 66–7.

Love, J.F. (1987) *McDonald's: Behind the Arches*. London: Bantam Books.

Lucas, R.E. and Laycock, J. (1991a) Developing an interactive personnel function for managing budget hotels: how the personnel function operates in Campanile. Paper presented to the Third International Journal of Contemporary Management Conference, October, Bournemouth.

Lucas, R.E. and Laycock, J. (1991b) An interactive personnel function for managing budget hotels. *International Journal of Contemporary Hospitality Management*, **3** (3), 33–6.

Lundberg, C.C. and Woods, R.H. (1990) Modifying restaurant culture: managers as cultural leaders. *International Journal of Contemporary Hospitality Management*, **2** (4), 4–12.

Macfarlane, A. (1982) Trade unionism and the employer in hotels and restaurants. *International Journal of Hospitality Management*, **1** (1), 35–43.

Machin, J.L.J. (1982) *The Expectations Approach*. Maidenhead: McGraw-Hill.

Machin, J., Stewart, R. and Hales, C. (1981) *Towards Managerial Effectiveness*. London: Gower.

Mackay, L. (1987) Personnel: changes disguising decline? *Personnel Review*, **16** (5), 3–11.

Mackay, L. and Torrington, D. (1986) *The Changing Nature of Personnel Management*. London: Institute of Personnel Management.

Mars, G. and Mitchell, P. (1976) *Room for Reform?* Milton Keynes: Open University Press.

Mars, G., Bryant, D. and Mitchell, P. (1979) *Manpower Problems in the Hotel and Catering Industry*. Farnborough: Gower.

Marshall, G. (1986) The workplace culture of a licensed restaurant. *Theory, Culture and Society*, **3** (1), 33–47.

Miller, P. (1989) Strategic HRM: what it is and what it isn't. *Personnel Management*, February, 46–51.

Miller, P. (1991) Strategic human resource management: an assessment of progress. *Human Resource Management Journal*, **1** (4), 23–39.

Millward, N. and Stevens, M. (1986) *British Workplace Industrial Relations*. Aldershot: Gower.

Mintzberg, H. (1973) *The Nature of Managerial Work*. New York: Harper and Row.

Mintzberg, H. (1979) *The Structuring of Organizations*. Englewood Cliffs, New Jersey: Prentice Hall.

Mintzberg, H. and Waters, J. (1985) Of strategies, deliberate and emergent. *Strategic Management Journal*, **6** (3), 257–72.

Monks, K. (1992/3) Models of personnel management: a means of understanding the diversity of personnel practices? *Human Resource Management Journal*, **3** (2), 29–41.

Nailon, P. (1968) A study of management activity in units of an hotel group. MPhil thesis, University of Surrey.

Nailon, P. (1982) Theory in hospitality management. *International Journal of Hospitality Management*, **1** (3), 135–43.

National Economic Development Council (1991) *Developing Managers for Tourism*. London: NEDC Tourism and Leisure Industries Sector Group, NEDO.

National Economic Development Office (1992) *UK Tourism: Competing for Growth*. London: NEDO.

Ogbonna, E. (1992/3) Managing organizational culture: fantasy or reality? *Human Resource Management Journal*, **3** (2), 42–54.

Ogbonna, E. and Wilkinson, B. (1988) Corporate strategy and corporate culture: the management of change in the UK supermarket industry. *Personnel Review*, **17** (6), 10–14.

Ouchi, W.G. (1981) *Theory Z: How American Business Can Meet the Japanese Challenge*. Reading, Massachusetts: Addison-Wesley.

Peters, T.J. and Waterman, S. (1982) *In Search of Excellence*. New York: Harper and Row.

Pizam, A. (1993) Managing cross-cultural hospitality enterprises. In P. Jones and A. Pizam (eds), *The International Hospitality Industry: Operational Issues*, pp. 205–25. London: Pitman.

Porter, M.E. (1980) *Competitive Strategy: Techniques for Analyzing Industries and Competitors*. New York: The Free Press.

Price, L. (1994) Poor personnel practice in the hotel and catering industry: does it matter? *Human Resource Management Journal*, **4** (4), 44–62.

Purcell, J. (1986) Employee relations autonomy within a corporate culture. *Personnel Management*, February, 38–40.

Purcell, K. (1993) Equal opportunities in the hospitality industry: custom and credentials. *International Journal of Hospitality Management*, **12** (2), 127–40.

Ralston, R. (1989) The changing nature of personnel management in the hotel and catering industry. MSc thesis, University of Manchester.

Riley, M. and Jauncey, S. (1990) Examining structure in decision making in hotels. *International Journal of Contemporary Hospitality Management*, **2** (3), 11–15.

Schaffer, J.D. (1984) Strategy, organization structure and success in the lodging industry. *International Journal of Hospitality Management*, **3** (4), 159–65.

Schein, E.H. (1985) *Organizational Culture and Leadership*. San Francisco, California: Jossey Bass.

Shamir, B. (1978) Between bureaucracy and hospitality – some organizational characteristics of hotels. *Journal of Management Studies*, **15** (3), 285–307.

Shortt, G. (1989) Work activities of hotel managers in Northern Ireland: a Mintzbergian analysis. *International Journal of Hospitality Management*, **8** (2), 121–30.

Snow, G. (1981) Industrial relations in hotels. MSc thesis, University of Bath.

Stewart, R. (1970) *Managers and Their Jobs*. London: Pan Piper.

Stewart, R. (1976) *Contrasts in Management*. London: McGraw-Hill.

Stewart, R. (1982) *Choices for the Manager: a Guide to Managerial Work*. London: McGraw-Hill.

Thompson, P. (1989) The end of bureaucracy. In M. Haralambos (ed.), *Developments in Sociology : an Annual Review, vol. 5*, pp. 51–72. Ormskirk: Causeway Press.

Torrington, D. and Hall, L. (1991) *Personnel Management: a New Approach*, 2nd edition. London: Prentice Hall.

Torrington, D., Weightman, J. and Johns, K. (1989) *Effective Management: People and Organisation*. London: Prentice Hall.

Tse, E. and Olsen, M.D. (1990) Business strategy and organisational structure: a case of US restaurant firms. *International Journal of Contemporary Hospitality Management*, **2** (3), 17–23.

Tyson, S. and Fell, A. (1986) *Evaluating the Personnel Function*. London: Hutchinson.

Webster, M. and Hudson, T. (1991) Strategic management: a theoretical overview and its application to the hospitality industry. In R. Teare and A. Boer (eds), *Strategic Hospitality Management – Theory and Practice for the 1990s*. London: Cassell.

Wood, R.C. (1983) Theory, management and hospitality: a response to Philip Nailon. *International Journal of Hospitality Management*, **2** (2), 103–4.

Wood, R.C. (1992) *Working in Hotels and Catering*. London: Routledge.

Wood, R. C. (1994) *Organizational Behaviour for Hospitality Management*. Oxford: Butterworth Heinemann.

Worsfold, P. (1989) Leadership and managerial effectiveness in the hospitality industry. *International Journal of Hospitality Management*, **8** (2), 145–56.

THREE

Contemporary Issues and Practices in Employment and Work

The purposes of this chapter are to:

- highlight past and predicted employment trends;
- identify key employment issues that centre on labour turnover, skills, qualifications and training;
- consider the main forms of employment;
- outline the nature of work;
- point to some of the implications established in relation to productivity, flexibility and other matters.

3.1 INTRODUCTION

The employment trends addressed in this chapter in part reflect the increasing importance of the service sector as Britain's major source of employment. Within private sector services, the hotel and catering industry is a significant player, and the industry is expected to sustain further growth to the end of the century. Thus the hotel and catering industry experience may provide pointers to the way in which the future nature and structure of employment in other sectors may develop. The industry employs high proportions of women, young people and ethnic minorities. The British experience is by no means untypical of practices observed elsewhere (see, for example, Timmo, 1993; Baldacchino, 1994). The extent to which such developments have been effectively 'managed' remains a moot point.

The nature of forecast employment trends in hotels and catering raises specific issues for the industry. Will the industry be able to meet the demands that have been predicted? Is training adequate to cope with changes in skills requirements or to produce appropriately qualified workers? How will above average labour turnover affect the industry's ability to cope with anticipated future demands for staff?

In relation to the above-mentioned points, work and employment can be arranged, constituted, organized and managed in an enormous variety of ways. It is probable that the variety of potential 'employment' arrangements in the hotel and catering industry is potentially more challenging than that in many other employment sectors. Therefore hotel and catering managers would appear to have a more difficult job of managing work organization (in its broadest sense). But the issue here is not simply about the needs of those managing labour; it must also take account of the needs of those offering their services for work. In other words, employment is dependent upon mutual reciprocal requirements (see Mumford, 1972): managers do not occupy a position of monopoly power. Employees can and do challenge managerial power through a variety of means, an area probed more fully in Chapters 4, 5 and 6.

Finally, the national 'flexibility' debate of the 1980s has suggested movement away from conventional full-time working towards more 'flexible' forms of employment that may include a greater degree of part-time working, variations in the pattern of hours worked and the contracting out of some work (see Atkinson, 1984), but too much of the focus has been on manufacturing industry. The hotel and catering industry was 'flexible' before flexibility became fashionable; a number of commentators subscribe to the view that many of the features of the 'flexibility' debate are rather too crude and hold little relevance to hotel and catering employment, even though the industry continues to strive for even greater flexibility (for example, Kelliher, 1989; Lockwood and Guerrier, 1989; Bagguley, 1990; Walsh, 1990). The issue of flexibility in its broadest sense underpins a substantial part of all the issues addressed in this chapter.

The employment characteristics of the establishments responding to the Workplace Industrial Relations Survey 1990 (WIRS3) are shown in Tables A1.4 to A1.11 (Appendix 1). These data, and those relating to non-standard employment, are mentioned at appropriate points in this chapter, but the data are not discussed separately in detail as in other chapters.

3.2 EMPLOYMENT TRENDS

Growth set to continue

Hotel and catering employment grew steadily in all sectors throughout the 1980s by some 300,000 jobs to a peak of 1,256,200 in 1990 (Lucas, 1992, 1993d, 1994). Between 1985 and 1990 there was a marked increase in the proportion of managerial and supervisory jobs (HCTC, 1992). More recent figures (1991–2) on managerial jobs in hotels are more difficult to assess because of the broadening of the standard occupational classification to include caravan site and holiday flat managers (HCTC, 1994a, p. 36).

The WIRS3 showed that 95 per cent of hotel and catering establishments had taken on new employees from outside the establishment for permanent posts in the previous 12 months. In the other sectors (all industries and services, service industries and private sector services) the proportion of establishments recruiting likewise was slightly less, at or around 90 per cent. The most frequently recruited levels of staff were unskilled, semi-skilled and skilled manual workers. The levels to which staff were recruited in the other sectors differ from that observed in hotels and catering, and this is most probably attributable to a different skills base in these other employment sectors.

The recession in 1991 and 1992 caused employment to fall, with hotels and contract catering being most seriously affected. Restaurants, public houses and clubs seemed to weather the recession most effectively. This may be a reflection of a rapidly expanding consumer demand for fast-food and the extension of opening hours in public houses and clubs (HCTC, 1993).

Industry and sectoral totals of employees in employment are shown in Table 3.1. These figures are based on a count of civilian jobs of employees paid by employers who run a PAYE scheme. Participants in government employment and training schemes are also included if they have a contract of employment. Employees holding two jobs with different employers will be counted twice. These figures do not include some part-timers and casual workers. HCTC (1994a) figures differ in detail, but not in substance, because they have been computed from a different base, in particular from the Census of Employment and the Labour Force Survey. Additionally, they do not go into as much detail as those from a separate, but related, source used to compile Table 3.1.

According to the HCTC (1993, p. 6), the industry is expected to regain 1990 levels of employment by 1995, and to continue with steady growth thereafter. Between 1995 and 2000, a further 105,000 jobs are expected to be created in the commercial sector (HCTC, 1994a, p. 41), of which 15,000 will be in the self-employed category. In the longer-term, the restaurant sector is likely to experience the strongest growth; public houses and hotels will show steady, but slow, growth. Contract catering will recover strongly as a consequence of the need to improve benefits to secure staff retention in a more buoyant economy.

In spite of employment growth, average establishment size is extremely small, at 4.5 employees (HCTC, 1994a, p. 44). There are sectoral differences in establishment size (see Table 1.1). The perpetuation of the dominance of small businesses may be a deliberate result of managerial resistance to growth in the size of the firm; as firms grow in size increasing problems of meeting labour market needs increase, although stability is reached at around 20 employees (Atkinson and Meager, 1994). Job quality is likely to be poor among small firms (Atkinson and Storey, 1994).

In addition to the need to recruit more staff as a result of continuing net job growth, the HCTC (1992, 1994a, b) has estimated that substantial recruitment is needed to maintain existing employment levels because of retirements and net losses from the industry (see below). These recruitment needs do not take account of the those people who move in and out of the different industry sectors. Most new recruits will be at operative level (the industry contains high proportions of manual workers, many of whom are unskilled – see Tables A1.5, A1.8 and A1.9, Appendix 1). Although the expected future demand for managerial and supervisory staff is likely to be less than in the 1980s, it will probably provide more of a challenge because of a more difficult recruitment climate.

The UK labour force will continue to grow in the 1990s, largely as a result of an increased participation rate among females. However, by 2001 the labour force will comprise a greater proportion of older workers than was the case in the 1980s and 1990s. The numbers of older people (aged 35–54) will increase by 1.9 million between 1992 and 2000, whereas the numbers of younger people (under age 35) will continue to decline by 1.3 million (see also Lucas and Jeffries, 1991). The number of students in higher education is expected to increase from the 1992 level of around one million to nearly 1.5 million in 2000 (Employment Department Group, 1992, p. 37). A squeeze on parental income and the freezing of student grants is expected to put increasing pressure on many such students to work part-time.

Table 3.1 *Numbers of employees (in thousands) employed (actual and forecast) in the hotel and catering industry for selected years (by sector).*

	Total	Male			Female		
		Full-time	Part-time	All	Full-time	Part-time	All
HCI[a]							
1980	959.3	–	–	321.7	208.2	429.4	637.6
1990	1256.2	258.7	186.6	443.3	244.4	566.5	810.9
1993	1177.1	241.9	203.2	445.1	214.8	517.2	732.0
2000	1338.0	–	–	–	–	–	–
Restaurants							
1980	196.4	–	–	69.4	41.6	85.3	126.9
1990	306.4	73.5	48.5	122.0	54.3	130.1	184.4
1993	298.4	74.2	50.2	124.2	50.1	123.9	174.0
2000	322.0	–	–	–	–	–	
Public houses							
1980	248.2	–	–	74.7	27.9	145.6	173.5
1990	337.2	41.2	61.4	102.6	37.2	197.4	234.6
1993	322.7	39.6	68.2	107.8	34.7	180.2	214.9
2000	357.0	–	–	–	–	–	–
Clubs							
1980	130.4	–	–	47.0	13.7	69.6	83.3
1990	142.3	19.2	35.8	55.0	13.2	74.3	87.5
1993	136.8	17.5	34.9	52.4	10.7	73.7	84.4
2000	142.0	–	–	–	–	–	–
Contract catering							
1980	104.2	–	–	24.0	40.4	39.8	80.2
1990	147.0	31.3	9.0	40.3	47.8	58.9	106.7
1993	113.1	28.1	10.7	38.8	37.7	36.6	74.3
2000	146.0	–	–	–	–	–	–
Hotels							
1980	241.7	–	–	88.4	73.3	80.0	153.3
1990	283.2	80.3	28.1	108.4	81.3	93.5	174.8
1993	262.5	68.9	35.1	104.0	70.0	88.5	158.5
2000	317.0	–	–	–	–	–	–

[a]Industry totals include other short-stay accommodation.

Any slight discrepancies between figures are due to the rounding of decimal points.

Source: Quarterly Estimates of Employees in Employment (June), Employment Department (published and unpublished data); HCTC, 1993, p. 6.

Gender issues

The hotel and catering industry employs high proportions of females (see also Tables A1.4 and A1.7, Appendix 1). As Table 3.2 shows, around three-quarters of part-timers are female. Male and female employment patterns vary across different industry sectors (see Lucas, 1992, 1994). Hotels, clubs and restaurants employ slightly higher proportions of males (about 40 per cent) than elsewhere (see also HCTC,

1994a, pp. 37–8). Gender, which is a societal construction designed to exaggerate the differences between men and women and to maintain sex inequality, is likely to be a feature of workforce construction policies (Kitching, 1994, pp. 35–139).

Table 3.2 *Distribution of employees in employment in the hotel and catering industry for selected years (by per cent).*

	Male			Female			Total
	All	Full-time	Part-time	All	Full-time	Part-time	
1971	38	27	11	63	30	32	100
1981	34	22	12	66	22	45	100
1990	35	21	15	65	19	45	100
1993	38	21	17	62	18	44	100

Any slight discrepancies between figures are due to the rounding of decimal points.

Source; Robinson and Wallace, 1983, Table 6, p. 270; Quarterly Estimates of Employees in Employment (June), Employment Department (published and unpublished data).

The gender implications of hotel and catering employment are important, and are manifested in three main ways outlined below (these issues are developed more fully in later chapters, particularly in relation to equality issues in Chapter 10).

Occupational segregation

The industry is 'an occupationally sex-segmented labour market ... women are horizontally and vertically segregated into particular jobs, grades and areas of operation' (Purcell, 1993, p. 127). As a general rule, females tend to be highly concentrated in the less skilled or sub-craft occupations. Over two-thirds of all women working in the industry are employed as counter and kitchen hands and domestic staff. Purcell (1993) also attributes the high concentration of women in lower level jobs to 'crowding'; they are seen as appropriate because 'they are disadvantaged workers, competing with other disadvantaged groups' (p. 128). There is a greater concentration of male employment in craft and semi-skilled occupations – nearly a quarter of males working in the industry are chefs or cooks. At more senior levels women are also proportionately under-represented; there are fewer female managers than males, but more female supervisors than males (for more details, see Bagguley, 1987; HCTB, 1987; Purcell, 1993). The proportion of women managers drops significantly after age 30 (HCIMA and Touche Ross Greene Belfield-Smith Division, 1992).

A broad overview of the presence of particular occupational groups (manual and non-manual) and data on skilled and unskilled manuals from the WIRS3 are shown in Tables A1.5, A1.8 and A1.9. The HCTC (1994a, p. 36) also provides data on occupational segregation, but this is for all catering employment (both the commercial and catering services sectors).

Age segregation

The commercial sector employs a sizeable minority of younger workers; as Table 3.3 shows, 30 per cent are aged between 16 and 24 years, compared to 20 per cent of all employees in the economy who are in this age band (HCTC, 1994a, p. 46). The proportion of 16 to 24 year olds working in the commercial sector increased from 22 per cent in 1984 to 30 per cent in 1992. Restaurants are more likely to employ such workers (44 per cent of the workforce) than other sectors (HCTC, 1994a, p. 46). The issue of 'young employment' is an area that has so far been inadequately addressed (Lucas and Jeffries, 1991; Lucas, 1993a, b, 1995) but is the subject of a current research project (Ralston and Lucas, 1994).

Table 3.3 *Age structure of the hotel and catering industry (1992)*

Age	Employees (%)
16–19	15.7
20–24	14.0
25–29	12.3
30–39	20.2
40–49	19.1
50–59	13.4
60–64	3.7
65+	1.5

Source: Adapted from Table 4.8, HCTC, 1994a, p. 46.

The age distributions of females and males in occupational groups differ markedly (see HCTB, 1987, pp. 24-8). Female employment patterns are related to child bearing: the majority of women working in the industry are between age 30 and 59. Male employment patterns show a higher concentration of employment among younger age groups, with numbers employed tailing off after age 44. Higher proportions of chefs and cooks, waiting and bar staff are found among the under-40s.

More recent evidence suggests that in recruitment terms the industry shows a bias towards the under-30s. Because of managerial perceptions related to the age of workers, age forms the basis of a structural construct in hotel and catering employment (Lucas, 1995). The tendency to favour the young at a time when they are in relative short supply suggests that organizations may not be gearing themselves for managing future changes in a labour market that will include larger proportions of older workers (Lucas and Jeffries, 1991; Lucas, 1993b). Young part-time labour is more likely to be female than male (Lucas and Bailey, 1993, p. 17; Ralston and Lucas, 1994).

Pay segregation

On average, full-time male earnings exceed full-time female earnings by a significant amount (for details see Table 5.1). A part explanation for this difference is related to occupational segregation. As a general rule, craft and semi-skilled occupations (males) will be more highly graded than unskilled occupations (females), although this is unlikely to be the outcome of any formal structural process or systematic job evaluation (Walsh, 1990, p. 525; Millward *et al.*, 1992, p. 275).

However, as the figures in Table 3.4 show, this male and female earnings gap is found among the same occupational groups, which suggests that men hold the more highly graded positions within particular occupations; for example, men are chefs, women are cooks (see Lashley, 1984). Such differences also reflect employers' perceptions that women's work is less valuable than men's (Walsh, 1990), and are potentially in breach of equal pay (equal value) legislation (see Chapter 10). Additionally, full-time female earnings exceed part-time earnings, but this smaller difference cannot necessarily be explained by the reasons advanced to explain the difference in male and female earnings, except to the extent that it reinforces the perception that part-time work is less valuable than full-time work (see Walsh, 1990, pp. 524–8). Youth pay is also lower on average than adult pay (Lucas and Bailey, 1993). The HCTC (1994a, pp. 84–7) also gives details of pay and earnings, but only in broad outline terms.

Table 3.4 *Gross hourly earnings[a] of manual occupations for selected years (catering and all industries and services).*

	Full-time males	Full-time females	Part-time females
Chefs/cooks			
1980	207.8	171.8	163.9
1991	482.3	394.6	352.8
1992	513.0	430.0	383.0
1993	509.0	444.0	398.0
Waiting staff			
1980	–	140.9	142.3
1991	402.9	335.0	324.0
1992	387.0	354.0	338.0
1993	444.0	358.0	351.0
Bar staff			
1980	162.1	137.3	136.5
1991	346.3	331.2	309.6
1992	398.0	333.0	309.0
1993	400.0	329.0	329.0
All manual occupations			
1980	240.5	170.4	153.9
1991	554.1	395.2	351.4
1992	589.0	421.0	376.0
1993	605.0	435.0	390.0

[a]Excluding overtime, in pence.

Source: New Earnings Survey, Tables 8, 9, 178.

Ethnic minority and migrant issues

Unfortunately, published information on ethnic minority and migrant workers in hotels and catering remains sparse. As a general rule, ethnic minority employment tends to be concentrated in particular cities and towns and the ethnic minority population is more concentrated among the under-30s than the white population (Lucas and Jeffries, 1991, p. 326; see also Chapter 10).

Byrne (1986) found that black workers (6.6 per cent of the hotel and catering labour force) were twice as likely to be found in the industry as in all industries. Black workers were more likely to be low-paid and clustered in more lowly graded work (see also CRE, 1991; Jameson and Hamylton, 1992). Data from the WIRS3 (see

Tables A1.4, A1.10 and A1.11) also confirm a relatively high extent of ethnic minority workers in hotel and catering workplaces. Hotels and catering has the highest representation of Malay, Chinese, Filipino, Far East and black African workers. There is a greater probability of ethnic minority workers being employed in larger establishments.

Migrant labour is also significant to the industry: 115,000 workers were foreign nationals, with two-thirds coming from non-European Community (EC) countries (Byrne, 1986, p. 14). While EC employees do not need work permits, non-EC employees, subject to certain qualifying requirements, do require work permits. As a general rule, permits are issued for a named overseas worker (between 23 and 54 years of age) for a specific job for a 12-month period, only if no suitable resident labour is available. 'Overseas students attending courses in the UK who seek part-time work also need a work permit' (HCTC, 1990, p. 32). Permits are also available for highly skilled and experienced workers to fill senior positions. More detailed guidance can be obtained from the Employment Department. For a discussion about the migrant contribution to the hotel industry, see Taylor (1983).

Summary propositions

- Recruitment will continue to be a major management activity that is likely to be most challenging in terms of finding managerial staff.
- Managers are likely to be more successful in their recruitment and staffing arrangements if they are aware of changes in the labour market.
- Traditional managerial attitudes that relate to gender may serve to compound negative employment experiences among women that may limit the chances of achieving a 'good fit' in the employment relationship.
- Ethnic minority and migrant labour is a significant issue in hotel and catering employment.

Discussion points

- How can managers make recruitment and selection more cost-effective?
- Have you come across workforce segregation by gender, occupation, age, pay or ethnicity? In what ways was this manifested and what was its effect on the workforce?

3.3 KEY EMPLOYMENT ISSUES

The HCTC has produced a number of research reports that highlight key employment and labour market issues for the 1990s and beyond (HCTC, 1992, 1993, 1994a, b). Some of these have already been mentioned above, including skills requirements and turnover problems, but this area requires a little more development because these issues impinge on a number of other issues discussed later on, including employment strategies, managing employee involvement and commitment, and health and safety. This part of the chapter summarizes some of the key managerial issues related to labour turnover, skills requirements, training and qualifications. Suffice it to say that they are all considered to be 'problems' for managers, and while there is nothing new in this (see Chapter 1 and Atkinson and Storey, 1994, p. 16), it is useful to identify the current situation.

Labour turnover

As noted in Chapter 1, high labour turnover has been an issue of much interest to hotel and catering researchers. High turnover may be a manifestation of industrial conflict (Analoui and Kakabadse, 1993). This is not always the case (for example, an employee may leave to secure promotion) but disaggregating turnover in this more instructive way has not always been possible (Kelliher, 1984, p. 5). Even if disaggregation were possible, it might be even more subjective to categorize industrial conflict as negative behaviour and seeking promotion as positive behaviour.

The subject of high labour turnover in hotels and catering has been viewed from two perspectives, which broadly view the phenomenon in either positive or negative terms. Briefly, to the disciples of 'conventional' management practice (for example, HCEDC, 1968, 1969; Mars *et al.*, 1979; Johnson, 1980, 1981, 1985, 1986; HCITB, 1984a; Bonn and Forbringer, 1992; Denvir and McMahon, 1992), high turnover is a major problem that should be managed and reduced because it has negative consequences for the organization, employers and employees. The 'deviant' view is that high turnover is inevitable, unavoidable and possibly even desirable (Bowey, 1976; Riley, 1980, 1981a, b; Simms *et al.*, 1988). A useful critique of these contrasting approaches is given by Wood (1992, pp. 95–103). More recently, Riley (1993) has suggested that it is time to change the paradigm by concentrating on individual decisions to leave, and to consider whether there is a relationship between commitment, job satisfaction and the intention to leave. Other recent contributions to the issue of retention and turnover include Riley (1990), Perrewe *et al.* (1991) and Boyd Ohlen and West (1993). Aspects of these differing perspectives are considered more fully later in the book; for example, in relation to managerial strategies to the employment relationship featured in Chapter 4.

The HCTC (1994a, b) has identified that employee turnover – 'the number of employees leaving their job within a certain period of time expressed as a proportion of the number of workers in that establishment at the start of the period' (1994a, p. 56) – remains too high and is, therefore, problematic. In 1991–2 average turnover was 27 per cent, much reduced from levels of 60 per cent in 1979–80 (HCITB, 1984a). Higher than average industry turnover was found in public houses and clubs (34 per cent), hotels (33 per cent) and restaurants (35 per cent). Turnover in 1992 was estimated to have cost the industry £430 million and turnover was higher among men than among women (HCTC, 1994b, p. 6). Only 6 per cent of managers monitored employee turnover.

Turnover continued to remain higher in the commercial sector than in the catering services sector (HCTC, 1994a, b), a similar position to that found over a decade earlier (HCITB, 1984a). Turnover was highest among operative staff and lowest among managers. Although 88 per cent of employers claimed to have 'done everything they could afford to make staff loyal to them' (HCTC, 1994a, p. 57), only 9 per cent of managers had made any attempt to manage turnover; training and multi-skilling were the most common measures used to reduce the extent of the problem.

The WIRS3 confirmed high turnover. The mean number of resignations in hotels and catering (26.6 per establishment) was substantially higher than elsewhere (around 15 per establishment in all industries and services, service industries and private sector services), and was most marked in establishments employing 50 or more employees. The HCTC (1994a, b) concluded that the recession was an important reason for reduced turnover levels, and that it had provided unforeseen benefits to employers in terms of savings in recruitment and training costs.

The HCTC (1994a) also found that few employees were seeking alternative employment. Stability was more marked among managerial positions (hotel and restaurant managers and publicans) than among bar staff, waiting staff and kitchen porters (HCTC, 1994a, p. 60). The main reason for seeking alternative employment was 'unsatisfactory' pay, and this was mentioned by a significant proportion of respondents in all occupational groups. For managers the most significant factor was 'other aspects of current job', but this was not connected to hours of work. A requirement for longer hours was a significant factor among all operative groups, presumably, in part, to compensate for their unsatisfactory pay.

Skills shortages

The HCTC (formerly the HCITB/HCTB) has undertaken a number of research exercises that have sought to assess skills needs against the provision of education and training (for example, HCITB, 1983a, b, c, d, e, 1984b; HCTB, 1988; HCTC, 1992, pp. 48–74, 1994a, pp. 58–82). Their findings, and those of others (Lucas and Jeffries, 1991; NEDC, 1991) point to a continuing major cause of concern.

Against a background of high levels of competition, demographic change and a greater emphasis on customer care and service quality and on health and safety issues (HCTC, 1992, pp. 48–9), hotels and catering face a more acute skills shortage than all other sectors of the economy (HCTC, 1994a, pp. 58–9). There were more catering and hospitality establishments (21 per cent) with hard-to-fill vacancies than anywhere else and the position had worsened during the recession (see also Lucas, 1995). The proportion of establishments with hard-to-fill vacancies in the previous 12 months was also highest in catering and hospitality (39 per cent).

The most difficult vacancies to fill appear to be managerial, and the least difficult to fill are for bar staff, catering assistants and kitchen porters. However, these conclusions are based on an analysis of vacancies registered at Job Centres, which is not regarded as giving accurate information on managerial jobs, which are probably more often recruited through other sources (HCTC, 1994a, p. 59).

The HCTC's most recent analyses (1992, 1994a) suggest that there are simply insufficient people on the job market with the requisite skills, but this observation should not be taken at face value. The employment decision is a two-way process. Employers argue that the number of applications is low and that they are of poor quality. This points both to an image problem for the industry, which is not new (see Ellis, 1981) and to a lack of clearly defined recruitment needs. Perhaps this latter point, the way managers recruit and select staff, is the more important. It cannot be assumed that managers approach recruitment in the most systematic way, or even systematically at all (for further discussion on selection criteria in small firms, including hotels and catering, see Kitching, 1994, pp. 109–21). Price (1993) found that only a minority of organizations had 'a recruitment policy or procedure to provide guidance on good practice' (p. 19), and that the use of job descriptions and person specifications in recruitment was a minority practice.

The absence of such formality in recruitment does not necessarily mean that managers have not defined their needs, but it can be inferred from the HCTC's (1994a) report that employers' requirement for 'motivated staff who are good with people' is rather more woolly than precise. For jobs with customer contact, particularly with male customers, there is often a presumption that women have better interpersonal skills and a better appearance than men (Kitching, 1994, p. 121). In other words, a rather more casual than systematic approach to recruitment may serve to exacerbate a poor recruitment outcome in terms of skills shortages because men are being overlooked. Low levels of investment generate low levels of return; this theme is now continued in relation to training and qualifications.

Qualifications and training

The issues of qualifications and training cannot be divorced from education, and here, as a general rule, the UK fares badly 'in a comparison of the number of qualified people with its major competitor countries' (HCTC, 1994a, p. 72). The difference is most marked at the level of GCSE and higher education entrance level qualifications. In other words, a low level of achievement in the education system will have a knock-on effect in terms of the quality of training (for trainers and trainees) and qualifications in the workplace. Even though the development of a system of more extensive vocational qualifications is a clearly stated government objective, Britain still has the highest proportion of workers with no such qualifications (63 per cent) (HCTC, 1994a, p. 72). In any event, vocational qualifications seem to be characterized by lack of rigour (Prais *et al.*, 1989).

Within hotels and catering, perhaps not surprisingly given some of the features of its employment profile noted above, the industry continues to remain relatively unqualified (HCITB, 1983a; HCTC, 1994a, p. 73), although there has been some improvement. For example, 30 per cent of all managers and professionals hold a degree but the proportion is a mere 6 per cent in catering and hospitality. In relation to all employment, 35 per cent of the catering and hospitality workforce is unqualified compared to 24 per cent of the workforce as a whole.

The WIRS3 data for hotels and catering used to inform this text and the evidence from Price's (1994) survey of hotels and restaurants discussed in Chapter 2 both indicate a higher than might be expected level of formal qualifications (including personnel management) among managers. However, Price (1994, p. 53) found that managers had limited access to continuing training. Almost all those with access worked within the relatively large multi-establishment PLCs. Hardly any respondents had taken advantage of HCTC courses on basic personnel management. Those most at risk through lack of qualification and training opportunities are the proprietors or partners in the smaller businesses.

The figures on qualifications and training for operatives shown in the HCTC's (1994a, pp. 73–4) report are interesting but cannot be fully developed here. Perhaps, not surprisingly, chefs and cooks were shown to be the most qualified occupational group (34 per cent held City and Guilds), but the extent of their qualifications levels is not exactly high. More interesting is the relatively high level of educational attainment of waiting and bar staff at GCSE and A-level standard. This suggests that many of these jobs are filled on a part-time basis by students, an area that has already been earmarked for more in-depth research (Lucas and Bailey, 1993; Ralston and Lucas, 1994). The opportunities for gaining further qualifications vary among the different sectors, and are highest in contract catering and lowest in clubs.

The range of provision of means to attain different types of qualification through education and training is changing rapidly and is best explained elsewhere (HCTC, 1992, pp. 54–74; HCTC, 1994a, pp. 64–82). Without allusion to detail, much of the thrust of 'new' training initiatives is that they will lead to some kind of recognized qualification. Thus, National (or Scottish) Vocational Qualifications (NVQs or SVQs), although no more than a recognition of achievements in the workplace (referred to as accreditation of prior learning (APL)) at a base level), constitute a formally recognized qualification.

It can be argued that the assumption that training must automatically lead to some kind of qualification is fundamentally flawed. Some training will do so, but this is not necessarily the case or even necessary. Qualifications without substance lack any real sense of meaning or value. To acknowledge a requisite level of achievement 'on-job' is not without purpose or value, but these elements should not be overstated or, more importantly, be used to compensate for inadequacies elsewhere. Training is about systematically developing knowledge, skills and attitudes. NVQs at Levels 1 and 2, the most common form of uptake of the 'new training', do nothing of this. The danger is that a qualification without substance demeans the whole range of qualifications.

In spite of this development towards more formally recognized vocational qualifications (this is not necessarily progressive, rigorous or standard raising; see Prais *et al.*, 1989), the issue of training remains problematic. The HCTC (1994a) considered that 'there is room for improvement in the industry's training record' (p. 64). Although a slightly higher proportion of catering and hospitality employers had a training budget than for all industries (HCTC, 1994a, p. 65), a smaller proportion of the industry's employees had recently received job-related training than for all industries.

Although there was a moderate-to-strong training need in areas underpinned by legislation, Price (1994) found the reluctance of organizations to pay for training 'disturbing'. In a follow-up survey of her sample of 'best' employers, she found that the majority would not pay more than £100 a day for training which was insufficient to cover the commercial cost of training offered by organizations like the HCTC. If more training is to be undertaken, it must be offered more cheaply (i.e. subsidized by public funds) or employers must be prepared to pay more. This finding raises important questions about the ability of the now 'privatized' HCTC to make any real impact on the training needs of most of the industry.

Additionally, the fact that 'training activity was rarely skills specific to the establishment' (HCTC, 1994a, p. 68) points to a low assessment of training need (this can be paralleled with the low assessment of recruitment need mentioned above). Lynch (1994) has identified a low take-up of training among bed and breakfast operators. He hypothesizes that training may not be perceived as relevant and that greater uptake might occur if training were to be marketed more effectively to those new to the operation. In any event, training is more likely to be more informal, proprietorial and 'on-job' in small firms (Atkinson and Meager, 1994). Training is a higher priority in larger establishments and most of it is 'in-house' and 'on-job'. Much training did not lead to formal qualifications but 'with the introduction of NVQs/SVQs the means are available to ensure all training can lead to the reward of qualifications' (HCTC, 1994a, p. 67).

Implications for the employment relationship

Much of the analysis above points to ways in which managers can choose, or choose not, to 'manage' key employment issues such as labour turnover, skills shortages and qualifications and training deficiencies, but the problems associated with these issues may lie much deeper than has been implied in that control is not entirely in managerial hands. The problem needs to be viewed in terms of the vagaries of the labour market, an issue that is addressed at various points throughout this text. As Riley (1991, p. 244) has observed, there is an inherent conflict between the economic interests of units and the skill development of the industry.

Managers, for a variety of reasons, may choose to recruit from among 'experienced' staff because suitably qualified individuals are not available. The point is that the labour market in hotels and catering is relatively 'closed', and not as 'open' as may have been suggested, that is, to the extent that everyone is ready and available to work in it (see also HCTC, 1994b). High turnover may exist, but where the source of joiners is known, 67 per cent come from within the industry (HCTC, 1994a, p. 56) and, therefore, the labour market is relatively self-contained and self-perpetuating – hence 'closed'. Where joiners come from the same sector, the majority are at managerial or craft level. Turnover is highest among operative staff and lowest among managers (HCTC, 1994a, p. 56), and men are more mobile than women (HCTC, 1994b). In other words, the workings of the labour market in relation to hotel and catering employment have not yet been fully explained and remain to be refined. It may well be segmented in ways that have yet to be defined.

Summary propositions

- The hotel and catering industry will continue to face more severe employment problems related to high labour turnover, skills shortages, and training and qualifications than the economy as a whole.
- Such problems are within managers' immediate powers to manage but some, most particularly in relation to education and training, are also related to deficiencies in government policy. Therefore, such problems can only be satisfactorily addressed through longer-term policy change.

Discussion points

- Is high labour turnover inevitable or should it be managed?
- How can the hotel and catering industry overcome skills shortages?
- How can the level of qualifications and training be raised?

3.4 FORMS OF EMPLOYMENT

The majority of people working in hotels and catering are employees employed under a contract of employment (contract of service), which affords a degree of legal protection that derives from common law, statute law and other sources. Although the more progressive employers with relatively sophisticated employee relations arrangements make fairly extensive use of written contracts (Price, 1993, p. 18), fewer than a quarter of employees have a written job offer or a formal contract (Parsons, 1992, p. 105). There are, however, sizeable numbers of workers who are contracted by an employer to work under a contract for services – they are not employees and are therefore not protected by most employment protection legislation. Additionally, a small, but significant, number of individuals opt for self-employment status, a sizeable number of whom are managers.

Atkinson's (1984) model of the flexible firm incorporates a variety of different working practices which can be constituted as particular forms of employment. An adaptation of this model is shown as Table 3.5. Atkinson's 'ideal' flexible firm, which comprises a core workforce of functionally flexible employees surrounded by a periphery of numerically flexible workers, is valid as a general principle but needs considerable revision in terms of how differing hotel and catering businesses function. For instance, small hotels may operate on high levels of functional flexibility but this is more difficult to achieve in large hotels (Guerrier and Lockwood, 1989b). Additionally, not all types of flexible work practices are appropriate to the hotel and catering industry (for instance, homeworking), but there is a clear predominance of some practices above others, the main examples of which are now outlined. Pay flexibility is featured in Chapter 5.

Part-time employment

The industry employs a much higher proportion of part-time workers (60 per cent) than the service sector (one-third) or the economy as a whole (a quarter) (see also Tables A1.4 and A1.6). (For a recent discussion on the role of part-time employment, see Dickens, 1992a, b.) As Table 3.1 shows, sectors like public houses and clubs rely heavily on part-timers. In employment statistics, part-time workers are people who normally work for not more than 30 hours a week, although patterns of part-time working vary considerably. In employment protection terms, part-time was also tantamount to working fewer than 16 hours a week (fewer than eight hours a week after five years' service) because, as a general rule, such workers were excluded from employment protection rights (now amended by the Employment Protection (Part-time Employees) Regulations (EPPTER) 1995 following *R* v *Secretary of State for Employment, ex parte EOC and another* [1994] IRLR 176 HL reversing [1993] IRLR 10 CA – see Chapter 10).

Employers may, however, define part-timers differently; for example, anyone who works fewer hours than the normal full-time 39 hours. In other words, part-time is a relative rather than an absolute measure and form of employment.

As Table 3.6 indicates, most part-timers in the hotel and catering industry, who happen to be women, do not work 30 hours. The proportions of female part-timers with no specified hours increased between 1980 and 1991, suggesting that employers used this device to maximize flexibility. Although figures after 1991 suggest that this practice may be declining, this may be linked to factors connected with the recession.

Table 3.5 *A framework for flexible working.*

Characteristics	Suggested examples
Functionally flexible working	
Multi-skilling	Manager; cross-department, e.g. operative able to work in restaurant, bar and housekeeping
Horizontal job enlargement	Room maids become responsible for checking own work
Vertical job enlargement	
Up	Substituting for a superior, e.g. departmental head
Down	Covering for a subordinate, e.g. manager, departmental head
Job rotation	Regular movement from bar to restaurant to reception
Career development	Manager
Autonomous work groups	Room maids in teams
Total retraining	Most occupational groups
Numerically flexible working	
(a) Employees	
Part-time	Restaurant, bar, housekeeping
Casual/temporary	Restaurant, bar, banqueting
Job sharing	Administrative jobs of fixed hours
Overtime	Most operative jobs
Sabbaticals, etc.	
Flexi-hours	
Daily	Some administrative jobs
Weekly	Some administrative jobs
Compressed working week	
Annual hours	
Committed hours	
Shiftwork	Kitchen, restaurant, bar, housekeeping, reception
Manipulating labour turnover	Any occupational group
(b) Contracted labour	
Agency staff	Banqueting
Contracting out/buying in	Laundry services, food preparation, cleaning
Homeworking	
Computer systems	
Government subsidized trainees	Youth Trainees

Source: Derived from Atkinson 1984; Lockwood and Guerrier, 1989.

Until 1992, average weekly part-time female hours also reduced slightly and increasing numbers of part-time jobs were at or below 16 hours a week. Whether the 1993 figures, which show a change in such hours' trends in hotels and restaurants, are symptomatic of new developments remains to be seen. Much part-time work is regular because the need for it is continuous, and therefore it is more core than peripheral in terms of organization structure (Guerrier and Lockwood, 1989b; Lockwood and Guerrier, 1989), although, in the main, it still constitutes a numerically flexible form of employment. The figures in Table 3.6 confirm that there is more numerically flexible employment in hotels and catering than in the economy as a whole.

Table 3.6 *Hours of part-time manual females for selected years (by former wages council sector[a] and all industries and services).*

	No specified hours (% in sample)	Of those with specified hours % below 8 hours	% 8–16 hours	Average weekly hours
Licensed residential establishment and licensed restaurant[b]				
1980	25	3.4	16.5	22.6
1991	49	11.3	29.3	19.1
1992	48	11.3	24.8	20.7
1993	42	9.0	18.1	22.1
Licensed non-residential establishment[c]				
1980	26	10.4	33.6	17.9
1991	39	16.2	44.9	15.5
1992	32	20.8	38.3	15.2
1993	30	18.8	46.4	15.1
All industries and services				
1980	11	7.6	23.2	19.9
1991	20	6.7	32.3	19.7
1992	20	7.7	32.0	19.4
1993	21	7.6	33.4	19.5

[a]These figures cannot be fully reconciled with those relating to employees in employment (see Table 3.1).

[b]Hotels and restaurants.

[c]Public houses and clubs.

Source: New Earnings Survey, Tables 176, 177, 181, 182.

Put simply, the high proportion of part-timers is a function of demand for this form of labour that is mirrored by a plentiful supply of labour, mainly women, that is prepared to work on this basis (for further development of this point, see Kitching, 1994, pp. 124–30). Most hotel and catering businesses are subject to widely varying patterns of demand which render the utilization of full-time labour uneconomic. In these circumstances, part-time employment is used as a control mechanism to manage fluctuating demand (for example, Walsh, 1990, pp. 519–21). Employers are not paying workers to stand idle. But most part-time labour is also cheaper in direct terms (lower wage and other employment costs) and, therefore, part-time employment offers considerable cost benefits to employers. To date there has been little evidence to support the view that indirect costs related to a perceived cost of employment protection rights influence the employment decision (see Lucas, 1993c, pp. 95–7).

The growth in part-time female employment was most marked in the 1970s and many attribute this to the availability of increasing numbers of women who were cheaper to employ (for example, Airey and Chopping, 1980; Robinson and Wallace, 1983). Bagguley (1990) sees the emergence of part-time employment as 'a combination of supply-side, demand-side and legislative factors' (p. 741). Since the 1980s the rate of part-time employment growth has been greater among males than among females, although the numbers of male part-time workers remain in the minority. Young workers have also constituted a sizeable part of the part-time labour force, and this group may become even more significant (Crompton and Sanderson, 1990; Lucas and Bailey, 1993).

Full-time employment

As 40 per cent of workers in hotels and catering work full-time (more than 30 hours a week) this is still a significant employment form. Table 3.1 shows that there are significant variations in the proportions of full-time and part-time jobs among the different sectors. As Table 3.7 shows, on average men work longer hours than women and there are also variations in the patterns of hours in the different sectors. But perhaps the most significant feature of the data shown in Table 3.7 is that smaller proportions of manual workers in hotels and catering work 40 hours or more than is the case in the economy as whole. In other words, these figures do not support one feature of hotel and catering mythology that alleges that many workers employed in the industry are unduly subjected to long hours of work, at least among manual workers (see for example, CIR, 1971; Mars *et al.*, 1979; Byrne, 1986; Gabriel, 1988).

Temporary employment

Temporary employment is a minority practice in Britain, but is used particularly in hotels and catering (Dickens, 1992a, pp. 11-13): an estimated 11.6 per cent of all hotel and catering employment was temporary in 1984 (Price, 1993, p. 22). It covers casual work, which may be hourly, daily, weekly, monthly or seasonal, fixed term contracts and where agencies supply labour on a casual basis. All of these types of temporary employment may 'disguise permanence' because some workers may have worked for the same employer for a number of years. However, regular casuals often have no employment rights. In *O'Kelly and others* v *Trusthouse Forte plc* [1983] IRLR 369 CA, the Court of Appeal ruled that there must be a mutual obligation to supply and perform work for a contract of employment to exist. Although the hotel in question had provided work on a regular basis, it was not obliged to do so. Mitchell (1988, p. 481) suggests that the *O'Kelly* case reveals how senior management may see even long-serving employees as peripheral. This judgement may yet be challenged by some future case (Price, 1993).

Table 3.7 *Hours of full-time manual workers for selected years (by former wages council sector[a] and all industries and services)*

	36 or less (%)	38–40 (%)	40+ (%)	Average weekly hours
Males				
Licensed residential establishment and licensed restaurant establishment[b]				
1980	5.3	64.6	29.7	42.8
1991	6.5	67.2	26.4	41.1
1992	6.0	72.5	21.5	40.5
1993	4.6	72.5	22.8	40.8
Licensed non-residential establishment[c]				
1980	7.3	37.3	55.2	47.8
1991	10.7	50.0	39.4	42.5
1992	13.3	50.8	35.7	42.3
1993	10.5	48.4	41.1	43.0
All industries and services				
1980	1.8	41.2	57.0	45.4
1991	2.4	44.8	52.8	44.4
1992	2.5	44.4	53.1	44.5
1993	2.7	45.0	52.3	44.3
Females				
Licensed residential establishment and licensed restaurant establishment[b]				
1980	20.9	56.4	23.0	39.8
1991	15.5	63.3	21.2	39.9
1992	16.4	64.9	18.6	39.7
1993	19.8	68.6	11.5	38.5
Licensed non-residential establishment[c]				
1980	26.1	41.5	32.4	40.7
1991	25.2	44.3	30.3	39.7
1992	27.8	52.5	19.6	39.4
1993	33.3	42.2	24.5	39.5
Unlicensed place of refreshment[d]				
1980	27.1	56.5	16.4	38.3
1991	19.6	58.8	21.5	39.5
1992	45.0	43.3	11.6	38.1
1993	25.5	56.9	17.7	38.7
All industries and services				
1980	17.3	65.5	17.2	39.6
1991	14.2	62.5	23.3	39.7
1992	15.3	60.5	24.2	39.8
1993	15.4	59.3	25.4	39.8

[a]See Table 3.6. [b]Hotels and restaurants.

[c]Public houses and clubs. [d]Cafes and snack bars.

Any slight discrepancies between figures are due to the rounding of decimal points.

Source: New Earnings Survey, Tables 1, 48 and 49 (1980), 43 and 44 (1991, 1992, 1993).

Some casual labour may not necessarily be cheaper in absolute terms. Some employers find it more expedient to contract an agency to muster 50 waiting staff for a banquet (see below) than to organize such an arrangement 'in house'. However, this will cost more because the agency's fee will probably amount to more than the 'going' employees' rate.

Data from the WIRS3 show that fewer employees worked on short-term fixed-term contracts in hotels and catering and private sector services than in all industries and services and all service industries. Such a practice was most likely to be found in workplaces employing 100 employees or more (see also Kitching, 1994, pp. 131–5).

Lee-Ross (1993) has identified that seasonal seaside jobs in hotels would appear to have a low skill content relative to that of other service jobs. Ball (1988) challenges the notion that such work is the poor relation because it provides valuable temporary work and job experience for 'voluntary' participants (for example, students) and respite from unemployment for 'involuntary' participants.

Self-employment

SIC Division 6, which includes hotels and catering, has the highest number of self-employed persons (Campbell and Daly, 1992, p. 271). There were an estimated 167,000 self-employed persons working in the hotel and catering industry in 1990 (HCTC, 1993). Around one-third of managers are self-employed proprietors (Purcell, 1993, p. 128). Sixty per cent of outlets in the commercial sector are owned or operated by self-employed people (HCTC, 1994a, p. 23). The numbers of self-employed actually fell by some 30,000 between 1988 and 1990, although the number of people in employment rose over the same period.

Self-employed workers contract their labour on a 'for services' basis and are not covered by most employment protection rights. The primary distinction for self-employed status is responsibility for payment of income tax and National Insurance contributions (Campbell and Daly, 1992, p. 292), including making personal pension arrangements. Employers have no obligation to provide employment terms and conditions (such as holidays and sick pay), but do have certain legal obligations (such as health and safety). For further commentary, see Kitching (1994, pp. 130–1).

Other employment variations and options

Shifts and unsocial hours

Much part-time and other employment is constituted as shiftworking. The hotel and catering industry has one of the highest incidences of shiftworking among all industries and services (Millward *et al.*, 1992, p. 346). As data from the WIRS3 in Table 3.8 show, 84 per cent of establishments in hotels and catering have shiftworking compared to around one-third of establishments in other sectors. Shiftworking was almost always found in hotel and catering establishments employing 50 or more employees.

Some full-time jobs are divided into split shifts (for example, much kitchen work), whereas other jobs may cover a complete unsocial hours shift (for instance, night portering). Although premium pay for some 'unsocial' hours arrangements were fixed by the wages councils until 1986, a subsequent study found little evidence of such premia being paid (Lucas, 1991, p. 279) and this position is supported by evidence from the New Earnings Survey discussed in Chapter 5 (see Table 5.2).

Table 3.8 *Shiftwork at establishment.*

	AIS	SI	PSS (by no. of employees) All	25–49	50–99	100+	HCI (by no. of employees) All	25–49	50–99	100+
Unweighted base	2061	1299	702	174	153	375	61	21	18	22
Weighted base	2000	1462	886	518	218	151	99	59	26	14
Yes (%)	37	37	33	26	38	49	84	77	95	95

Any slight discrepancies between figures are due to the rounding of decimal points.
Source: WIRS3 (1990).

Contracting out

Many catering businesses contract out some activities, the most common of which are laundry services and the preparation of convenience foods, such as cook–chill and cook–freeze. Other potential areas of contracting out include cleaning services, the use of agency staff and the management of on-site facilities such as shops and leisure centres.

Data from the WIRS3 show that hotels and catering are less likely (11 per cent) to utilize individuals operating on a freelance basis than elsewhere (around 15 per cent). There are no cases of outworkers or homeworkers being used in hotels and catering. There are some differences between hotels and catering and elsewhere in terms of services carried out by people not employed at the establishment, shown in Table 3.9. The very high incidence of in-house catering in hotels and catering does not necessarily mean that many establishments do not use cook–chill or cook–freeze methods. The WIRS3 also shows that a smaller proportion of hotel and catering establishments 'contracted out' cleaning, building maintenance, printing, photocopying and pay roll to another part of the organization compared to elsewhere.

Table 3.9 *Services carried out by people not employed at establishment*

	AIS	SI	PSS (by no. of employees) All	25–49	50–99	100+	HCI (by no. of employees) All	25–49	50–99	100+
Unweighted base	2061	1299	702	174	153	375	61	21	18	22
Weighted base	2000	1462	886	518	218	151	99	59	26	14
Cleaning (%)	48	51	53	50	51	67	34	40	25	26
Security (%)	27	26	32	33	28	37	21	23	16	22
Catering (%)	25	27	18	16	18	26	1	1	–	[a]
Maintenance (%)	67	72	70	74	66	61	69	75	59	64
Printing/ photocopying (%)	31	31	36	38	35	29	29	22	46	27
Pay roll (%)	43	49	38	40	37	31	46	48	44	45
Transport (%)	49	47	45	44	43	51	31	32	25	35

[a]Less than 0.5 per cent.
Any slight discrepancies between figures are due to the rounding of decimal points.
Source: WIRS3 (1990).

Government-subsidized trainees

The potential cost benefits of utilizing youth trainees are not inconsiderable because the government effectively pays their wages for two years through the training allowance. However, they seem to be less likely to be found in small businesses which would seemingly accrue most benefit from such an arrangement (Lucas, 1991, pp. 278–9).

Summary proposition

- Managers have the option to exercise wide discretion over the forms of employment that are used. The resultant diversity and complexity of such arrangements may create more difficulties in terms of managing staff.

Discussion points

- What factors influence managers to choose particular employment forms?
- What are the benefits and drawbacks of each type of employment form for managers and for workers?
- Identify an appropriate mix of employment forms for a unit of your choice. Justify your decision.

3.5 THE NATURE OF WORK

Forms of employment are inextricably linked to the nature of work and the way in which it is organized. Perhaps most appropriate to the subject of employee relations is the labour process approach, which broadly argues that managerial strategies have deliberately sought to deskill work in order to increase managerial control (see for example, Thompson, 1987). Work organization and working in hotels and catering are more comprehensively dealt with by others (for example, Bagguley, 1987; Gabriel, 1988; Wood, 1992). Wood (1992) speaks for most commentators by noting that 'For many employees – probably the majority – labouring in hotels and catering is a last ditch option characterised by exploitative and conflictual relationships' (p. 16). This section attempts only to outline some of the key issues in order to illustrate certain points that are made about work at appropriate junctures throughout the book, and to present evidence about work organization from the WIRS3.

Wood (1992) asserts that 'the preoccupations of mainstream industrial sociologists have left the analysis of hotel and catering work relatively untouched' (p. 8) (see also Wood, 1993; 1994, pp. 41–53). Even so, the orthodox paradigm that hotels and catering is unique and beset by special problems that largely relate to food service staff has grown out of a very narrow occupation-based approach to the study of hotel and catering employment (for a discussion of occupations see Wood, 1992, pp. 61–91). As noted in Chapter 2, there is little research that has addressed the issue of organizational structure. Thus Wood identifies problems in attempting to synthesize an account of hotel and catering work, for reasons that are similar to the factors that inhibit a synthesis of employee relations in the industry (noted in Chapter 1).

Chivers (1973) and Bagguley (1987) probably offer some of the best accounts of the restructuring of work, primarily among food preparation staff. Chefs' work has been restructured through technical change, and is considered to have been subject to systematic deskilling since the early 1950s; a similar trend has also been identified among dishwashing staff. Bagguley (1987) thus argues that a strategy of economic restructuring has driven the movement towards increased flexibility, a recurrent theme in contemporary discourse that is neatly encapsulated by the phrase 'the McDonaldization of the economy' (Gabriel, 1988, p. 4). The HCTC has hinted that employment restructuring in the fast-food industry, rather than a growth in numbers employed, is responsible for increased fast-food turnover in the late 1980s (HCTC, 1992, pp. 23–4) (see also the section on managerial strategies in Chapter 4).

Evidence from the WIRS3 shows that work reorganization is more an outcome of managerial prerogative in hotels and catering than elsewhere. Rather more workplaces in the industry (81 per cent) responded that management was able to organize work as it wished among non-managerial employees than was the case in the rest of the economy (all industries and services, 67 per cent; service industries, 63 per cent; private sector services, 74 per cent). The higher incidence of trade union organization in sectors outside hotels and catering would seem to be the main factor to explain this difference. In answer to a question on specific limits to the way in which management can organize work, union constraints (for example, formal agreements with trade unions and opposition from shop stewards) were the main reasons given. Hotel and catering workplaces reported unanimously that there was no union constraint of any kind in this regard. The main factor limiting the way in which management could organize work in hotels and catering was the lack of skills among the workforce, but this only applied in 8 per cent of cases (see also Bagguley, 1987, p. 41).

Interestingly, the WIRS3 shows that in the three years prior to 1990 changes in working practices designed to reduce job demarcation or increase flexibility occurred in a smaller proportion of workplaces in hotels and catering (18 per cent) than in other sectors (typically 36 per cent or slightly above). Where working practices had changed, the sections of the workforce most affected in hotels and catering were semi-skilled manual (8 per cent of establishments) and skilled manual (36 per cent of establishments).

The WIRS3 provides further evidence that hotels and catering was 'flexible' before flexibility became fashionable in the 1980s, a point supported by Kitching (1994, p. 140), who found no evidence of a new approach to flexibility in small firms, including those in hotels and catering. It is conceivable that the opportunities to become even more flexible may be more limited in such firms.

Summary proposition

- Lack of constraining forces and influences may have given hotel and catering managers more freedom to organize and reorganize work than managers in other employment sectors.

Discussion points

- What evidence of deskilling have you come across? Did this affect particular occupations or age groups? Were women more likely to be affected by deskilling than men?

- Is further deskilling inevitable and, if so, where is this likely to occur?
- Should managers have considerable freedom to organize work with little input from the workforce? If you agree, what are the consequences of this for the workforce? If you disagree, how can the workforce be more involved?

3.6 PRODUCTIVITY AND FLEXIBILITY

Even though the hotel and catering industry is more 'flexible' than most sectors elsewhere, at least in a numerically flexible sense, businesses need continually to evaluate, shape and reshape employment forms and working practices in order to adapt, survive and compete. But the means by which the industry does so may, in reality, militate against the achievement of increased efficiency, productivity, quality, commitment and motivation. Some of these issues are now explored very briefly, although they recur in later chapters.

The state of the art

Lockwood and Guerrier (1989) have considered the alternative strategies which organizations (15 major hotel groups) may pursue in order to achieve flexibility, have related these to methods of flexible working and have considered the potential problems that may result from management actions. What stands out from this matching exercise is an apparent lack of functionally flexible employment forms that are designed to improve workforce development, improve job satisfaction and motivation, and reduce labour turnover and recruitment problems. Furthermore, this is likely to result in poorer product and service quality, reduced staff commitment and poorer terms and conditions for staff. The main reasons for the lack of functional flexibility stem from implementation difficulties and problems related to established cultural norms such as departmentalization (see also Riley, 1992). Bagguley (1990) has suggested that the development of 'quite significant levels of functional flexibility during the 1960s ... raises the question of whether or not there is any room left for further functional flexibility on a significant scale' (p. 739) in the hotel and catering industry.

Lockwood and Guerrier (1989) identified that more extensive use of numerically flexible employment forms, such as part-time working, temporary/casual employment, shiftworking, using labour turnover and agency staff, was associated with strategies of managing peaks and troughs, economy and risk avoidance. There was a high reliance on *ad hoc* unplanned methods of achieving this type of employment flexibility, which should come as no surprise given the observations made about managing in Chapter 2, except that their study centred on businesses where more sophisticated practice could expected to be found. Interestingly, where flexible employment strategies have been pursued in public sector catering, the emphasis has been on numerical and pay flexibility (for example, contractors are no longer obliged to observe national agreements) rather than on functionally flexible means (Kelliher and McKenna, 1987, 1988; Kelliher, 1989). Thus the presence of trade unions does not appear to be a factor inhibiting the development of more functionally flexible work patterns.

Productivity implications

Measuring productivity in much of the hotel and catering industry is problematic (Ball *et al.*, 1986), although it is concerned with increasing the conversion rate of inputs, such as labour and materials, to outputs, such as value added per employee or net profit. There are simply too many unpredictable variables involved. A National Economic Development Council (1992) survey of productivity in hotels found a large degree of unsophistication in the way management attempted to use various techniques of operational management and productivity measurement, a point also noted by Witt and Witt (1989). For example, in response to a question about the use of various ratios to measure productivity, the most common ratio used was revenue : wage costs or wage costs : revenue, but productivity cannot be measured by reference to a single ratio alone (see also McMahon, 1994). Conversely, in more standardized operations, such as fast-food, productivity measurement ought to be more straightforward. However, the failure of a fast-food chain to incorporate quality into productivity assessment was deemed to be a significant omission given the presence of customers and their integral involvement in the service process (Ball and Johnson, 1989).

Although Prais *et al.*'s (1989) comparative study of labour productivity in hotels in Britain and Germany is probably the most comprehensive, a weakness is the omission of food and beverage operations. This was, however, deliberate, largely because of the difficulties in isolating unpredictable variables mentioned above. Instead the study focused on housekeeping and front office operations, which are not typical of most hotel and catering operations. Nevertheless, German hotels utilized half the labour per guest night of comparable British hotels. The main differentiating factor lay in the superior level of vocational training in Germany, and was not attributable to physical layout or superior equipment.

Guerrier and Lockwood (1989a), using the simple and simplistic definition that productivity is 'the efficient use of human resources', suggest that it will be affected by job design, motivation, skills and the ability to do the work and scheduling. The HCTC (1992) sees that 'Productivity in service industries is very largely a function of the quality of the workforce at all levels, especially skilled people, supervisors and managers' (p. 8), and that high productivity is closely related to effective training provision. Both definitions beg a number of questions about how labour productivity can be improved in the hotel and catering industry, seemingly ignoring, among other things, the effects of technology on output.

There is persuasive evidence to suggest that the industry's approach to flexible working, coupled with a poor training record, is not conducive to productivity improvement (Guerrier and Lockwood, 1989a; Prais *et al.*, 1989; NEDC, 1992; NEDO, 1992). Numerical flexibility is more representative of short-term cost cutting and a lack of strategic thinking (Hakim, 1990). These views are not shared by others, who view the utilization of part-time jobs as being central to efficiency (Walsh, 1990).

> In the hotel and catering sector over the period 1979–85 output barely increased although major growth occurred in employment figures and as a result productivity has actually declined. There is an obvious need for the service sector to make substantial improvements in productivity and

> a different approach to flexible working from that already adopted may provide the answer. (Guerrier and Lockwood, 1989a, p. 407)

Guerrier and Lockwood's thesis is that greater functional flexibility would improve productivity in hotels, but that the traditional management culture and style of 'being there' (with little or no emphasis on planning) and *ad hoc* decision making militate against its development. Additionally, managers are not productive if they trade their skills downwards, a criticism also levelled by Prais *et al.* (1989). Furthermore, part-time and casual staff are said to lack commitment and this results in lower output and service standards. Riley (1992) suggests that 'A realistic view might be that functional flexibility will evolve only when there is a constant qualitative demand from consumers which forces management to retain their staff' (p. 367).

Summary propositions

- Most hotel and catering businesses need to be managed on the basis of a combination of functionally flexible and numerically flexible means.
- Some businesses may have a propensity to operate on the basis of more functionally flexible means than others, but this may not always be appropriate for all types of business.
- It is questionable whether an appropriate balance between functional and numerical flexibility is being 'managed' effectively in some sectors of the industry, such that it works towards maximizing labour productivity.

Discussion points

- How can managers take steps to improve labour productivity?
- How can staff help an organization to become more productive?

REFERENCES

Airey, D.A. and Chopping, B.C. (1980) The labour market. In R. Kotas (ed.) *Managerial Economics for Hotel Operations*, pp. 45–69. Guildford: University of Surrey Press.

Analoui, F. and Kakabadse, A. (1993) Industrial conflict and its expressions. *Employee Relations*, **15** (1), 46–62.

Atkinson, J. (1984) Manpower strategies for flexible organizations. *Personnel Management*, August, 28–31.

Atkinson, J. and Meager, N. (1994) Running to stand still: the small firm in the labour market. In J. Atkinson and D. Storey (eds), *Employment, the Small Firm and the Labour Market*, pp. 28–102. London: Routledge.

Atkinson, J. and Storey, D. (1994) Small firms and employment. In J. Atkinson and D. Storey (eds), *Employment, the Small Firm and the Labour Market*, pp. 1–27. London: Routledge.

Bagguley, P. (1987) *Flexibility, Restructuring and Gender: Changing Employment in Britain's Hotels*. Lancaster Regionalism Group, University of Lancaster.

Bagguley, P. (1990) Gender and labour flexibility in hotel and catering. *The Service Industries Journal*, **10** (4), 737–47.

Baldacchino, G. (1994) Peculiar human resource management practices? A case study of a microstate hotel. *Tourism Management*, **5** (1), 46–52.

Ball, R.M. (1988) Seasonality: a problem for workers in the tourism labour market? *The Service Industries Journal*, **8** (4), 501–13.

Ball, S.D. and Johnson, K. (1989) Productivity management within fast food chains – a case study of Wimpy International. *International Journal of Hospitality Management*, **8** (4), 265–70.

Ball, S.D., Johnson, K. and Slattery, P. (1986) Labour productivity in hotels: an empirical investigation. *International Journal of Hospitality Management*, **5** (3), 141–8.

Bonn, A.N. and Forbringer, I.R. (1992) Reducing turnover in the hospitality industry: an overview of recruitment, selection and retention. *International Journal of Hospitality Management*, **11** (1), 47–63.

Bowey, A. (1976) *The Sociology of Organizations*. London: Hodder and Stoughton.

Boyd Ohlen, J. and West, J.J. (1993) An analysis of fringe benefit offerings on the turnover of hourly housekeeping workers in the hotel industry. *International Journal of Hospitality Management*, **12** (4), 323–36.

Byrne, D. (ed.) (1986) *Waiting for Change*. London: Low Pay Unit.

Campbell, M. and Daly, M. (1992) Self-employment into the 1990s. *Employment Gazette*, June, 269–92.

Chivers, T.S. (1973) The proletarianisation of a service worker. *Sociological Review*, **21**, 633–56.

Commission for Racial Equality (1991) *Working in Hotels*. London: Commission for Racial Equality.

Commission on Industrial Relations (1971) *The Hotel and Catering Industry, Part 1. Hotels and Restaurants, Report No. 23*. London: HMSO.

Crompton, R. and Sanderson, K. (1990) *Gendered Jobs and Social Change*. London: Unwin Hyman.

Denvir, A. and McMahon, F. (1992) Labour turnover in London hotels and the cost effectiveness of preventative measures. *International Journal of Hospitality Management*, **11** (2), 143–54.

Dickens, L. (1992a) *Whose Flexibility? Discrimination and Equality Issues in Atypical Work*. London: The Institute of Employment Rights.

Dickens, L. (1992b) Part-time employees: workers whose time has come? *Employee Relations*, **14** (2), 3–12.

Ellis, P. (1981) *The Image of Hotel and Catering Work*. London: HCITB.

Employment Department Group. (1992) *Labour Market and Skills Trends*. London: Employment Department.

Gabriel, Y. (1988) *Working Lives in Catering*. London: Routledge & Kegan Paul.

Guerrier, Y. and Lockwood, A. (1989a) Managing flexible working. *The Service Industries Journal*, **7** (3), 406–19.

Guerrier, Y. and Lockwood, A. (1989b) Core and peripheral employees in hotel operations. *Personnel Review*, **18** (1), 9–15.

Hakim, C. (1990) Core and periphery in employers' workforce strategies: evidence from the 1987 ELUS survey. *Work, Employment and Society*, **4** (2), 157–88.

Hotel and Catering Economic Development Committee (1968) *Why Tipping?* London: HMSO.

Hotel and Catering Economic Development Committee (1969) *Staff Turnover*. London: HMSO.

Hotel and Catering Industry Training Board (1983a) *Hotel and Catering Skills – Now and in the Future. Part 1, Summary. Report Prepared for the Education and Training Advisory Council.* Wembley: HCITB.

Hotel and Catering Industry Training Board (1983b) *Hotel and Catering Skills – Now and in the Future. Part 2, Jobs and Skills. Report Prepared for the Education and Training Advisory Council.* Wembley: HCITB.

Hotel and Catering Industry Training Board (1983c) *Hotel and Catering Skills – Now and in the Future. Part 3, Review of Vocational Education and Training. Report Prepared for the Education and Training Advisory Council.* Wembley: HCITB.

Hotel and Catering Industry Training Board (1983d) *Hotel and Catering Skills – Now and in the Future. Part 4, Meeting Future Needs. Report Prepared for the Education and Training Advisory Council.* Wembley: HCITB.

Hotel and Catering Industry Training Board (1983e) *Hotel and Catering Skills – Now and in the Future. Part 5, Questionnaires. Report Prepared for the Education and Training Advisory Council.* Wembley: HCITB.

Hotel and Catering Industry Training Board (1984a) *Manpower Flows in the Hotel and Catering Industry*. Wembley: HCITB.

Hotel and Catering Industry Training Board (1984b) *Manpower Forecasts for the Hotel and Catering Industry: Supplementary Report to Hotel and Catering Skills – Now and in the Future. Report Prepared for the Education and Training Advisory Council.* Wembley: HCITB.

Hotel, Catering and Institutional Management Association (HCIMA) and Touche Ross Greene Belfield-Smith Division (1992) *Salaries and Benefits in the Hotel and Catering Industry 1991 Survey, Volume 1, Survey*. London: Touche Ross Greene Belfield-Smith Division.

Hotel and Catering Training Board (1987) *Women in the Hotel and Catering Industry*. Wembley: HCTB.

Hotel and Catering Training Board (1988) *New Employment Trends and Forecasts 1988–1993*. Wembley: HCTB.

Hotel and Catering Training Company (1990) *Employee Relations in the Hotel and Catering Industry*, 7th edition. London: HCTC.

Hotel and Catering Training Company (1992) *Meeting Competence Needs in the Hotel and Catering Industry – Now and in the Future*. London: HCTC.

Hotel and Catering Training Company (1993) *Employment Forecasts Update 1992–2000 – Meeting Competence Needs in the Catering and Hospitality Industry and Licensed Trade*. London: HCTC.

Hotel and Catering Training Company (1994a) *Catering and Hospitality Industry – Key Facts and Figures*. London: HCTC.

Hotel and Catering Training Company. (1994b) *Employment Flows in the Catering and Hospitality Industry*. London: HCTC.

Jameson, S.M. and Hamylton, K. (1992) The CRE's investigation into the UK hotel industry. *International Journal of Contemporary Hospitality Management*, **4** (2), 21–6.

Johnson, K. (1980) Staff turnover in hotels. *Hospitality*, February, 28–36.

Johnson, K. (1981) Towards an understanding of labour turnover? *Service Industries Review*, **1** (1), 4–17.

Johnson, K. (1985) Labour turnover in hotels – revisited. *The Service Industries Journal*, **5** (2), 135–52.

Johnson, K. (1986) Labour turnover in hotels – an update. *The Service Industries Journal*, **6** (3), 362–80.

Kelliher, C. (1984) An investigation into non-strike conflict in the hotel and catering industry. MA thesis, University of Warwick.

Kelliher, C. (1989) Flexibility in employment: developments in the hospitality industry. *International Journal of Hospitality Management*, **8** (2), 157–66.

Kelliher, C. and McKenna, S. (1987) Contract caterers and public sector catering. *Employee Relations*, **9** (1), 8–13.

Kelliher, C. and McKenna, S. (1988) The employment implications of government policy: a case study in public sector catering. *Employee Relations*, **10** (1), 8–13.

Kitching, J. (1994) Employers' work-force construction policies in the small service sector enterprise. In J. Atkinson and D. Storey (eds), *Employment, the Small Firm and the Labour Market*, pp. 103–146. London: Routledge.

Lashley, C. (1994) Is there any power in empowerment? Paper presented to the Third Annual CHME Research Conference, Napier University, April.

Lee-Ross, D. (1993) An investigation of 'core job dimensions' amongst seaside hotel workers. *International Journal of Hospitality Management*, **12** (2), 121–6.

Lockwood, A. and Guerrier, Y. (1989) Flexible working in the hospitality industry: current strategies and future potential. *Contemporary Hospitality Management*, **1** (1), 11– 16.

Lucas, R.E. (1991) Remuneration practice in a wages council sector: some empirical observations in hotels. *Industrial Relations Journal*, **22** (4), 273–85.

Lucas, R.E. (1992) Employment trends in the hotel and catering industry in the 1980s. Paper presented to the International Association of Hotel Management Schools Conference, Manchester, May.

Lucas, R.E. (1993a) Some age-related issues in restaurant employment in Greater Manchester. Paper presented to the CHME Second Research Conference, Manchester Metropolitan University, April.

Lucas, R.E. (1993b) Ageism and the UK hospitality industry. *Employee Relations*, **15** (2), 33–41.

Lucas, R.E. (1993c) The Social Charter – opportunity or threat to employment practice in the UK hospitality industry? *International Journal of Hospitality Management*, **12** (1), 89–100.

Lucas, R.E. (1993d) Hospitality industry employment – emerging trends. *International Journal of Contemporary Hospitality Management*, **5** (5), 23–6.

Lucas, R.E. (1994) Trends in British hotel and catering employment in the 1980s. *Tourism Management*, **15** (2), 145–50.

Lucas, R.E. (1995) Some age-related issues in hotel and catering employment. *The Service Industries Journal*, **15**(2), 234–50.

Lucas, R.E. and Bailey, G. (1993) Youth pay in catering and retailing. *Personnel Review*, **22** (7), 15–29.

Lucas, R.E. and Jeffries, L.P. (1991) The 'demographic timebomb' and how some hospitality employers are responding to the challenge. *International Journal of Hospitality Management*, **10** (4), 323–37.

Lynch, P. (1994) Demand for training by bed and breakfast operators. *International Journal of Contemporary Hospitality Management*, **6** (4), 25–31.

McMahon, F. (1994) Productivity in hotel industry. In A.V. Seaton, C.L. Jenkins, R.C. Wood, P.U.C. Dieke, M.M. Bennett, L. R. MacLellan and R.Smith (eds), *Tourism: the State of the Art*, pp. 616–25. Chichester: John Wiley & Sons.

Mars, G., Bryant, D. and Mitchell, P. (1979) *Manpower Problems in the Hotel and Catering Industry*. Farnborough: Gower.

Millward, N., Stevens, M., Smart, D. and Hawes, W. R. (1992) *Workplace Industrial Relations in Transition*. Aldershot: Dartmouth Publishing Company.

Mitchell, P. (1988) The structure of labour markets in the hotel and catering industry: what do employment law cases indicate? *The Service Industries Journal*, **8** (4), 470–87.

Mumford, E. (1972) Job satisfaction: a method of analysis. *Personnel Review*, **1** (3), 48–57.

National Economic Development Council (1991) *Developing Managers for Tourism*. London: NEDC Tourism and Leisure Studies Sector Group.

National Economic Development Council (1992) *Costs and Manpower Productivity in Hotels*. London: NEDO.

National Economic Development Office (1992) *UK Tourism: Competing for Growth*. London: NEDO.

Parsons, D. (1992) Developments in the UK tourism and leisure labour market. In R. Lindley (ed.) *Women's Employment: Britain and the Single European Market*, pp. 101–12. London: Equal Opportunities Commission, HMSO.

Perrewe, P.L., Brymer, R.A., Stepina, L.P. and Hassell, B.L. (1991) A causal model examining the effects of age discrimination on employee and psychological reactions and subsequent turnover intentions. *International Journal of Hospitality Management*, **10** (3), 245–60.

Prais, S.J., Jarvis, V. and Wagner, K. (1989) Productivity and vocational skills in services in Britain and Germany: hotels. *National Institute Economic Review*, November, 52–74.

Price, L. (1993) The limitations of the law in influencing employment practices in UK hotels and restaurants. *Employee Relations*, **15** (2), 16–24.

Price, L. (1994) Poor personnel practice in the hotel and catering industry: does it matter? *Human Resource Management Journal*, **4** (4), 44–62.

Purcell, K. (1993) Equal opportunities in the hospitality industry: custom and credentials. *International Journal of Hospitality Management*, **12**, (2), 127–40.

Ralston, L.M. and Lucas, R.E. (1994) Youth employment in hotels and catering – beyond demographic phenomena. Paper presented to the Third CHME Research Conference, Napier University, April.

Riley, M. (1980) The role of mobility in the development of skills for the hotel and catering industry. *Hospitality*, March, 52–3.

Riley, M. (1981a) Recruitment, labour turnover, and occupational rigidity: an essential relationship. *Hospitality*, March, 22–5.

Riley, M. (1981b) Labour turnover and recruitment costs. *Hospitality*, September, 27–9.

Riley, M. (1990) The labour retention strategies of UK hotel managers. *The Service Industries Journal*, **10** (3), 614–18.

Riley, M. (1991) An analysis of hotel labour markets. In C.P. Cooper (ed.), *Progress in Tourism, Recreation and Hospitality Management, Vol. 3*, pp. 232–46. London: Belhaven Press.

Riley, M. (1992) Functional flexibility in hotels – is it feasible? *Tourism Management*, **13** (4), 363–7.

Riley, M. (1993) Labour turnover: time to change the paradigm? *International Journal of Contemporary Hospitality Management*, **5** (4), Viewpoint, i–iii.

Robinson, O. and Wallace, J. (1983) Employment trends in the hotel and catering industry in Great Britain. *The Service Industries Journal*, **3** (3), 260–78.

Simms, J., Hales, C., and Riley, M. (1988) Examination of the concept of internal labour markets in UK hotels. *Tourism Management*, **9** (1), 3–12.

Taylor, D. (1983) The migrant contribution to the British hotel industry. *International Journal of Hospitality Management*, **2** (2), 61–8.

Thompson, P. (1987) *The Nature of Work: an Introduction to Debates on the Labour Process*, 2nd edition. London: Macmillan.

Timmo, N. (1993) Employment relations and labour markets in the tourism and hospitality industry. *International Journal of Employment Studies*, **1** (1), 33–50.

Walsh, T. (1990) Flexible labour utilization in the private service sector. *Work, Employment and Society*, **4** (4), 517–30.

Witt, C.A. and Witt, S.F. (1989) Why productivity in the hotel sector is low. *International Journal of Contemporary Hospitality Management*, **1** (2), 28–34.

Wood, R.C. (1992) *Working in Hotels and Catering*. London: Routledge.

Wood, R.C. (1993) Status and hotel and catering work. *Hospitality Research Journal*, **16** (3), 3–15.

Wood, R. C. (1994) *Organizational Behaviour for Hospitality Management*. Oxford: Butterworth Heinemann.

PART TWO
THE EMPLOYMENT RELATIONSHIP

FOUR

The Employment Relationship: Perspectives and Participants

The purposes of this chapter are to:

- identify the nature of the employment relationship in hotels and catering;
- consider the perspectives from which the employment relationship may be managed;
- discuss the role and attitudes of the main institutions and parties that influence the employment relationship.

For the purpose of this chapter, the terms employee relations and industrial relations are treated as broadly synonymous because both are concerned with the employment relationship, in one way or another. Their respective presence largely reflects their usage in the appropriate source material.

4.1 WHAT IS THE EMPLOYMENT RELATIONSHIP?

Employee relations and the employment relationship are defined in different ways by different commentators, which largely reflect the particular perspective of the author. Nevertheless, the working definition of employee relations cited at the beginning of Chapter 1 – employee relations in hotels and catering is about the management of employment and work relationships between managers and workers and, sometimes, customers; it also covers contemporary employment and work practices – has provided a useful starting point because it acknowledges a broader view of the employment relationship than implied in many other definitions.

However, it can be said that employee relations is also about control over the employment relationship and the organization of work (Gospel and Palmer, 1993, p. 3). The existence of a wide variety of work patterns (identified in Chapter 3), such that some workers are not employees, requires us to use the term employment in a looser sense than as a reference to the conventional employer–employee relationship. In some service occupations the influence of customers on the employment relationship is much more direct than virtually anywhere else in the economy.

Gospel and Palmer's (1993) definition of the employment relationship as 'an economic, social and political relationship for which employees provide manual and mental labour in exchange for rewards allotted by employers' (p. 3) is thus too narrow in absolute terms, although it acknowledges the complexity of the relationship that centres on the reward-effort bargain or exchange. The *rewards* from employment may be economic, social and psychological. While the job of a top chef may be highly paid, powerful and prestigious, the job of kitchen porter is usually degrading, dead-end and badly paid. Rewards, both extrinsic and intrinsic, may be heavily influenced by customers (for example, affecting staff in restaurants, bars and front office). Individual *effort* may be rigidly specified and controlled (most operational jobs in fast-food) or be open to individual interpretation and initiative (chefs, managers and proprietors).

Within the reward-effort bargain, it is customary to distinguish between two sets of issues or rules. The *substantive issues* are about pay and conditions of employment, and about levels of performance in an absolute and relative sense. The *procedural issues* are about how the substantive issues are decided – they are about power, both legitimate and illicit, and administrative arrangements that may be formal and informal.

Rose (1988) argues that conflict and cooperation are always found together (a bipolar approach) in the employment relationship (see pluralist perspectives below). Here conflict has traditionally been viewed as taking two discrete forms. Conflicts of right are disputes about the application of existing rules. Conflicts of interest are disputes about the development of new rules. Successful conflict resolution must build from a basis of some consent and cooperation. More recently there has been more emphasis of the role of cooperation or consent in the employment relationship, particularly in HRM (see, for example, Guest, 1987) and elsewhere (see, for example, Sturdy *et al.*, 1992).

The employment relationship may exist where there is a contract of employment or a contract for services and in circumstances of self-employment. The job – managerial, professional, technical, skilled, semi-skilled or unskilled – may be full-time, part-time, fixed-term, seasonal, temporary or casual. The 'employing' firm may be a large public company, such as McDonald's, with a bureaucratic structure, or a small self-managed business, such as a boarding house, with an entrepreneurial structure. The employment relationship is multifactorial and multifarious and is important to those associated with it, leading Gospel and Palmer (1993) to note that 'it is hardly surprising that there will be many people and groups interested in influencing the employment relationship and the way work is organized, or that the processes that shape these phenomena should form a complex area for study' (p. 4).

Summary proposition

- The employment relationship in the hotel and catering industry is diverse and complex. The influence of customers at workplace level has the potential to differentiate the industry from many other employment sectors.

Discussion points

- Are conflict and cooperation always found together?
- In what ways can conflict and cooperation be manifested?

4.2 PERSPECTIVES ON THE EMPLOYMENT RELATIONSHIP

The way we perceive (our frame of reference for) the nature of employee relations is related to how we approach and analyse specific issues and situations, how we expect others to behave and our response to their behaviour, and the means we deploy to influence or modify that behaviour (Salamon, 1992, p. 29). Additionally, precise definitions of the subject are impossible, given that it is 'concerned with subjective, value judgements about concepts for which there are no universally accepted criteria' (Salamon, 1992, p. 58). Questions of what is right or wrong in any given situation depend on concepts of fairness and equity, which in turn are linked to the exercise of power and authority, notions of individualism and collectivism, integrity and the establishment of trust (see Fox, 1985).

In spite of these difficulties, it is generally agreed that there are three major approaches to managing employee relations (the author's perspective is contained in Appendix 2):

- the *unitary perspective* views the organization as a coherent and integrated team 'unified by a common purpose' (Fox, 1966, p. 2);
- the *pluralist perspective* views the organization as 'a miniature democratic state composed of sectional groups with divergent interests over which the government tries to maintain some kind of dynamic equilibrium' (Fox, 1966, p. 2);
- the *Marxist perspective* 'emphasises the organization as a microcosm and replica of the society within which it exists' (Salamon, 1992, p. 31).

In reality, the variation within each perspective and the differences between them are considerable. The unitary perspective can involve an authoritarian or paternalistic style of management; cooperation and conflict are inherent in pluralism; and Marxism may imply either evolutionary or revolutionary social change. Salamon (1992) asserts that the three major perspectives have manifested themselves differently into four types of approach that inform the analysis of the employment relationship. The unitarist perspective underpins the HRM approach; the pluralist perspective has influenced the input–output and systems approaches; and the Marxist perspective has driven the 'control of the labour process approach'. He argues that the HRM and labour process approaches 'represent opposite forms of analysis of the same phenomenon – control of the human activity of paid work' (p. 39). Gospel and Palmer (1993) have proposed five broad theoretical perspectives on industrial (employee) relations, derived from these three major approaches.

Unitary perspectives

The unitary perspective assumes that the organization is an integrated group of people who espouse a single authority/loyalty structure, a set of common values, goals and objectives. In other words cooperation is the natural order of things. Managerial prerogative is absolute and any opposition to it is irrational. Thus conflict is unnecessary and must derive from deviance. Trade unions are seen as intrusions into the organization that compete with management for employee loyalty.

Gospel and Palmer (1993) suggest that the unitary perspective has 'a simple ideological appeal' that is subscribed to by many managers and often surfaces in media comment as a simple way of explaining things. A more sophisticated variant of unitarism is associated with the Human Relations School, influenced by Elton Mayo's development of the workplace as a community, and is associated with job enrichment, employee involvement and participation. Some, such as Keenoy (1990), see HRM, which is reminiscent of the Human Relations School, as unitarism by another name. While others also consider that HRM is unitary in nature (Guest, 1987; Purcell, 1987; Salamon, 1992; Gospel and Palmer, 1993), this view is not a universally held one.

Pluralist perspectives

Pluralism assumes that individuals in organizations form distinct sectional groups which espouse particular interests, objectives and leadership styles. Thus the organization is structured on a variety of levels and is competitive in terms of its constituent elements, which give rise to a 'complex of tensions and competing claims which have to be "managed" in the interests of maintaining a viable collaborative structure' (Fox, 1973, p. 193).

Three variations of the pluralist perspective proposed by Gospel and Palmer (1993), although discrete, can be traced in the historical development of post-war British industrial relations (see Kessler and Bayliss, 1992, pp. 1–65). These variations also encompass the input–output and systems models.

Liberal collectivism and collective bargaining

This dimension of pluralism derives from the work of Sydney and Beatrice Webb and the 'Oxford School' (Allan Flanders, Hugh Clegg, Otto Kahn-Freund, Alan Fox and Bill (now Lord) McCarthy). Collective bargaining between employers and trade unions is the best method of institutionalizing conflict at the workplace and establishing employment rules, with government playing a limited role. The process, which seeks to regulate the labour market, requires a political system that allows employers and employees to associate freely, a mutual recognition of divergent interests between the parties and a willingness to accept compromise. The negotiating positions of both sides are backed by economic sanctions: the employees' right to take industrial action, often in the form of strike action, and the employers' right to dismiss employees in dispute as a last resort. Collective bargaining has been under attack in Britain since the 1980s.

Attempts to develop a more formal academic theory of industrial relations within this dimension of pluralism have been manifested in the input–output and systems models. The input– output model regards industrial relations as a process for converting conflict into regulation but does not provide an adequate framework for understanding the integrative nature of parts of industrial relations and its relationship to wider society (Salamon, 1992). A more useful approach, underlying Gospel and Palmer's approach and that of others (Wood *et al.*, 1975), derives from the influential work of Dunlop (1958), who, through the development of a general theory of industrial relations, proposed an industrial relations system which, 'at any one time in its development is regarded as comprised of certain actors, certain contexts, an ideology which binds the industrial relations system together, and a body of rules created to govern the actors at the work place and work community' (p. 7). Here the actors (managers, workers and government agencies) interact in an environmental context (technology, the market and societal power) to produce workplace and work community rules. In spite of its limitations of not fully explaining industrial relations issues or events, Dunlop's approach is a useful analytical tool to organize facts. Dunlop's model has been modified to produce an industrial relations system for hotels (Mars and Mitchell, 1976, p. 5). To replicate such a picture for hotels and catering is more problematic because of the complex range of variables involved and the unknown ways in which they interplay and react; more empirical work is needed here.

Corporatism

This dimension, and the third, 'give more attention to the questions of the relative power of different interest groups and the role of the state' (Gospel and Palmer, 1993, p. 19). Corporatism implies more state intervention and tripartite decision making, involving government, employer and employee representatives. This occurred in Britain in the 1970s, most notably at national level, when state intervention (for example, incomes policies) was necessary to regulate the activities of powerful groups for the national economic good. Variants of corporatism are more widely accepted in Europe, and have been strongly implied in some of the developments associated with the single European market. This approach has been vehemently rejected by British Conservative governments of the 1980s and 1990s.

Liberal individualism and the neo-laissez-faire

This contemporary dimension recognizes that organizations are best led by strong managers but acknowledges that there are conflicts of economic interest between employers and employees. Collective bargaining, organized groups (employers or trade unions) and any form of corporatism are to be deplored. The linchpin is 'the individual contract of employment as determined by market forces and common law notions of the rule of law' (Gospel and Palmer, 1993, p. 23). In pursuit of a deregulated labour market, trade unions have been marginalized (Marchington and Parker, 1990, pp. 203–28), protective legislation has been reduced and state interventions in the employment relationship, such as wages councils, have been removed. Marchington and Parker (1990) have identified two main themes in the recent management of employee relations as 'macho management' and HRM. HRM is more likely to be found in unionized firms (Sisson, 1993). Such developments, often referred to as the 'new industrial relations', are said to have generated a renaissance of unitarist ideas among employers and have placed more emphasis on individual relations in organizations. Evidence from the WIRS3 is ambiguous to the extent that it can be used to confirm or deny the existence of the 'new industrial relations' (Millward, 1994).

Radical perspectives

Although radicalism can be traced back to Robert Owen and the Chartists in the early nineteenth century, it is now most closely associated with variations of Marxism. The Marxist perspective (radical but also pluralist) views the employment relationship not in organizational terms but in social, political and economic terms (Salamon,

1992, p. 39). It has a wide following and provides a radical critique of the other perspectives (see, for example, Hyman, 1975, 1989) and has also served as a basis from which to review the employment relationship and the labour process in capitalist society (for example, Thompson, 1987). The labour process approach has underpinned a number of industry studies on work discussed in this book (for example, Bagguley, 1987, 1990; Gabriel, 1988).

Implications for hotel and catering

It is probable that if you asked hotel and catering managers to conceptualize their approach to managing employee relations, most would probably not have thought about it in this way but would nevertheless articulate a number of statements about management's right to manage, with perhaps an acknowledgment that staff have a right to some kind of involvement, most probably through management-initiated means. In other words, managers do adopt a perspective but may not consciously recognize they do so. What more direct evidence we have from Chapters 2 and 3 suggests that the theoretical perspectives of unitarism and the liberal-individualist and the neo-*laissez-faire* variants of pluralism have most practical application in the industry. What is interesting here is that while this variant of pluralism is associated with the 1980s, this approach, or at least a close relation of it, has underpinned managerial thinking in the hotel and catering industry since at least the 1960s.

Summary propositions

- Managers may not be consciously aware of how their beliefs, values and expectations of behaviour influence and guide the way they manage employee relations.
- Most managers in the hotel and catering industry probably subscribe to the unitary or the pluralist liberal-individual and the neo-*laissez-faire* approaches to managing employee relations.

Discussion points

- What are the positive and negative effects of managing from a unitary perspective for managers and the workforce?
- What are the positive and negative effects of managing from a pluralist perspective for managers and the workforce?
- Which perspective do you most identify with? Why?

4.3 WHO HAS AN INTEREST IN THE EMPLOYMENT RELATIONSHIP?

Before the role and attitudes of the institutions and parties interested in influencing the employment relationship can be considered, the relevant actors need to be identified. For this purpose, the broad approach implied in Dunlop's systems model has been used. Riley (1994) holds the view that Dunlop's systems approach is the only approach that can adequately capture the determinants of the industry and workplace rules in hotels and catering, although he does not fully substantiate this claim.

For the remainder of the chapter, institutions and parties are differentiated as follows. Institutions are defined as the state and its agencies, including the Employment Department, Training and Enterprise Councils (TECs) – named Local Enterprise Companies (LECs) in Scotland – ACAS, the Health and Safety (HSE) Executive, the Health and Safety Commission (HSC), the Commission for Racial Equality (CRE), the Equal Opportunities Commission (EOC), pressure groups such as the Low Pay Network, industry-specific bodies such as the HCTC and the HCIMA, the media and the legal system (the HCTC (1990) provides a useful summary of institutions (pp. 10–20), and a list of their addresses (pp. 117–25)). The parties are employers and workers, who may be constituted as institutions such as trade unions, and customers.

4.4 INSTITUTIONS: ROLE AND ATTITUDES

The state

'The state can be described as the institutional system of political government with a monopoly over taxation and the legitimate use of force in a society. The state is not synonymous with government' (Gospel and Palmer, 1993, p. 154). The state is the guardian of the national interest and the ultimate custodian of the economic well being of the nation. The state is not a cohesive and single body; it comprises a number of institutions that impinge on the employment relationship, whose objectives do not necessarily coincide. These include the legislature (Parliament), the executive (government ministers), the judiciary, central administration (the civil service), local government and specialist agencies in the employment field.

Since the late nineteenth century, the state has been involved in regulating the employment relationship to varying extents. State-sponsored machinery to resolve industrial disputes (1896) and to fix legally enforceable minimum wages in selected industries (1909) are examples of early measures of involvement. Broadly speaking, the role of the state was relatively passive until the 1960s, which witnessed the creation of the industrial training boards, the beginnings of a mass of employment laws and more comprehensive incomes policies. State involvement became even more active during the 1970s in areas of incomes policy and employment law (see Chapters 9, 10 and 11) and through the creation of tripartite institutions such as ACAS. For a more detailed historical review of the role of the state, see Kessler and Bayliss (1992, pp. 4–39) and Gospel and Palmer (1993, pp. 154–73).

While the government has been keen to stress that it has been attempting to reduce its 'domestic' involvement in the employment relationship since the 1980s to enable employers better to 'call the tune', the extent to which this has been achieved remains questionable. The government has at the same time been forced to accept increased regulation in some areas as a result of Britain's European Community (EC) membership. On the one hand, the scope of some employment laws has been deliberately reduced; for example, in protection against unfair dismissal the qualifying service period was increased from six months to two years, and the wages councils were reformed (1986) and abolished (1993). On the other hand, domestic regulation has been increased in some areas, particularly in relation to trade unions (see Chapter 9), while European developments have forced changes in the fields of equal pay, sex discrimination, pensions and retirement and the transfer of undertakings (see Chapters 9 and 10). The practical effects of ostensibly differing roles for the state may come to the same thing. To some, the difference between a statutory incomes policy (corporatist Labour administrations of the 1970s) and the use of pay cash limits in the public sector (associated with the neo-*laissez-faire* Conservative governments of the 1980s and 1990s) is questionable. Kessler and Bayliss (1992, pp. 40–66) offer an excellent analysis and review of the economic background and government values and policies of the 1980s.

Other institutions

The Employment Department is the executive and administrative arm of the government's employment policy responsible to the Secretary of State for Employment. Its many functions include drafting employment legislation, commissioning and carrying out research, the compilation and publication of labour market statistics and running Job Centres. Training and Enterprise Councils (Local Enterprise Companies in Scotland) administer government-funded training programmes such as Youth Training and Employment Training.

ACAS provides three main types of service to employers and workers in furtherance of its aim of improving industrial relations. An advisory service (by telephone, publications or through a personal visit by an ACAS official) provides assistance on any employment matter, from writing a simple procedure to devising a job-evaluated payment system (Armstrong and Lucas, 1985; Kessler and Purcell, 1993/4). An arbitration service provides a panel of arbitrators who are available to provide final stage arbitration in disputes (Brown, 1992). A two-fold conciliation service offers individual conciliation to all industrial tribunal applicants and collective conciliation to the parties involved in major industrial disputes. Of its publications, *Discipline at Work: the ACAS Advisory Handbook* (ACAS, 1987) is probably the most influential (see Chapter 7).

The Health and Safety Executive, the Health and Safety Commission, the Commission for Racial Equality and the Equal Opportunities Commission exercise specific responsibilities for specialist employment legislation (see Chapters 10 and 11). Their roles include the publication of codes of practice, monitoring the effect of legislation and mounting investigations into specific complaints or problems.

Many other institutions, some of which have a political bias, offer a mixture of fact and opinion on employment matters. These include the Confederation of British Industry, the Institute of Directors, the Trades Union Congress, the Institute of Personnel and Development, the Low Pay Network, the Institute of Employment Rights, Labour Research, the Policy Studies Institute and the National Federation of the Self-employed. The HCTC's role is primarily about training. The media are often thought to portray a simplistic and unbalanced (too pro-employer) view of employment issues. The legal system, including the industrial tribunal system and other related institutions, is considered more fully in Chapters 9, 10 and 11.

Summary proposition

- In spite of pursuing deregulatory policies for over 15 years, the state still regulates the employment relationship directly and indirectly through a substantial framework of laws and institutions.

Discussion point

- What is the role of the state in the employment relationship in hotels and catering?

4.5 EMPLOYERS: ROLE AND ATTITUDES

Earlier chapters have dealt with some aspects of the role and attitudes of employers but some of these features now need to be considered more specifically in terms of the employment relationship. Employers are the prime movers and shapers of the employment relationship, regardless of whether the workforce is organized or unorganized, and are practically and theoretically more powerful than even the best mustered organized labour. Employers' institutions, such as the British Hospitality Association (formerly the British Hotels and Restaurants Association) and the Brewers' Society, have not played a major part in employee relations at an industry level, although it had been thought that a more active role would be of particular benefit to small hotels (CIR, 1971; HCEDC, 1975). Although such organizations held seats on the employers' side of the wages councils, they are not constituted as employers' associations and therefore have no negotiating mandate on behalf of the industry. Their role is largely limited to providing information and advice on employment matters to their members.

Although the hotel and catering industry contains some examples of large corporations that may operate multinationally or internationally, most hotel and catering businesses operate in what is termed the secondary economy, which is comprised of smaller and more fiercely competitive organizations. There is relative ease of entry into and exit from the market. Small, independently owned and managed businesses are less likely to provide good terms and conditions of employment and to follow systematic employment policies and procedures. This is, in part, related to the needs of such firms to minimize labour costs and maximize effort in order to remain price competitive and survive in highly turbulent conditions. For further discussion on the small firm and the labour market, see Atkinson and Storey (1994).

Management strategies

Cost and control

Although in sociological analysis the classical and radical theories of capitalist production work from the premise that employers treat their labour forces as a factor of production or a cost, employers are unlikely to state publicly that their objectives are 'to achieve maximum effort and production and maximum subordination to managerial objectives from their employees, in return for minimum costs' (Gospel and Palmer, 1993, p. 37). Central to the labour process debate (deriving from Braverman, 1974), is that employers have set out to deskill employees to gain closer control over employee behaviour and costs. Using a labour process approach, Bagguley (1987, 1990), Gabriel (1988) and McKenna (1990) have provided interesting evidence on employers' deskilling strategies and their effects on hotel and catering employment from both a macro- and a micro-perspective across different parts of the industry. Among other things, increased control has been secured through flexible labour strategies (Bagguley, 1987, 1990; Walsh, 1990), confrontation and technological means (Gabriel, 1988). Control in small establishments is likely to be personal, direct and *ad hoc*, typifying Edwards's (1979) 'simple' control.

Resource

An alternative strategy of HRM views the workforce as a resource or asset which is to be retained and developed in the longer-term through the deployment of enlightened and sophisticated employment policies. Although HRM has become an increasingly widely used term, it remains loosely defined around notions of individualism, involvement, commitment and the integration of personnel policies. If companies wish to remain competitive or improve performance they are exhorted to take a strategic approach to people.

A broadly consistent observation about HRM (that differentiates it from personnel management) is that it is strategic (Armstrong, 1987; Purcell, 1987; Thomas, 1990; Torrington and Hall, 1991), although there may be different strategies for different situations (Thomason, 1991). Some see two types of HRM, one that is strategic and the other that is non-strategic (Miller, 1987, 1989; Monks, 1992–3). Others see HRM as either 'soft', embracing a more caring attitude towards staff, or 'hard', where a more ruthless management regards employees as simply another resource to be managed (Storey, 1992). Keenoy (1990) is more sceptical and argues that 'HRM is best regarded as a patch-work quilt concept stitched together from the diverse currents of change that presently impinge upon the management of the employment relationship' (p. 4) and that 'the "practice" of strategic HRM has little or nothing to do with the "practice" of managing people in the employment relationship' (p. 5).

Studies on HRM and employee relations strategy in the hotel and catering industry have found little evidence that any kind of HRM approach is being followed, even among the larger organizations. In a study of four hotel groups, Croney (1988b) found that the management of labour did not resemble anything approaching HRM. Ralston (1989) concluded that trends in personnel management in the industry were more a response to changes in the labour market than a strategic approach. Kelliher (1989) observed that many private sector catering firms did not have a 'consciously strategic approach to managing their human resources or at least not in the sense of a formalised written policy' (p. 9). Hiemstra's study (1990) of strategic personnel management policies in the lodging industry found them to be far from adequate because they focused on the short-term rather than the long-term. Ishak and Murrmann (1990, p. 151) 'noted that the two top HRM issues related to restaurant managers are their retention and recruitment', and that their preoccupation with dealing with labour turnover left executives limited time to deal with strategic issues.

Externalization and the market

An economic approach which addresses the extent to which the market influences employers' strategies on employee relations is best explained using Gospel's (1992) application of the concepts of internalization and externalization to decisions related to the employment relationship in hotels and catering. Gospel argues that strategic externalization is most likely to exist where product markets are small, fragmented and competitive, where labour markets provide an ample supply of workers with general skills that are easily acquired in the market, where divisions of labour are simple and organizations lack the capacity to sustain strong internal systems.

Thus, in regard to work organization, hotel and catering firms rely on the external labour market, both to subcontract work out (laundry, pre-prepared foods) and to subcontract work in (the use of casual labour). They do little or no training, recruiting labour that others have trained or relying on others to provide skills training (City and Guilds courses, to be replaced by National Vocational Qualifications, for kitchen skills). In terms of the employment relationship, recruitment and lay-off are demand driven, higher positions can be filled by external or internal candidates, and wages are determined by external market signals. The employment contract is likely to be minimal, and the internal labour market will be poorly developed. In short, many hotel and catering organizations rely heavily on the external labour market and internalize little within the firm, a concept which has also been developed by Simms *et al.* (1988) in relation to explaining a perceived problem of high labour turnover. 'High turnover is seen as a mechanism by which the size of the labour force can be regulated in relation to current demand' (Kelliher, 1984, p. 57). A strategy that is simple, direct and personal, and reflects a low degree of bureaucratization and high reliance on the market (Gospel and Palmer, 1993, p. 53), would therefore seem to reflect one appropriate typology of employee relations strategy in the hotel and catering industry.

Individualism and collectivism

The last approach to employee relations strategy, expressed in terms of management style, draws on the work of Purcell and Sisson (1983) and, mainly, Purcell (1987). It also incorporates the cost and control and resource strategies outlined above. In aiming to move employee relations beyond 'the frame of reference debate' (Fox, 1966), Purcell (1987) identifies two dimensions of style – individualism and collectivism – the adoption of which (separately or together) are deliberate choices related to business policy.

Three variants of individualism are identified. Low individualism stems from the more 'ruthless' employers, who view employees as a commodity and as a cost concern (akin to the labour process approach). The middle variant – paternalism – centres on notions of caring and welfare. Finally, organizations with highly individualistic policies demonstrate strong leadership, and centre on the individual as a resource. Such organizations are most likely to be associated with HRM.

Both aspects of collectivism are concerned with 'the extent to which the organization recognizes the right of employees to have a say in those aspects of management decision-making which concern them' (Purcell, 1987, p. 338). In the first type, democratic structures representing employees must exist. The second type is based on the degree of management legitimation of collectivism, which is reflected in the extent to which legitimacy is accepted or opposed.

To avoid confusion with the way in which the term management style in the hotel and catering industry was used in Chapter 2 (for instance, expressed as reactive and *ad hoc* management), Purcell denies that those who dabble in short-term activities have a management style because their approach is characterized by 'pure short-run commercial logic'. Management style must be 'capable of reproduction', which would seem to imply something more deliberate than accidental, more Porter than Mintzberg (see, for example, Porter, 1980; Mintzberg and Waters, 1985; Mintzberg, 1987,

1990). Using Purcell's (1987) model, Kelliher (1989) has identified that private catering firms do not have a consciously strategic approach to employee relations, although management style is 'typically individualistic'. By contrast, in public sector catering, a clearly identifiable pattern of employee relations had been pursued that had moved from collectivism towards individualism through policies of competitive tendering and reducing the bargaining agenda.

Summary propositions

- Managerial strategy in the hotel and catering industry can be understood within a number of different theoretical frameworks.
- Strategies of cost and control, externalization and the market, and low individualism provide much more persuasive evidence of the dominant management styles in employee relations than HRM and collectivism.

Discussion points

- What are the pros and cons of each type of management strategy for the organization and the workforce?
- Is any particular framework more appropriate and effective for managing the employment relationship in hotels and catering?

4.6 WORKERS: ROLE AND ATTITUDES

Orientations to work

All theoretical analyses of workers' orientations to work assume that workers have demands that go beyond economic rewards and impinge upon managerial issues, including how work is organized and how the business operates. Empirical evidence shows that work is valued in different ways by different kinds of workers. In other words, there can be no simple set of assumptions about orientations to work; for example, that casual workers seek less job satisfaction from their work or are less committed than regular employees, or that women demand less control over their work than men. Different types of workers may have different priorities in relation to their work, such that equal opportunities and working time arrangements are more important considerations to women than to men. Some individuals will have more interest than others in demanding some kind of involvement in decision making in the workplace. Most of the attachment to work is 'instrumental' in the sense that it is a means to an end to secure the best financial reward. Nevertheless, workers do look for other things, including status, satisfying work and social relationships, and some degree of control over the way they perform their work (see also Marchington, 1992).

A variety of outlooks and expectations to work is confirmed in hotel and catering studies. In hotels, although Mars (1973) suggested that hotel employees were typically 'instrumental' (money-oriented) in their orientation to work, Shamir (1981) found 'many hotel employees whose orientation to work was primarily expressive or social, i.e. they expected to satisfy their "higher order" needs at work' (p. 55). Snow (1981) proposed that hotel workers have a *profile* of orientations which they seek to maximize in the labour market. Those aspects of the 'orientation profile' which are not satisfied have to be accepted because the choice is between the job held or no job at all. Such 'orientation profiles' encourage an individualistic perception of workplace relations. Marshall (1986) suggested that restaurant workers were not wholly 'instrumental' because they were engaged in activities that they would have readily accepted as 'leisure'. Dodrill and Riley (1992) found that hotel workers' orientations to work were based on strong positive attitudes to ambition and scope, but that there was no significant attitude to security, mobility and autonomy.

From a broader sample of catering businesses, Gabriel (1988) found an even wider variety of orientations, although job protection was the common primary preoccupation of all workers. For example, owners of small restaurants were prepared to put up with long hours and hard work as 'a price for independence', whereas in a gentlemen's club the informal and intimate atmosphere compensated for poor pay. Some workers, particularly casual workers, may hold more of an instrumental orientation to work in the sense that work is purely for short-term economic gain, but Gabriel (1988) argues that many hotel and catering workers have 'an instrumental orientation *thrust upon them* by unrewarding jobs and lack of alternatives' (p. 157).

Conversely, workers may develop negative orientations to work. Although the explanations for some of the behaviours associated with such negativism are complex, they are often a response to unfair rules, role ambiguity, arbitrary management and other deficiencies in the way the organization is run. Analoui and Kakabadse's (1989) fascinating study of a night club provides a wealth of data on such industrial conflict.

The view that 'work in a leisure industry is not completely similar to other types of work in spite of the fact it is done for money and under restricting circumstances' (Shamir, 1981, p. 54), would seem to merit closer examination (see also Marshall, 1986).

Workers as individuals

The employer has the upper hand from the beginning of the employment contract and has considerably more experience in 'negotiating' it. So the bargain is not being struck between equals, but it is proposed that there are three main ways in which workers can act individually, termed here as the conformist, the deviant and the terminator. These are similar to Marchington's (1992) forms of employee response ('getting on', 'getting by' and 'getting back') and have parallels in the negative behaviours of facilitative, inhibitive and futile defiance identified by Analoui and Kakabadse (1989, pp. 52–5).

Gospel and Palmer (1993) are sceptical about the extent to which the first two types of such action (conformist and deviant) actually change the basic employer features of the employment relationship, but do not consider the third (terminator) as an explicit feature of individual action. Analoui and Kakabadse (1989, 1993) consider behaviours such as pilferage (deviant) and labour turnover (terminator) to be expressions of industrial conflict that have received insufficient analysis and evaluation in mainstream industrial relations; they are, however, not always individual manifestations (see below).

The conformist

The conformist works within the rules; for example, by working hard to achieve promotion. However, this form of individual action can only be effective where there is a strong internal labour market, which is not a characteristic of most hotel and catering organizations.

The deviant

The deviant evades or seeks to manipulate the rules and regulations. Mars and Mitchell (1976) and Mars *et al.* (1979) identified this in relation to the reward system whereby workers supplement low pay with 'fiddles' and 'knock offs' (see also Marshall, 1986). Kelliher (1984, pp. 33–4) offers an interesting critique of Mars's reformist attitude towards such illicit practices. Mars argued that these practices should be translated into 'official means' of operation by franchising or subcontracting activities such as dispense bars and food lets to employees, who would, in effect, channel their illicit practices into entrepreneurial activities. Kelliher argues that Mars appears to accept such behaviours as inevitable rather than tackling the question of conflict in the employment relationship.

Macaulay and Wood (1992) noted how an emphasis on individual 'survival' skills used to gain access to informal rewards manifested itself as expressions of 'belligerent individualism' and 'evidence of some degree of instrumentalism, or at least self-reliance among employees' (p. 27). Wynne (1993) has also shown how employees can take control of an employer's empowerment initiative to their own advantage.

The terminator

The terminator can no longer work for the organization and 'votes with the feet' by leaving altogether. This expression of individual dissatisfaction has received considerable attention in hotel and catering research as labour turnover (see Chapters 1 and 3, and Wood, 1992, pp. 95–103), but not in mainstream industrial relations (see, for example, Analoui and Kakabadse, 1989, 1993). The predominant perspective (championed by Johnson, 1980, 1981, 1985, 1986) views high labour turnover as damaging to the industry and the organization through its creation of, among other things, high recruitment and selection costs, disruption to the business and unstable work relations. Shamir (1981) also maintains that the notion of the workplace as a community tends to promote high worker mobility that could be harmful for the individual and the work organization. An alternative perspective (see Simms *et al.*, 1988) perceives high labour turnover as a deliberate and necessary benefit, both to organizations seeking to maintain flexibility and to the individual who aspires to promotion in an industry where strong internal labour markets are not the norm.

Analoui and Kakabadse (1993, p. 57) suggest that the absence of legitimized and effective channels of communication for conflict expression results in isolated employees who are only left to choose from among the less desirable choices of actions available. Marchington (1992), in relation to the labour process debate, notes 'that more attention needs to be given to the missing subject: that is the employee/workers ... Of particular interest here is the notion of "tacit skills", those attributes which all workers ... learn by virtue of their experience and then utilize in order to assist or thwart the achievement of managerial objectives' (p. 150).

Such attention is justified given that deviant and terminator behaviour in hotels and catering has been seen by many as a function of the labour force – women, part-timers and casual workers. Kelliher (1984) believes that the problem stems from management attitudes. 'Employers in the industry are prepared to tolerate and, in some case, even to encourage individualistic forms of conflict expression in order to defend themselves against collective resistance' (Kelliher, 1984, p. 66). However, Kelliher's observations about the threat of trade unions have rather less credence a decade later. The propensity of hotel and catering workers to behave in 'negative' ways, compared to others, remains to be explained more fully than simply in terms of individualism or collectivism.

Workers as groups

Informal groups in organizations are extremely common and may, or may not, match formally constituted departments, sections or teams. Groups in the workplace became the focus of attention in national industrial relations during the 1970s and 1980s. Such attention was concentrated on unofficial trade union activity in organized industries rather than on informal groups in industries where workers were unorganized. Predominant workplace issues included restrictive practices and unofficial industrial action. It would be naive to assume that such collective 'practices' only occur in unionized workplaces, and are manifested through formally recognized means (the lesser extent of industrial action in hotels and catering is mentioned in Chapter 9). Analoui and Kakabadse (1989) report how night club employees acted in a collective but covert manner to repay the manager for insensitivity to a colleague in distress.

The existence of different groups in the workplace is indicative of a plurality of different interests within the organization. Much of the history of management has been concerned with formally controlling, breaking up or institutionalizing these conflicting sets of interests. This may have had the effect of 'missing' the much less visible, but by no means less important, informal network of practices, processes and relationships (see Kolb and Bartunek (1992) for an invaluable insight into 'hidden conflict' and Analoui and Kakabadse (1989, 1993) for an alternative perspective on industrial conflict and its expressions).

Although most of the major studies on group action have been centred on unionized manufacturing industry (for example, Sayles, 1958; Batstone *et al.*, 1977), group action is not inevitable, but it is more likely to occur, and groups are more likely to be cohesive, under certain conditions. Dann and Hornsey (1986) have identified that the interdependence of departmental groups may generate conflict. Differences in group solidarity that produce different value systems, territorial issues and gender differences between waiting and kitchen staff may aggravate conflict. This view also provides some support to the contention that 'differences in workgroup activity and power relate to the workgroup's position in the organizational structure and to the extent to which it occupies an important position in the production process' (Gospel and Palmer, 1993, p. 108). Thus it could be expected that chefs would be in a position to exert more pressure than accommodation staff. Analoui and Kakabadse (1993, p. 56) suggest that:

> For managers of work organizations, it is essential to bear in mind that the kind of actions which are specifically chosen by the employees to express their discontent are in fact responses to the situations with which they are confronted. Management can influence the process of choice making by altering, modifying and even changing the structure of work relationships.

If the shop floor is the 'frontier of control' between managerial and workgroup interests, then the view that 'the most alienated, oppositional workgroups have emerged as the unintended consequences of externalisation and Taylorist policies' (Gospel and Palmer, 1993, p. 115) also merits further development in terms of the hotel and catering industry.

Summary propositions

- Different individuals value their work in a wide variety of different ways.
- Different approaches to work can lead to conflict between individual workers and to conflict between worker and manager.
- Individual workers adopt modes of deviant or terminator behaviour which are potentially more damaging than helpful to the organization.
- Hotel and catering workers align into formal and informal groups. Such groups may adopt restrictive practices that conflict with the organization's objectives.

Discussion points

- What does work mean to you?
- What orientations to work have you observed in others? Are there differences in work orientations that are based on a person's sex, occupation, age, type of contract or other factors?
- Are positive and negative individual orientations to work inevitable? If so, how can these be managed?
- What evidence have you observed of group behaviour? Has this occurred under particular circumstances or manifested itself in particular forms?

4.7 TRADE UNIONS

While workers in the hotel and catering industry do align themselves into groups, this has not been translated into the more formal means of trade unionism. The questions 'Why have the vast majority of workers in the hotel and catering industry not joined trade unions?' or, put another way, 'Why have trade unions failed to organize workers in the hotel and catering industry to any significant extent?' have probably been among the most frequently posed questions in employee relations terms. Although accurate figures are difficult to come by, WIRS3 estimates suggest that only 3 per cent of the industry's workforce is unionized and that this may have declined from around 5 per cent (Byrne, 1986). Put simply, arguments about causes of the national decline in trade union membership and density since the end of the 1970s are largely irrelevant to an industry where trade unionism is at best a marginal practice. Nevertheless, if trade unions are to recover their position nationally, they must recruit in areas where they have traditionally been weak.

Before we consider some of the broader factors that underlie these questions and the reality that trade unions are not a force in hotel and catering employment, a brief overview of the main forms of trade union organization is presented. Gospel and Palmer (1993) have identified four main types of trade union in Britain, that is, unions that have been formed by workers themselves rather than by governments and employers. Since one type, organization-based skills unions (for example, as in the civil service or the Post Office), has no obvious relevance to the hotel and catering industry, it has been omitted from the summary.

Market-based skill: occupational institutions

The earliest form of worker organization derived from the medieval feudal guilds that were established to protect and enhance the value of an occupational skill through policies of 'closure' and monopoly control. These organizations extended across the hierarchy of occupations from master to apprentice. Modern day equivalents include professional associations and craft trade unions. The success of such organizations is contingent upon their own power resources and the countervailing power of any rival groups, the most powerful of which are employers and governments.

Occupational controls predominated in the nineteenth century when most employers were small and competitive, and since these are features that have continued to characterize most of the hotel and catering industry over a century later, it is not unreasonable to ask why skilled occupational groups such as chefs have not organized formally in this way, particularly when a propensity to organize has been found (for example, Chivers, 1973).

Gospel and Palmer (1993) maintain that it is the emergence of large organizations that has enabled employers to gain control through alternative strategies of deskilling and creating an internal labour market. Only the deskilling strategy has real relevance to the hotel and catering industry: to avoid being deskilled, labour must be difficult to replace or substitute. Since the deskilling of skilled kitchen work can be effected through a variety of means (mainly technological) by employers of all sizes, such labour can be replaced and substituted. The deployment of deskilling is not necessarily a function of organizational size, as Gospel and Palmer (1993) have suggested.

Unskilled unions

The second phase of trade union development occurred during the late nineteenth century and the early twentieth century, and was based on the belief that the mass mobilization of labour could influence industrial and societal arrangements. This manifested itself in two forms. *General unions* constituted a 'general class of workers' from among different occupations and industries. The two main unions, the Transport and General Workers Union (TGWU) and the GMB, are set to amalgamate. *Industrial unions* were based on significant industrial groups who could exercise more power against their employer and government. The National Union of Mineworkers is the most well known example, although the extent of its power and influence has been greatly reduced by long-term governmental policies of phasing out the coal industry, and more recent employment legislation.

Attempts to deal specifically with an 'appropriate' union within the hotel and industry came late and have had the effect of creating an industrial union within a general union. The Hotel and Catering Workers' Union (HCWU) – now the Hotel and Catering Union (HCU), part of the Food and Leisure Section of the GMB – was 'a specialist union spawned by the GMB in 1980' (Macaulay and Wood, 1992, p. 22). The HCU has only around 30,000 members (Lucas, 1991b), out of an estimated 57,000 union members (Byrne, 1986). Although the TGWU has a national officer for the hotel and catering industry, the industry is integrated into the broader based food and drink group.

White collar

Trade union growth in Britain between the end of the Second World War and the late 1970s was largely attributable to clerical, technical and managerial staff joining trade unions, particularly in the unorganized service sector. In other words, trade unions are not just a function of male manual workers employed in manufacturing industries. One such union, the National Association of Licensed House Managers (NALHM), has been relatively successful in terms of recruiting members in hotels and catering.

Gospel and Palmer (1993) suggest that white collar unions can fit all of the other three types identified. The most significant union of this type is UNISON, born in 1993 from a merger of the National Association of Local Government Officers (NALGO), the Confederation of Health Service Employees (COHSE) and the National Union of Public Employees (NUPE). UNISON has become the largest single union, with 1.4 million members, one million of whom are women. The membership is mainly, but not exclusively, white collar. UNISON undoubtedly has the highest concentration of catering membership among trade unions, although its sphere of influence extends predominantly to the catering services sector, where organization has been less problematic than in most of the commercial sector. Even so, with a high part-time female membership, UNISON may be better placed than others to offer ways of increasing hotel and catering membership in other sectors of the industry.

The WIRS3

In hotels and catering, aggregate trade union density is substantially lower than among all industries and services, all service industries and, perhaps more importantly, private sector services, as shown in Table 4.1.

The significance of this disparity between hotels and catering and private sector services is even more interesting given that the workplaces in both sectors share a number of important similarities in the WIRS3 sample base, including workplace size,

Table 4.1 *Union density and recognition among all workers.*

	AIS	SI	PSS (by no. of employees) All	25–49	50–99	100+	HCI (by no. of employees) All	25–49	50–99	100+
Unweighted base	2061	1299	702	174	153	375	61	21	18	22
Weighted base	2000	1462	886	518	218	151	99	59	26	14
No members (%)	36	34	56	61	52	44	92	99	85	76
1–59% (%)	22	21	19	17	22	23	7	1	15	19
60–100% (%)	35	36	20	19	17	24	[a]	–	[a]	1
Missing (%)	8	9	5	3	9	7	[a]	–	–	3
Aggregate density[b] (%)	48	47	25	19	22	31	3	[a]	1	7
Any recognition	53	55	34	29	39	45	5	1	10	15
Aggregate % employees covered by bargaining groups[c]	80	81	77	75	72	79	51	36	6	67
Total employees covered by bargaining groups (millions)	8.402	5.699	1.676	0.316	0.318	1.042	0.206	0.001	0.001	0.024

[a]Fewer than 0.5 per cent. [b]Excluding missing cases.
[c]Where recognized.
Any slight discrepancies between figures are due to the rounding of decimal points.
Source: WIRS3 (1990).

status of establishment, level of specialist personnel expertise and experience and proportions of female employees. There are also a number of obvious differences, apart from the level of part-time employment, which is higher in hotels and catering than in private sector services. Differential factors such as the skill composition and the ethnic mix of the workforce do not seem to have been examined seriously as factors to explain the extremely low level of unionization in hotels and catering. Put another way, the data suggest that smaller workplace size (see Abbott, 1993; Beaumont and Harris, 1994) and high levels of female employment (Gabriel, 1988) are not factors that can be positively correlated with low levels of union density with any degree of certainty.

What follows from a low level of trade union density in hotels and catering is that there is very little trade union recognition. Even where trade unions are recognized, fewer employees in hotels and catering are covered by bargaining groups than elsewhere (see Table 4.1). But this is not to say that the trade unions are standing idle to a greater extent in hotels and catering than elsewhere. What Table 4.2 shows represents something of a damning indictment of trade union recruitment efforts everywhere. From the low levels of recruitment activity that are identified, it follows that recognition requests will also be low. What is perhaps surprising is that there have been more attempts to gain recognition in hotels and catering than elsewhere among non-manual workers.

Table 4.2 *Recruitment and recognition attempts during the past six years.*

	AIS	SI	PSS (by no. of employees) All	25–49	50–99	100+	HCI (by no. of employees) All	25–49	50–99	100+
Manual workers[a]										
Unweighted base	704	526	412	136	99	177	51	21	15	15
Weighted base	1014	792	624	401	141	83	92	59	22	11
Yes attempt to recruit (%)	11	8	7	5	5	21	2	–	6	2
Yes attempt to gain recognition (%)	4	4	5	1	7	20	2	–	6	–
Non-manual workers[b]										
Unweighted base	712	400	391	122	96	173	55	20	17	18
Weighted base	991	611	596	368	142	86	98	59	26	13
Yes attempt to recruit (%)	10	11	11	9	14	15	5	–	19	4
Yes attempt to gain recognition (%)	4	5	5	3	10	8	8	3	17	8

[a]Base: manuals present and no manual members.

[b]Base: non-manuals present and no non-manual members.

Any slight discrepancies between figures are due to the rounding of decimal points.

Source: WIRS3 (1990).

An equivalent level of apathy towards positively advancing trade union membership, perhaps implied rather than explicit, can be discerned among managers, as shown in Table 4.3. The marginal proportions of those 'in favour' can hardly be taken as a positive endorsement of trade unionism. Paradoxically, positive endorsement of union membership is slightly more pronounced, and opposition to membership is less forceful, in hotels and catering than elsewhere. From an analysis of WIRS3 data across all sectors, Beaumont and Harris (1994) have shown that management opposition to trade unions is not disproportionately associated with small firms.

Is there a role for trade unions?

In spite of a number of different approaches to this question, none has come up with a wholly satisfactory explanation. In a review of published research, Wood (1992, pp. 103–24) has identified four categories of explanation, which are summarized briefly here.

The ethos of hotel and catering work

Low pay encourages individual 'instrumentalism' and illicit behaviour to gain access to informal rewards (Mars and Nicod, 1984; Mars *et al.*, 1979). The emphasis of trade unions on formal regulation is perceived to threaten such 'total rewards' (Johnson,

Table 4.3 *Management attitude towards trade union membership.*[a]

	AIS	SI	PSS (by no. of employees) All	25–49	50–99	100+	HCI (by no. of employees) All	25–49	50–99	100+
Unweighted base	493	331	326	106	80	140	49	20	15	14
Weighted base	720	506	498	315	115	68	92	59	22	11
In favour (%)	2	2	2	3	3	–	4	6	–	–
Not in favour (%)	31	26	26	26	24	29	17	8	33	33
Neutral (%)	64	67	67	66	71	63	74	81	67	48
Other (%)	3	5	5	4	3	8	6	6	–	19

[a]Where no members present at establishment.

Any slight discrepancies between figures are due to the rounding of decimal points.

Source: WIRS3 (1990).

1983). An individualistic worker attitude also reinforces the idea that trade unions are incompatible with 'service' work (Riley, 1985, 1991).

Structure of the workforce

Many part-timers, particularly women, are seen as difficult to recruit (Gabriel, 1988). The combination of various working time arrangements and shiftworking undermines any sense of workplace cohesion. The presence of casual workers, the existence of high labour turnover and the wide geographical spread of small units serve to undermine effective organization.

Employer and management attitudes to trade unions

Although examples of explicit and implicit hostility to trade unions have been well documented (CIR, 1971; Wood and Pedler, 1978; Forte, 1982; Macfarlane, 1982a, b), there are also examples of positive attitude (Mars *et al*., 1979; Croney, 1988a; Aslan and Wood, 1993). While such positivism has been espoused at corporate level, it would appear to have been rendered ineffective by hostile unit management actions.

The role of trade unions

Trade unions have fought and lost major recognition disputes (Palmer, 1968; Wood and Pedler, 1978; Macfarlane, 1982a, b). Although a 'spheres of influence' dispute between the TGWU and the GMB (then GMWU) was resolved in 1973, and both unions took steps to address membership problems in the industry (Airey and Chopping, 1980), little real recruitment success followed. The extent to which failure can be attributed to the lack of an effective union strategy (Johnson and Mignot, 1982) or ineffectual shop steward organization (Jameson and Johnson, 1985, 1989) remain moot points.

Wood's classification of explanatory factors overlooks some other significant points, some of which he has himself developed. Shamir (1981) found a degree of solidarity among hotel employees rather than exceptionally individualistic behaviour. Individualistic behaviours may be more characteristic of some occupations than others

(Lennon and Wood, 1989). Writing later, Macaulay and Wood (1992) have rejected the notion of a 'dominant research paradigm' that depicts employees as conservative and powerless, a view that is strongly supported by Gabriel (1988).

Aslan and Wood (1993) also noted a more positive managerial attitude to trade unions than observed in the earlier studies, which have seemingly proved deficient in explaining the totality of hotel managers' attitudes to trade unions. The hotel and catering industry has remained poorly organized regardless of whether public policy has been supportive of, or hostile to, trade unions (Lucas, 1991a) and, as the WIRS3 data have confirmed, trade union membership in hotels and catering is a very marginal practice.

Summary propositions

- Trade unionism is a very marginal practice in the hotel and catering industry for a complex mix of reasons.
- It is difficult to find a holistic explanation for the consistently low levels of unionization.
- It is hard to envisage the development of trade unionism on any significant scale in the foreseeable future.

Discussion points

- Is there a role for trade unions in hotels and catering, now and in the future?
- What form of unionization is most appropriate in hotels and catering?

4.8 CUSTOMERS

The influence of customers on the employment relationship can be significant but may have been overstated. Lennon and Wood (1989) contend that much of the analysis of work and industrial relations in the hotel and catering industry has stemmed from a 'dominant research paradigm' that is inadequate because, among other things, it has been unduly focused on 'waiting at table' jobs to the neglect of other occupational groups, especially those with little or no customer contact. Other groups, including kitchen staff serving at a carvery or room maids, may have more customer contact than has been formally acknowledged.

While Hornsey and Dann (1984) estimated that 40 per cent of hotel staff held service jobs, the ratio of service jobs to production and other jobs will vary across the different sectors of the industry according to business type and the standards of service provision. Butler and Snizek (1976) noted that customers lead service workers to use a variety of manipulative ploys to control the work's reward structure, but there are qualitative and quantitative differences between customers in terms of their potential effect on the employment relationship. Thus the clientele of an exclusive high-class restaurant has potentially more influence on staff reward through tipping than the hordes of teenagers who regularly patronize the local hamburger restaurant. Tipping does not 'benefit' service workers alone because there are different methods of tipping practice. In a pool system, all the tips are shared out among staff in all departments (see Lucas, 1991b, 1992).

Since the 1980s there has been a rash of customer service initiatives in service industries, in the form of attitude training for employees in how to handle customers to give competitive advantage. Customers may be both a source of stress (Shamir, 1983) and a source of reward and satisfaction (Marshall, 1986), although some workers may be able to devise coping mechanisms that allow them to control their workflow – a form of 'self-empowerment' (Wynne, 1993). Those unable to 'cope' will simply leave. It is difficult to know the extent to which customer stress causes deviant and terminator individual behaviours, or whether such behaviours are triggered by other factors, such as arbitrary management, poor workplace relationships and other general work-related pressures. Dann and Hornsey (1986) have noted that the differing demands of customers and colleagues may create a situation of conflict for some staff. The issue of customers is further developed in Chapters 5 and 6.

Summary proposition

- While customers have a potentially greater effect on the employment relationship in hotels and catering than in many other industries, the extent of their influence may have been overstated; nevertheless, it merits further study.

Discussion point

- Evaluate the role of the customer in the employment relationship.

REFERENCES

Abbott, B. (1993) Small firms and trades unions in services in the 1990s. *Industrial Relations Journal*, **24** (2), 308–17.

Advisory, Conciliation and Arbitration Service (1987) *Discipline at Work: the ACAS Advisory Handbook*. London: HMSO.

Airey, D.A. and Chopping, B.C. (1980) The labour market. In R. Kotas (ed.), *Managerial Economics for Hotel Operations*, pp. 45–69. Guildford: University of Surrey Press.

Analoui, F. and Kakabadse, A. (1989) Defiance at work. *Employee Relations* **11** (3), 1–62.

Analoui, F. and Kakabadse, A. (1993) Industrial conflict and its expressions. *Employee Relations*, **15** (1), 46–62.

Armstrong, E.G.A. and Lucas, R.E. (1985) *Improving Industrial Relations: the Advisory Role of ACAS*. London: Croom Helm.

Armstrong, M. (1987) Human resource management: a case of the Emperor's new clothes? *Personnel Management*, August, 31–5.

Aslan, A.H. and Wood, R.C. (1993) Trade unions in the hotel and catering industry: the views of hotel managers. *Employee Relations*, **15** (2), 61–70.

Atkinson, J. and Storey, D. (eds) (1994) *Employment in the Small Firm and the Labour Market*. London: Routledge.

Bagguley, P. (1987) *Flexibility, Restructuring and Gender: Changing Employment in Britain's Hotels*. Lancaster Regionalism Group, University of Lancaster.

Bagguley, P. (1990) Gender and labour flexibility in hotel and catering. *The Service Industries Journal*, **10** (4), 737–47.

Batstone, E., Boraston, I. and Frenkel, S. (1977) *Shop Stewards in Action: the Organization of Workplace Conflict and Accommodation*. Oxford: Basil Blackwell.

Beaumont, P. B. and Harris, R.I.D. (1994) Opposition to unions in the non-union sector in Britain. *The International Journal of Human Resource Management*, **5** (2), 457–71.

Braverman, H. (1974) *Labor and Monopoly Capital*. New York: Monthly Review Press.

Brown, A. (1992) ACAS arbitration: a case of consumer satisfaction? *Industrial Relations Journal*, **23** (3), 224–34.

Butler, S.R. and Snizek, W.E. (1976) The waitress–diner relationship. *Sociology of Work and Occupations*, **3** (2), 209–22.

Byrne, D. (ed.) (1986) *Waiting for Change*. London: Low Pay Unit.

Chivers, T.S. (1973) The proletarianisation of a service worker. *Sociological Review*, **21**, 633–56.

Commission on Industrial Relations (1971) *The Hotel and Catering Industry, Part 1. Hotels and Restaurants, Report No. 23*. London: HMSO.

Croney, P. (1988a) An investigation into the management of labour in the hotel industry. MA thesis, University of Warwick.

Croney, P. (1988b) An analysis of human resource management in the UK hotel industry. Paper presented to the International Association of Hotel Management Schools Autumn Symposium, Leeds.

Dann, D. and Hornsey, T. (1986) Towards a theory of interdepartmental conflict. *International Journal of Hospitality Management*, **5** (1), 23–8.

Dodrill, K. and Riley, M. (1992) Hotel workers' orientations to work: the question of autonomy and scope. *International Journal of Contemporary Hospitality Management*, **4** (1), 23–5.

Dunlop, J. (1958) *Industrial Relations Systems*. New York: Henry Holt and Company.

Edwards, R. (1979) *Contested Terrain: the Transformation of the Workplace in the Twentieth Century*. London: Heinemann.

Forte, R. (1982) How I see the personnel function. *Personnel Management*, August, 32–5.

Fox, A. (1966) *Industrial Sociology and Industrial Relations. Research Papers 3, Royal Commission on Trade Unions and Employers' Associations*. London: HMSO.

Fox, A. (1973) Industrial relations: a social critique of pluralist ideology. In J. Child (ed.), *Man and Organization: the Search for Explanation and Social Relevance*, pp. 185–233. London: Allen and Unwin.

Fox, A. (1985) *Man Mismanagement*, 2nd edition. London: Hutchinson.

Gabriel, Y. (1988) *Working Lives in Catering*. London: Routledge & Kegan Paul.

Gospel, H.F. (1992) *Markets, Firms, and the Management of Labour in Modern Britain*. Cambridge: Cambridge University Press.

Gospel, H.F. and Palmer, G. (1993) *British Industrial Relations*, 2nd edition. London: Routledge.

Guest, D. (1987) Human resource management and industrial relations. *Journal of Management Studies*, **24** (5), 503–21.

Hiemstra, S.J. (1990) Employment policies and practices in the lodging industry. *International Journal of Hospitality Management*, **9** (3), 207–21.

Hornsey, T. and Dann, D. (1984) *Manpower Management in the Hotel and Catering Industry*. London: Batsford Academic and Educational.

Hotel and Catering Economic Development Committee (1975) *Manpower Policy in the Hotel and Catering Industry – Research Findings*. London: HMSO.

Hotel and Catering Training Company (1990) *Employee Relations in the Hotel and Catering Industry*, 7th edition. London: HCTC.

Hyman, R. (1975) *Industrial Relations: a Marxist Introduction*. London: Macmillan.

Hyman, R. (1989) *The Political Economy of Industrial Relations: Theory and Practice in a Cold Climate*. Basingstoke: Macmillan.

Ishak, N.K. and Murrmann, S.K. (1990) An exploratory study of human resource management practices and business strategy in multi-unit restaurant firms. *Hospitality Research Journal*, **14** (2), 143–55.

Jameson, S.M. and Johnson, K. (1985) The hotel shop steward – an emerging role in British industrial relations. *International Journal of Hospitality Management*, **4** (3), 131–2.

Jameson, S.M. and Johnson, K. (1989) Hotel shop stewards – a critical factor in the development of industrial relations in hotels? *International Journal of Hospitality Management*, **8** (2), 167–77.

Johnson, K. (1980) Staff turnover in hotels. *Hospitality*, February, 28–36.

Johnson, K. (1981) Towards an understanding of labour turnover? *Service Industries Review*, **1** (1), 4–17.

Johnson, K. (1983) Trade unions and total rewards. *International Journal of Hospitality Management*, **2** (1), 31–5.

Johnson, K. (1985) Labour turnover in hotels – revisited. *The Service Industries Journal*, **5** (2), 135–52.

Johnson, K. (1986) Labour turnover in hotels – an update. *The Service Industries Journal*, **6** (3), 362–80.

Johnson K. and Mignot, K. (1982) Marketing trade unionism to service industries: an historical analysis of the hotel industry. *Service Industries Review*, **2** (3), 5–23.

Keenoy, T. (1990) HRM: a case of the wolf in sheep's clothing? *Personnel Review*, **19** (2), 3–9.

Kelliher, C. (1984) An investigation into non-strike conflict in the hotel and catering industry. MA thesis, University of Warwick.

Kelliher, C. (1989) Management strategy in employee relations: some changes in the catering industry. *Contemporary Hospitality Management*, **1** (2), 7–11.

Kessler, I. and Purcell, J. (1993/4) Joint problem solving and the role of third parties: an evaluation of ACAS advisory work. *Human Resource Management Journal*, **2** (2), 1–21.

Kessler, S. and Bayliss, F. (1992) *Contemporary British Industrial Relations*. London: Macmillan.

Kolb, D.M. and Bartunek, J.M. (eds) (1992) *Hidden Conflict in Organizations*. Newbury Park, California: Sage Publications.

Lennon, J.J. and Wood, R.C. (1989) The sociological analysis of hospitality labour and the neglect of accommodation workers. *International Journal of Hospitality Management*, **8** (3), 227–37.

Lucas, R.E. (1991a) Promoting collective bargaining: wages councils and the hotel industry. *Employee Relations*, **13** (5), 3–11.

Lucas, R.E. (1991b) Remuneration practice in a wages council sector: some empirical observations in hotels. *Industrial Relations Journal*, **22** (4), 273–85.

Lucas, R.E. (1992) Minimum wages and the labour market – recent and contemporary issues in the British hotel industry. *Employee Relations*, **14** (1), 33–47.

Macaulay, I.R. and Wood, R.C. (1992) Hotel and catering industry employees' attitudes towards trade unions. *Employee Relations*, **14** (3), 20–8.

Macfarlane, A. (1982a) Trade union growth, the employer and the hotel and restaurant industry: a case study. *Industrial Relations Journal*, **13** (1), 29–43.

Macfarlane, A. (1982b) Trade unionism and the employer in hotels and restaurants. *International Journal of Hospitality Management*, **1** (1), 35–43.

McKenna, S. (1990) The business ethic in public sector catering. *The Service Industries Journal*, **10** (2), 377–98.

Marchington, M. (1992) Managing labour relations in a competitive environment. In A. Sturdy, D. Knights and H. Willmott (eds), *Skill and Consent: Contemporary Studies in the Labour Process*, pp. 149–84. London: Routledge.

Marchington, M. and Parker, P. (1990) *Changing Patterns of Employee Relations*. Hemel Hempstead: Harvester Wheatsheaf.

Mars, G. (1973) Hotel pilferage: a case study in occupational theft. In M. Warner (ed.), *The Sociology of the Workplace*, pp. 200–10. London: Allen & Unwin.

Mars, G. and Mitchell, P. (1976) *Room for Reform?* Milton Keynes: Open University Press.

Mars, G. and Nicod, M. (1984) *The World of Waiters*. London: Allen and Unwin.

Mars, G., Bryant, D. and Mitchell, P. (1979) *Manpower Problems in the Hotel and Catering Industry*. Farnborough: Gower.

Marshall, G. (1986) The workplace culture of a licensed restaurant. *Theory, Culture and Society*, **3** (1), 33–47.

Miller, P. (1987) Strategic industrial relations and human resource management – distinction, definition and recognition. *Journal of Management Studies*, **24** (4), 347–61.

Miller, P. (1989) Strategic HRM: what it is and what it isn't. *Personnel Management*, February, 46–51.

Millward, N. (1994) *The New Industrial Relations?* London: Policy Studies Institute.

Mintzberg, H. (1987) Crafting strategy. *Harvard Business Review*, **65** (4), 66–75.

Mintzberg, H. (1990) The design school: reconsidering the basic premises of strategic management. *Strategic Management Journal*, **11** (3), 271–96.

Mintzberg, H. and Waters, J. (1985) Of strategies, deliberate and emergent. *Strategic Management Journal*, **6** (3), 257–72.

Monks, K. (1992/3) Models of personnel management: a means of understanding the diversity of personnel practices? *Human Resource Management Journal*, **3** (2), 29–41.

Palmer, G. (1968) Inter-union dispute in the Torquay hotel industry. *British Journal of Industrial Relations*, **6** (2), 250.

Porter, M.E. (1980) *Competitive Strategy: Techniques for Analyzing Industries and Competitors*. New York: The Free Press.

Purcell, J. (1987) Mapping management styles in employee relations. *Journal of Management Studies*, **24** (5), 533–48.

Purcell, J. and Sisson, K. (1983) Strategies and practices in the management of industrial relations. In G.S. Bain (ed.), *Industrial Relations in Great Britain*, pp. 95–120. Oxford: Basil Blackwell.

Ralston, R. (1989) The changing nature of personnel management in the hotel and catering industry. MSc thesis, University of Manchester.

Riley, M. (1985) Some social and historical perspectives on unionisation in the UK hotel industry. *International Journal of Hospitality Management*, **4** (3), 99–104.

Riley, M. (1991) Technological differentiation and union associate choice. *The Service Industries Journal*, **11** (1), 47–62.

Riley, M. (1994) Industrial relations in the hotel and catering industry. In C.P. Cooper and A. Lockwood (eds), *Progress in Tourism, Recreation and Hospitality Management Volume 5*, pp. 242–7. Chichester: John Wiley and Sons (published in association with the University of Surrey).

Rose, M. (1988) *Industrial Behaviour*, 2nd edition. Harmondsworth: Penguin.

Salamon, M. (1992) *Industrial Relations*, 2nd edition. London: Prentice Hall.

Sayles, L.R. (1958) *The Behavior of Industrial Work Groups*. New York: Wiley.

Shamir, B. (1981) The workplace as a community: the case of British hotels. *Industrial Relations Journal*, **12** (6), 45–56.

Shamir, B. (1983) A note on tipping and employees' perceptions. *Journal of Occupational Psychology*, **56**, 255–9.

Simms, J., Hales, C., and Riley, M. (1988) Examination of the concept of internal labour markets in UK hotels. *Tourism Management*, **9** (1), 3–12.

Sisson, K. (1993) In search of HRM. *British Journal of Industrial Relations*, **31** (2), 201–10.

Snow, G. (1981) Industrial relations in hotels. MSc thesis, University of Bath.

Storey, J. (1992) HRM action: the truth is out at last. *Personnel Management*, April, 28–31.

Sturdy, A., Knights, D. and Willmott, H. (1992) *Skill and Consent: Contemporary Studies in the Labour Process*. London: Routledge.

Thomas, M.A. (1990) What is a human resources strategy? *Employee Relations*, **12** (3), 12–16.

Thomason, G.F. (1991) The management of personnel. *Personnel Review*, **20** (2), 3–10.

Thompson, P. (1987) *The Nature of Work: an Introduction to Debates on the Labour Process*, 2nd edition. London: Macmillan.

Torrington, D. and Hall, L. (1991) *Personnel Management: a New Approach*, 2nd edition. London: Prentice Hall.

Walsh, T. (1990) Flexible labour utilization in the private service sector. *Work, Employment and Society*, **4** (4), 517–30.

Wood, R.C. (1992) *Working in Hotels and Catering*. London: Routledge.

Wood, S. and Pedler, M. (1978) On losing their virginity: the story of a strike at the Grosvenor Hotel, Sheffield. *Industrial Relations Journal*, **9** (2), 15–37.

Wood, S.J., Wagner, A., Armstrong, E.G.A., Goodman, J.F.B. and Davies, J.E. (1975) The 'industrial relations system' concept as a basis for theory in industrial relations. *British Journal of Industrial Relations*, **13** (3), 291–308.

Wynne, J. (1993) Power relationships and empowerment in hotels. *Employee Relations*, **15** (2), 42–50.

FIVE

Managing the Effort–Reward Bargain

The purposes of this chapter are to:

- identify and discuss the main elements of substance in the effort–reward bargain, which centre on pay and conditions of employment;
- evaluate the extent to which the effort–reward bargain represents a 'fair' and effective deal for employers, workers and customers.

5.1 INTRODUCTION

The most important substantive issue in the effort–reward bargain is pay. Pay is highly visible and tangible and is undoubtedly the most problematic term of employment. It forms an important element of employers' fixed and variable costs, and determines the standards of living of workers. Evidence from official surveys (including annual New Earnings Surveys and the WIRS3) and empirical studies shows that, as a rule, hotel and catering employment embodies poor pay practice (for example, Robinson and Wallace, 1984; Byrne, 1986; Gabriel, 1988; Lucas, 1991, 1992; Macaulay and Wood, 1992). As will be shown later, pay is low in both an absolute sense (in real terms related to amount of buying power) and a relative sense (compared to most other workers in Britain).

Workers in hotels and catering are also less likely to enjoy other benefits of 'good' employment (Johnson, 1983; Lucas, 1991, 1992). Such benefits can be placed into two broad categories, although these tend to overlap in practice. For the purposes of this book, 'fringe benefits', include incentives, transport to and from work and subsidized meals and accommodation. They tend to be associated with fairly immediate, often indirect, pecuniary gain, although fringe benefits like profit-sharing schemes are more long-term. 'Other conditions of employment' include paid holiday entitlements, sick pay schemes and the opportunity to join a company pension scheme. Although, since *Barber* v *Guardian Royal Exchange Assurance Group Ltd* 262/88 [1990] IRLR 240

ECJ, pensions are now technically pay (discussed in Chapter 10 under equal pay), these arrangements can be said to fit more easily with employer welfare and social responsibility policies and are therefore also discussed in more detail in Chapter 11. The net result of poor practice in both these areas serves to widen the 'earnings gap' between hotel and catering workers and other workers, with the consequence that hotels and catering can be said to be at the forefront of pay or financial flexibility (Atkinson, 1984). A list of potential fringe benefits and terms of employment is given later in this chapter.

The more thorny point is, perhaps, the worth of pay in relation to the level of effort that is demanded of the worker by the employer. Workers may claim that their pay is low or unreasonable compared to that of other workers, inside or outside the organization, but a more common feeling among hotel and catering workers may well be that their pay is unreasonable in relation to what they have been asked to do by their manager (see, for example, Lucas and Bailey, 1993). For instance, an employee may be asked to work overtime without being paid anything extra (Mars *et al.*, 1979; Macaulay and Wood, 1992, pp. v–vi), or asked to 'mind reception' although it is an official rest period. This may well explain why low pay is often cited as a reason for high staff turnover (for example, Byrne, 1986); it is not the absolute level of pay but rather the extent of perceived unfairness in the effort–reward exchange that triggers this employee response. As Gabriel (1988, p. 159) observes

> For most of the catering workers in this book there was no question of a choice between well-paid intrinsically unpleasant jobs and more intrinsically rewarding but less well-paid ones; their jobs were generally seen as both intrinsically unrewarding *and* badly paid.

As Gabriel notes later, problems with management, supervision and the work itself were more likely to be the main issues in the work situation than low pay. Higher job satisfaction was associated with minimal interference by management and lower job satisfaction was associated with work that had been deskilled, such as where cook–freeze was utilized and in fast-food operations.

One important consequence of perceived unfair reward is that some workers find ways of increasing their total remuneration ('the total reward system') through illicit practices and behaviours, termed the 'hidden items' because they are almost impossible to estimate (Mars and Mitchell, 1976). Such illicit practices may even be condoned by their immediate managers, although this would not necessarily be in line with official company policy. This more individualistic approach seems to have been preferred to that of seeking the help of trade unions in order to press for higher pay, a point which will be considered more fully later in the chapter.

A statement by the former Chancellor of the Exchequer, Nigel Lawson (cited in Smith, 1993, p. 47), that 'Pay is now deemed to be a reward and as such is a key part of the so-called Enterprise Culture', is turned on its head in terms of practices that are found in hotels and catering. Pay is an unfair reward and drives workers towards entrepreneurship in the form of pilfering and fiddling, behaviour that is in reality a manifestation of non-strike conflict (Kelliher, 1984) and defiance at work (Analoui and Kakabadse, 1989, 1993).

5.2 THE CONTRACT FOR PAYMENT

Torrington and Hall (1991) have identified sets of employer and employee objectives that underpin the contract for payment. Although these may be criticized for verging on the idealistic, they do at least offer a framework that acknowledges the legitimacy of attempting to strike something of a bargain in fixing one of the most important aspects of the effort–reward exchange.

The employer or administrator's objectives are related to: being prestigious (it is a good thing to pay well); remaining competitive (the need to sustain a given quantity and quality of workforce); controlling operations; motivating staff and generating increased performance; and controlling costs. Employees' or recipients' objectives relate to: purchasing power; the principle of felt-fairness (a personal evaluation that pay is fair); rights to a share in company success (such as profit sharing); relativities with others inside and outside the organization; recognition of personal contribution; and the composition of total payment or remuneration.

In terms of control of the employment relationship, a key question seems to be 'To what extent is the employer able to call the tune and achieve his or her objectives, or are employees able to hijack the payment system successfully to achieve their ends?' The extent to which both sets of objectives appear to have been met in hotels and catering is a recurrent theme throughout the first part of this chapter.

Managers have to decide how much to pay workers, determine pay levels and set pay differentials. To manage the effort–reward bargain, employers need to administer payment systems and to develop methods of control and discipline in order to maintain performance (see Torrington and Hall, 1991, pp. 571–90). Smith (1993) maintains that 'the management of remuneration in the UK has for too long been seen as an *ad hoc* process lacking direction and integration, relying on short-term methods and "quick-fix" solutions' (p. 46). This observation is supported by Ralston's study (1989, p. 288), which found that only one-quarter of hotel and catering managers recognized the importance of payment administration as a contributor towards organizational effectiveness. Half the respondents did not know the percentage contribution of manpower costs or salaries and wages to total costs.

The WIRS3

Evidence from the WIRS3 shows that in hotels and catering the location of the decision about pay is most likely to be at establishment level, for both manual and non-manual workers. However, managers in hotels and catering are also characterized by a much higher degree of 'interference' than managers in businesses in the other comparator groups because a sizeable number of decisions about pay are made at a higher point in the organization (see Table 5.1). Surprisingly the wages councils, although more important than elsewhere, were not perceived as being a locus for decision making, although they may have been used to guide managerial decisions. While the locus of decision-making was less diverse for non-manuals than manuals, in both categories a locus outside the organization was perceived to be of little consequence.

One important basis for determining pay systems is job evaluation, which is a method that seeks to determine systematically the value of job content, not the job

Table 5.1 *Location of decision when pay last considered (where no recognized trade unions) (percentages)*

	AIS	SI	PSS (by no. of employees) All	25–49	50–99	100+	HCI (by no. of employees) All	25–49	50–99	100+
Manual workers										
Weighted base[a]	51	50	70	78	64	54	96	100	90	88
At this establishment	31	27	45	55	34	33	47	52	35	53
Higher in organization	15	18	23	22	28	21	42	36	59	34
Wages council	2	2	3	3	3	1	10	12	5	9
Other[b]	4	4	1	1	4	–	–	–	–	–
Multi-response	2	1	2	2	4	1	3	–	8	8
Non-manual workers										
Weighted base[a]	56	51	73	77	70	64	95	93	100	93
At this establishment	37	30	50	56	41	41	57	67	35	58
Higher in organization	17	18	26	24	32	24	44	32	73	38
Wages council	[c]	1	1	1	1	1	1	–	–	8
Other[b]	5	4	[c]	–	1	1	–	–	–	–
Multi-response	2	2	3	3	5	3	7	6	8	12

[a]Percentage of cases with no recognized trade unions.
[b]By an employers' association or national negotiating body or by some other body.
[c]Fewer than 0.5%.
Table does not include percentage figures for missing cases.
Source: WIRS3 (1990).

holder (the value of the job holder is assessed from an appraisal of performance). Ranking and grading are methods of non-analytical job evaluation. Analytical job evaluation concentrates on measuring factors such as effort, skill and decision-making in a systematic way, although it can never be wholly objective. Some studies have found that job evaluation is not used to any extent in hotels and catering (Croney, 1988; Ralston, 1989). The WIRS3 also shows that job evaluation is a minority practice, but only to a slightly lesser degree than in other comparator groups (see Table 5.2). The extent of job evaluation schemes among the smaller hotel and catering industry establishments (25 to 49 employees) is, perhaps, surprising given the general tendency towards informal workplace practices. This lack of job evaluation suggests that pay differentials may not be founded upon any sound basis, and that employers may well be leaving themselves wide open to equal pay claims under the equal value provisions of the Equal Pay Act, which is discussed more fully in Chapter 10.

As a general rule, pay is likely to be reviewed annually (Croney, 1988; Ralston, 1989; Lucas, 1991), although the absence of the annual wages council order, which has traditionally provided the basis for reviewing manual workers' pay, may conceivably lead to a change in this practice in the future. According to the WIRS3, the three factors that are most likely to influence the level of manual workers' pay –

Table 5.2 *Presence and type of job evaluation schemes.*

			PSS (by no. of employees)				HCI (by no. of employees)			
	AIS	SI	All	25–49	50–99	100+	All	25–49	50–99	100+
Unweighted base	2061	1299	702	174	153	375	61	21	18	22
Weighted base	2000	1462	886	518	218	151	99	59	26	14
Scheme present (%)	26	26	25	22	22	37	18	18	11	35
Type of scheme										
Points rating[a] (%)	45	45	46	49	30	53	42	49	–	50
Factor comparison[a] (%)	15	17	12	14	11	10	19	32	–	–
Ranking whole job[b] (%)	21	19	25	23	35	19	26	18	49	29
Grading[b] (%)	15	15	14	15	12	13	6	–	2	21
Other (%)	5	5	4	–	11	5	8	–	49	–

[a]Analytical.
[b]Non-analytical.
Any slight discrepancies between figures are due to the rounding of decimal points.
Source: WIRS3 (1990).

increased cost of living, labour market considerations and economic performance/ability to pay – were stronger influences in sectors other than hotels and catering. However, individual employee performance was a more significant factor in hotels and catering than elsewhere. For non-manual workers, there was much closer correspondence in terms of the most important factors influencing pay across all sectors. The exception was increased cost of living, which was of lesser importance in hotels and catering than elsewhere. These differences are shown in Table 5.3.

Evidence from the WIRS3, summarized in Table 5.4, shows that in hotel and catering establishments manual and non-manual workers are most likely to be paid on a weekly basis. Monthly payment is a much more marked practice in all the other sectors. Cash payments, rather than the direct transfer of pay into a bank or building society, are considerably more common in hotels and catering than elsewhere. The incidence of weekly pay paid in cash is much more marked in smaller workplaces in hotels and catering, for both manual and non-manual workers. This lack of security may leave the way open for illicit practices, a point which is developed later in the chapter.

What is pay?

Pay can be referred to in different ways, such as the hourly rate, the weekly wage, the annual salary or remuneration, which is the term usually used to denote the 'total package' of basic pay and other elements such as bonuses, tips and incentive schemes. The European definition of pay for equal pay purposes now includes pensions (see Chapter 10). One of the problems about discussing pay in hotels and catering is deciding what it means because of the variety of permutations that can be combined to constitute total remuneration. Some pay elements, monetary and in kind, including the receipt of tips and provision of meals and accommodation, are, in a way, 'special' to the industry. But not all workers are entitled to all, some or indeed any of these

Table 5.3 *Most significant factors influencing the level of pay.*[a]

	AIS	SI	PSS (by no. of employees) All	PSS 25–49	PSS 50–99	PSS 100+	HCI (by no. of employees) All	HCI 25–49	HCI 50–99	HCI 100+
Manual workers										
Unweighted base	556	364	319	89	76	154	48	17	16	15
Weighted base	789	544	463	285	110	68	82	49	22	11
Increased cost of living (%)	45	41	44	40	46	57	29	29	28	31
Labour market (%)	36	37	42	42	36	52	30	34	25	25
Economic performance (%)	31	27	31	35	14	42	17	21	10	17
Individual performance (%)	20	20	24	22	27	26	26	27	16	38
Non-manual workers										
Unweighted base	874	471	456	134	104	218	52	18	17	17
Weighted base	1111	696	660	404	154	103	90	52	26	12
Increased cost of living (%)	45	41	44	39	47	57	28	27	24	42
Labour market (%)	30	34	35	33	33	48	29	25	40	23
Economic performance (%)	32	30	32	33	21	42	28	26	33	21
Individual performance (%)	31	29	30	30	31	33	27	32	14	36

[a]These are the most frequently mentioned factors given in response to a multi-response question. These factors relate to decisions about pay taken by management alone or through discussions/consultations with employees or representatives shown in Table 6.1.
Any slight discrepancies between figures are due to the rounding of decimal points.
Source: WIRS3 (1990).

elements. Therefore to state that hotel and catering workers enjoy a plethora of pay elements, which is taken by some to mean that workers are adequately paid, would be misleading. The provision of poor quality meals and substandard accommodation do little to compensate for a low basic wage and tips that may vary from nothing one week to a few pounds the next.

Although low pay in hotels and catering is not unique to Britain (Wood, 1992a, b), the pay of many, particularly part-time workers, is likely to be little more than a low rate of basic hourly pay. As Wood (1992a, p. 38) has noted, 'The majority of academic evidence concurs in suggesting both that basic rates of pay in hotels and catering are inadequate and employers are frequently ruthless in pursuing low-pay strategies.' This point applies to managers (see HCIMA and Touche Ross Greene Belfield-Smith Division, 1992a, b), non-manual and manual workers. The annual New Earnings Surveys provide the most consistent source of data on hotel and catering workers' pay that can be compared with that of other workers. As the figures in Table 5.5 show, hotel and catering workers, the majority of whom are manual workers, fare less well than workers elsewhere (see also Table 2.3).

Table 5.4 *Frequency and method of payment.*

	AIS	SI	PSS (by no. of employees) All	PSS 25–49	PSS 50–99	PSS 100+	HCI (by no. of employees) All	HCI 25–49	HCI 50–99	HCI 100+
Manual workers										
Unweighted base	1831	1090	563	122	129	312	58	19	18	21
Weighted base	1697	1170	689	383	184	121	95	55	26	13
Weekly (%)	74	67	62	66	60	51	79	86	76	57
Monthly (%)	23	28	36	33	38	44	13	14	8	18
Cash (%)	27	22	34	39	27	26	50	70	20	27
Cheque (%)	13	15	8	11	5	4	3	2	5	–
Direct transfer (%)	59	62	58	50	68	69	47	28	75	72
Non-manual workers										
Unweighted base	1831	1090	563	122	129	312	58	19	18	21
Weighted base	1697	1170	689	383	184	121	95	55	26	13
Weekly (%)	17	14	22	26	17	19	55	67	47	21
Monthly (%)	81	84	76	73	81	78	36	27	43	55
Cash (%)	8	7	12	14	9	8	34	51	11	8
Cheque (%)	9	7	10	13	5	3	3	2	5	–
Direct transfer (%)	83	85	78	72	86	89	60	41	84	90

Table excludes figures for fortnightly payment and missing cases.
Any slight discrepancies between figures are due to the rounding of decimal points.
Source: WIRS3 (1990).

Data from the WIRS3 also show that hotels and catering had a substantially higher incidence of low paid employees (less than £3.28 per hour) and very low paid (less than £2.70 per hour) than elsewhere, shown in Table 5.6. One-quarter of smaller workplaces in hotels and catering had very low paid employees. The statutory minimum hourly rate set by the Licensed Residential Establishment and Licensed Restaurant Wages Council (hotels and restaurants) in force for most of 1990 was £2.33; the rate set by the Licensed Non-residential Establishment Wages Council (public houses and clubs) that was in force for 1990 was £2.50; the rate in force for the Unlicensed Place of Refreshment Wages Council (cafes and snack bars) was £2.38 until June, after which it increased to £2.58.

In the annual (1992) survey of managerial salaries and benefits, undertaken by the HCIMA in conjunction with Greene Belfield-Smith (the specialist hotel, catering, tourism and leisure division of Touche Ross Management Consultants), teachers and lecturers were the highest paid managerial group, with 87 per cent earning over £15,000, whereas managers in hotels were the lowest paid group, with only 48 per cent earning over £15,000 (HCIMA and Touche Ross Greene Belfield-Smith Division, 1992a, b). The survey also found significant earnings differences between males and females, although the position of females was improving, higher salaries in London and the South East, and increasing salary with age.

Croney (1988) found a certain sensitivity about the issue of pay in the four hotel companies featured in his study. Although these companies made corporate statements about the need to provide good overall reward packages, this was contradicted by the type of systems in operation. There was no formal job evaluation, subjective

Table 5.5 *Gross hourly earnings[a] of manual males and females for selected years (by former wages council sector and all industries and services).*

	Full-time males	Full-time females	Part-time females
Licensed residential establishment and licensed restaurant establishment[b]			
1980	183.4	142.1	136.3
1991	400.4	339.0	289.0
1992	416.0	361.0	299.0
1993	444.0	353.0	320.0
Licensed non-residential establishment[c]			
1980	159.4	134.1	130.7
1991	386.0	322.2	303.3
1992	428.0	318.0	316.0
1993	419.0	334.0	327.0
Unlicensed place of refreshment[d]			
1980	–	146.6	–
1991	–	322.2	292.3
1992	–	318.0	–
1993	–	339.0	370.0[e]
All industries and services			
1980	240.5	170.4	153.9
1991	554.1	395.2	351.4
1992	589.0	421.0	376.0
1993	605.0	435.0	390.0

[a]Excluding overtime (in pence).
[b]Hotels and restaurants.
[c]Public houses and clubs.
[d]Cafes and snack bars.
[e]Non-manual workers.
Source: New Earnings Survey, Tables 1, 2, 3, 176, 177..

performance assessment criteria were used and pay rates were close to, or at, the level of the wages council rate.

Basic pay is also a more important component of gross pay than is the case in the economy as a whole. This is shown in Table 5.7 using another set of figures drawn from selected New Earnings Surveys. The low incidence of overtime pay is consistent with the findings of others that normative premium overtime rates are not paid (Croney, 1988; Lucas, 1991; Macaulay and Wood, 1992), often because management prefers to give time off in lieu. In particular, Macaulay and Wood (1992) found what appeared to be widespread law breaking by employers in terms of the overtime rates set by the wages councils (one and a half times the basic statutory minimum rate). The low incidence of shift work pay is also surprising. Evidence from the WIRS3 presented in Chapter 3 showed a very high incidence of shiftworking in hotel and catering establishments. Therefore the figures in Table 5.7 (hours of work figures are shown in Tables 3.6 and 3.7) seem to suggest that overtime and shiftwork are not compensated for separately.

Table 5.6 *Presence of 'low pay' for any group of employees working 18 or more hours per week.*

			PSS (by no. of employees)				HCI (by no. of employees)			
	AIS	SI	All	25–49	50–99	100+	All	25–49	50–99	100+
Unweighted base[a]	1898	1151	634	153	136	345	58	19	17	22
Weighted base	1728	1199	796	465	192	139	93	55	25	14
Any low-paid employees[b] (%)	27	29	30	30	34	26	70	68	83	54
Any very low-paid[c] (%)	7	6	8	10	9	3	20	25	15	9

[a]Base is cases where any of five groups (unskilled manual; semi-skilled manual; skilled manual; clerical, administrative and secretarial; supervisors) has typical pay and hours data for workers doing 18 hours or more.
[b]Less than £3.28 per hour.
[c]Less than £2.70 per hour.
Any slight discrepancies between figures are due to the rounding of decimal points.
Source: WIRS3 (1990).

The 'total reward system'

The term that is probably the most frequently used to describe total remuneration in hotels and catering is 'the total reward system' (Mars and Mitchell, 1976; Mars *et al.*, 1979), defined as

> basic pay + subsidized lodging + subsidized food + tips or service charge (where applicable) + 'fiddles' and 'knock offs'.

Employees in hotels and restaurants are presumed to receive these terms. Fiddles are usually about obtaining certain monetary benefits by devious means which can be at the employers' or customers' expense. Knock-offs are usually the theft of small items like soap, toilet rolls and food (Mars, 1973). Johnson (1983) has rightly criticized the limited scope of this definition because it excludes other monetary and in-kind elements, such as bonuses, incentive schemes, free uniforms and free transport to and from work. Lennon and Wood (1989) have also criticized the work of Mars and others for taking too restrictive a view of the occupational groups to whom such a package might apply.

In spite of such reservations, the total reward system does illustrate some of the potential variations that constitute the substance of the employment relationship in hotels and catering. Furthermore, the three related concepts that are crucial to the analysis of the total reward system, noted by Wood (1992a, p. 37), are consistent with some of the points about managing in hotels and catering that featured in Chapter 2. As Wood notes, the system is based on the need for managers to respond to unpredictable demand in an *ad hoc* and flexible manner. They do so by forming individual contracts based on secrecy with key staff, which has the effect of creating core and peripheral staff (similar but not identical to Atkinson's (1984) core and periphery). Core staff benefit substantially from individual contracts; because of their specific

Table 5.7 *Contribution of other pay to average gross weekly earnings (full-time manual males and females).*

	Males			Females		
	O/T (%)	PBR etc. (%)	Shift etc. (%)	O/T (%)	PBR etc/ (%)	Shift etc. (%)
Licensed residential establishment and licensed restaurant wages council						
1982	4.5	4.2	0.6	5.7	1.4	0.5
1989	4.8	1.6	0.2	3.3	2.7	0.5
1992	1.9	1.7	1.1	2.7	1.0	0.1
1993	2.6	4.1	0.4	1.4	1.4	0.2
Licensed non-residential establishment wages council						
1982	9.7	2.4	0.4	4.7	1.0	0.5
1989	7.5	2.9	0.4	no data	no data	no data
1992	7.1	1.0	0.1	5.5	0.2	0.5
1993	7.8	1.4	0.4	5.7	0.2	0.2
All industries and services						
1982	12.9	7.6	3.3	3.6	6.8	2.4
1989	16.0	7.1	3.4	6.0	8.1	2.4
1992	14.1	5.5	3.7	6.1	6.4	2.8
1993	13.7	5.2	3.6	6.0	6.0	3.0

O/T, overtime; PBR etc., payment by results etc.; Shift etc., shift premia etc.
Source: Selected New Earnings Surveys, Table 1 (all years), Tables 25 and 26 (1982), Tables 41 and 42 (1989, 1992, 1993).

skills or technical knowledge, they are vital to the organization. The presumption that must follow from Wood's analysis is that peripheral workers must be treated less favourably. For example, although they may be able to take advantage of some 'fiddles' or 'knock offs', these may be less lucrative than those available to core workers. Perceived unfair access to rewards may also generate a situation of potential conflict (Dann and Hornsey, 1986).

Meals and accommodation

Although some employers have argued that workers are not low paid because they receive payments in kind, the true value of benefits such as subsidized meals and lodging is both difficult to quantify and highly questionable. Yet Johnson (1983) found that employees placed more value on these 'technological' benefits (those related to the technology of hotels and catering) than other 'institutionalized' terms (common across industry in general), including sick pay and pensions (see also Johnson, 1982).

The provision of free and, to a lesser extent, subsidized meals while on duty is certainly a widespread practice, particularly in hotels (HCEDC, 1975; Lucas, 1991; Lucas and Bailey, 1993). The WIRS3 shows that 89 per cent of hotel and catering establishments made free or subsidized meals available to employees, and that this provision was found in all establishments with 50 employees and above. This provision is more widespread in hotels and catering than in the other sectors: 47 per cent of establishments in all industries and services and 54 and 52 per cent of establishments in all service industries and private sector services respectively. However, until

August 1993 (when wages councils were abolished) the Wages Inspectorate could include the value of free meals (provided the arrangements had been agreed) in order to determine whether a worker had been statutorily underpaid. The effect of this was that a deduction could be made from the basic statutory rate of pay to cover the cost of such meals (Lucas, 1991, p. 280). which rendered the meals of some rather less of a benefit than first appeared to be the case. Additionally, the quality of meals and the circumstances in which they are taken are often poor (Wood, 1992a, p. 42).

The issue of subsidized lodging as a positive benefit is similarly problematic. It is likely to be of potentially greater benefit to hotel workers than those in other sectors (Shamir, 1981; HCIMA and Touche Ross Greene Belfield-Smith Division, 1992a) and it may generate harmful consequences for the individual and work organization (Shamir, 1981). Live-in accommodation has often been shown to be of a variable standard and usually involves room sharing (HCEDC, 1975; Saunders and Pullen, 1987). Employees living in may also work longer hours because they are on call (Knight, 1971; Brown and Winyard, 1975; Shamir, 1981). As Jordan observed,

> A change of attitude is imperative in the hotel and catering industry which tends to regard itself as exceptional in the value of the non-monetary benefits it provides. This is nonsense. There are many other industries and services whose employees live 'on the job' and even more who provide free or inexpensive meals, or perks associated with the products sold or manufactured by the enterprises concerned. These companies provide the accommodation or perks because it is in their interests to do so and because it is economically viable. So too does the hotel and catering industry. In most industries the provision of these benefits is not offered as a justification for low pay. Nor should it be in hotel and catering. (Jordan, 1978, p. 11).

Tips

The issue of tipping has been a topic of considerable interest for some time (for example, HCEDC, 1968; CIR, 1971; Mars and Mitchell, 1976; Nailon, 1978; Butler and Skipper, 1981; Shamir, 1983; Ralston, 1989; Lucas, 1991; Wood, 1992a, pp. 44–7; Lucas and Bailey, 1993). Some, but not all, workers may benefit from tips (Mars and Mitchell, 1976; Lucas, 1991) and the receipt of tips is not simply distinguished between those with direct customer contact and those with little or none, implied by Dronfield and Soto (1980); who benefits is very much a function of the type of tipping system in force (Lucas, 1991). Where a tronc is in operation, tips and the service charges are allocated on a points system; points are 'frequently a trade-off between status and length of service and bear little necessary relationship to the effort–reward nexus' (Wood, 1992a, p. 46).

To an employer, tips provide a free subsidy to low basic pay. Customer service staff, in effect, become entrepreneurs rather than employees (Mars *et al.*, 1979). Management may, however, be able to reinforce and support this system to an extent through a reasonably sophisticated system of customer 'prompts'. Those staff collecting high tips would be most likely to oppose trade unions who would seek to increase basic pay and seek the elimination of informal rewards (Mars and Mitchell, 1977).

The adverse consequences of tipping have been emphasized rather more forcefully than their supposed benefits. Tips are largely outside the control of workers because they are customer driven and the amounts collected may be subject to considerable variations. Tips have an adverse effect on productivity, leading managers to abdicate some responsibility for standards and involvement because of dependency on the customer (HCEDC, 1968); they do not act as a motivator because the amount of effort and skill put in may not be matched by commensurate reward (Nailon, 1978). Tips serve to reinforce competition among workers, which increases the potential for disharmony (CIR, 1971; Snow, 1981; Dann and Hornsey, 1986).

Tips are not wages for the purpose of being detailed on a pay slip and cannot be included in remuneration for the purposes of calculating a week's pay for compensation in industrial tribunals (see Chapter 9) because they are a discretionary payment by a third party, as in *Hall* v *Honeybay Caterers Ltd* (1967) 2 ITR 538 and *Palmanor Ltd (t/a Chaplins Night Club)* v *Cedron* [1978] IRLR 303 EAT. A fixed service charge shared out between employees does count as part of a week's pay, as in *Tsoukka* v *Potomac Restaurants Ltd* (1968) 3 ITR 259 and *Keywest Club Ltd* v *Choudhury and others* [1988] IRLR 51 EAT.

The tax position is complex and depends on the system operating. Individual tips should be declared to the Inland Revenue by workers themselves. Where the employer operates a pool system, s/he should make the appropriate deductions. Where the employees pool the tips among themselves (a tronc), the person who organizes the tronc must make the appropriate deduction, but establishing whether this system exists may be problematic. In *Figael* v *Fox (Inspector of Taxes)* [1990] (cited in Peters, 1992, p. 198), the company contended that a director in charge of the restaurant who divided the tips was the tronc master and, therefore, responsible for payment of the tax. This was not upheld and the company was liable for making the appropriate deductions. Although there is no obligation on an employer to pay out a service charge to employees, where it is paid out it is liable to tax.

Illicit rewards: fiddles and knock-offs

Fiddles and knock-offs are not exclusive to hotels and catering, and similarly deviant practices of 'getting back' (Marchington, 1992) and defiance (Analoui and Kakabadse, 1989) undoubtedly exist elsewhere. This could include the purloining of miscellaneous items like pens, paper and sellotape, using the company telephone and putting the odd letter through the company mailing system (see Mars, 1983). Such actions are deemed to be covert actions of discontent, and may be individual or collective (Analoui and Kakabadse, 1993, pp. 54–6). However, in some cases such practices may simply reflect the cultural norms of the workplace that employees conform to subconsciously rather than consciously.

In a hotel and catering study, Kelliher (1984) found that the majority of fiddles were of low unit value on a regular basis, although the 'cumulative effect over a period of time is undoubtedly substantial' (p. 39). Big fiddles were usually 'the domain of only very experienced staff or management, who find it easier to "account" for a case of wine and usually rely on a high degree of collusion from other staff' (Kelliher, 1984, p. 38).

The real issue for managers is about setting boundaries within which pilferage can be tolerated, but as noted by Mars and Mitchell (1976), this gives managers considerable power to sack a particularly troublesome worker because theft is normally unacceptable behaviour outside the industry. As Kelliher (1984) noted in relation to the Wimbledon tennis championship, management used the process of pilfering as a control mechanism. At the end of the first week fewer staff were needed. In order to reduce numbers and reassert managerial control, a rigorous security check was made at the staff exit. Twenty staff were sacked for being in possession of strawberries which would otherwise have been thrown away, a practice which management had closed its eyes to earlier in the week. Staff who had worked there in previous years were well aware of this 'control' mechanism and were more selective about the days when they took home goods.

As with many of the features of total reward or remuneration already outlined, it is difficult to make any accurate assessment of the extent and value of illicit rewards. Wood (1992a, p. 49) argues that 'access to type, quantity and quality of fiddle is highly stratified'. Fiddling mechanisms can be extremely complex (see Mars, 1973, 1983; Mars *et al.*, 1979; Mars and Nicod, 1981, 1984). To suggest that fiddling and pilfering is only undertaken to supplement poor pay (for example, Marshall, 1986) overlooks the complexities in the workplace that drive workers to become involved in such behaviours (Analoui and Kakabadse, 1989, 1993). Since so many workers are paid in cash, the opportunities for managerial fiddles, including the avoidance of paying tax and paying non-existent workers, are also rife unless organizations operate effective pay control procedures.

Incentives

There is some evidence to show that incentive schemes are used in hotels and catering to varying extents (Johnson, 1982, 1983; Ralston, 1989, p. 291; Lucas, 1991, p. 280), but for manual workers the payments that derive from them are lower than elsewhere (see Table 5.7). Managers are more likely to benefit from profit sharing schemes than operatives (Ralston, 1989). The WIRS3 found that hotels and catering had lower levels of any form of profit sharing (38 per cent of establishments) than was the case elsewhere: all industries and services, 43 per cent; all service industries, 49 per cent; private sector services, 47 per cent. This was most likely to be in the form of profit-related payments or bonuses.

One foreign-owned hotel company, Campanile UK, is known to remunerate hotel managers with a high incentive element through a managers' bonus scheme and an annual managerial competition, which is considered to be generally successful (Lucas and Laycock, 1991a, b, 1992). Even so, at a more general level, there are considerable questions about the validity and effectiveness of incentives, put neatly by Torrington and Hall (1991, p. 629) as follows:

> Incentive payments remain one of the ideas that fascinate managers as they search for the magic formula. Somewhere there is a method of linking payment to performance so effectively that their movements will coincide ... This conviction has sustained a continuing search for this elusive formula, which has been hunted with all the fervour of the Holy Grail or the crock of gold at the end of the rainbow.

Incentive pay and payment by results schemes are designed to stimulate performance, relate to manual workers, are collective and are output related. As a general rule, managers expect incentive schemes to motivate workers, control output and improve cost-effectiveness. This is largely supported by Ralston's (1989, p. 293) finding that the most common reason for having incentives was to increase productivity; other reasons included to increase flexibility, to improve levels of pay, to reduce unit labour costs, to improve quality of work and to change work methods.

Employees with a strong instrumental (financial) orientation to work will welcome incentives, which can give scope for greater employee autonomy and can increase employee interest. However, even if managers believe that they control the scheme, employees are often able to manipulate the scheme to their advantage. Consequently, all schemes need to be reviewed constantly and changed frequently to circumvent fiddling and provide fresh stimulus. Apart from the temptation to fiddle, operational inefficiencies, fluctuating earnings, threats to quality standards and quality of working life are other problems associated with incentives (see Torrington and Hall, 1991, p. 630–40).

Performance-related pay is designed to reward input, usually by managerial, professional and technical staff, and is more individual and personal. It can be divisive and ineffective in practice because of factors outside the individual performers' control which hamper individual performance. As Smith (1993) notes, in both white and blue collar employment there is a lack of evidence to link performance-related pay with a positive effect on standards of performance. Performance-related pay schemes almost always turn out to be more costly than envisaged. The way in which performance is assessed – performance appraisal – is briefly mentioned in Chapter 7.

As noted earlier (see Table 5.3), the WIRS3 found that individual performance was as significant a factor in influencing pay levels in hotels and catering as it was in other sectors. Yet, ironically, it does not lead to levels of pay that constitute good reward in hotels and catering. The industry appears to conform to Smith's (1993) 'lean and mean' approach, where pay policy is no more than an execution of a low-wage, low-salary strategy in pursuit of low labour costs. To paraphrase Nolan (1989, p. 84), the more pay is kept down in order to solve the industry's problems, the weaker is the incentive for firms to innovate and modernize, which must have adverse consequences for the industry.

Commission schemes, usually sales-related, are not an area that has been studied widely (Torrington and Hall, 1991, p. 641).

Fringe benefits

In general, the incidence and extent of fringe benefits are more marked at managerial level but, even so, they are 'seldom managed in a positive way with a sense of purpose about why they are provided and what they are to achieve' (Torrington and Hall, 1991, p. 646).

In comparison with other employment sectors, hotels and catering provide a poorer range and level of fringe benefits (Johnson, 1982, 1983; Ralston, 1989; Lucas, 1991). Although managers tend to fare better (see HCIMA and Touche Ross Greene Belfield-Smith Division, 1992a), a large number of managers receive no fringe benefits (Ralston, 1989, p. 295). Johnson (1983) found that employers saw no reason to alter this situation, because employees 'only leave anyway, so what's the point?', which is consistent with policies that place reliance on using the external labour market rather than those which seek to create a more stable internal labour market. Boyd Ohlen and West (1993) suggest that the provision of increased fringe benefits may not necessarily reduce turnover and aid retention. The benefits offered by low--

turnover employers were related to fostering and enhancing a long-term relationship, using retirement plans, for example.

Conversely the WIRS3 did find some evidence of longer-term benefits in the form of share ownership schemes in hotels and catering. Higher levels of any kind of share ownership (43 per cent of establishments) were found in hotels and catering than elsewhere (all industries and services, 32 per cent; all service industries, 37 per cent; private sector services, 39 per cent), particularly in establishments employing 100 employees or more (76 per cent). However, it should be remembered that 78 per cent of hotel and catering establishments in the WIRS3 sample were private or public limited companies and 77 per cent were one of several establishments (see Tables A1.2 and A1.3), so they were more likely to be part of a larger concern that is in a position to offer benefits such as share option schemes – proprietor-run establishments simply do not have shares to offer.

The main examples of the potential range of benefits available to hotel and catering workers are shown in the following list.

basic pay;
performance-related pay;
incentive bonus;
payment by results scheme;
Christmas or seasonal bonus;
overtime paid at premium rates;
shift premia;
unsocial hours premia;
service pay;
tips/service charge;
subsidized meals;
subsidized accommodation;
paid annual/statutory holiday entitlement;
sick pay above SSP;
maternity pay/leave above statutory minimum;
compassionate leave;
medical facilities
- private health insurance;
- subsidized medical facilities;

company pension scheme
- contributory
- non-contributory;

life insurance scheme;
personnel accident insurance;
transport to and from work;
provision of uniforms/footwear;
subsidized laundering of uniforms;
clothing allowance;
employee discount on products;
long service awards
- shareholdings;
- monetary;

relocation expenses;

company car;
car servicing;
petrol credit card;
free petrol;
profit-sharing schemes
- cash bonus
- approved deferred share trust plan
- profit-related pay;

subscription to professional bodies;
subsidized holidays and travel;
subsidized personal loans;
mortgage facilities;
employee introduction bonus;
suggestion scheme;
bar allowance;
entertainment allowance;
company savings scheme;
language proficiency award;
employee of the month/year award;
social club.

Summary propositions

- Although hotel and catering managers may have a broader range of remuneration elements with which to reward workers, most workers do not enjoy good levels of remuneration overall.
- Pay systems that generate poor rewards, which are also unlikely to be administered formally, are likely to result in reduced managerial control because they afford greater incentives and opportunities for workers to seek extra rewards; for example, by fiddling.
- Pay levels and differentials that are not objectively or formally determined are likely to generate feelings of perceived unfairness among individuals within and across departments.
- Reward systems are more likely to be geared to short-term rather than long-term benefits.

Discussion points

- As a manager, what would be your objectives in the contract for payment? Justify your choice.
- Outline the payment/reward system you would utilize to enable you to fulfil your objectives?
- As a worker, what are your objectives in the contract for payment?
- Outline the payment/reward system that would satisfy those objectives.
- Do incentives motivate? Discuss the pros and cons of different types of incentives for different categories of hotel and catering workers.

5.3 A FAIR AND EFFECTIVE DEAL?

Although in an ideal world the effort–reward bargain should always satisfy all the parties, the bargain is not absolute and is subject to constant adjustment and refinement. Thus while pay may be reviewed formally on an annual basis, interim adjustments in terms of standards of performance expected for a given rate of pay will occur if, for example, working methods alter or fewer staff are required to undertake the same amount of work. Such adjustments may be made on the basis of informal and formal negotiations (see Chapter 6). But adjustments to the effort–reward bargain may give rise to dissatisfactions among managers and/or workers, which may need be dealt with by other means, such as the discipline or grievance procedures (see Chapter 7). Given the general picture of low levels of remuneration and informal pay systems in hotels and catering, the final part of the chapter considers the extent to which this scenario represents a fair and effective deal for those involved: employers, workers and customers.

Employers

Much of what has been presented in this chapter does not square with the employer objectives in the contract for payment identified by Torrington and Hall (1991), which were set out earlier. Hotel and catering employers may claim, and some do, that their pay is good enough to give them prestige, secure competitive advantage, motivate staff and increase performance, but the facts simply do not bear this out. The one aspect of Torrington and Hall's framework that does seem to 'fit' is in the area of controlling operations and costs, where relatively low pay and the flexible forms of employment that are utilized in the industry are closely interrelated (see Chapter 3).

While it is probably unlikely that most managers set out to be deliberately unfair in terms of rewarding staff for their efforts, it is understandable that they want to strike what they perceive to be the best deal for their business. Stereotypical perceptions about the value of workers' worth – women and part-timers are worth less – and the nature of the work – much is unskilled – undoubtedly serve to keep pay rates at the lower end of the spectrum.

Furthermore, if a manager can recruit an equivalent replacement worker at a lower rate, he or she will probably feel that this is justifiable. Whether this is really effective in some circumstances is questionable because the new recruit may find that he or she is paid less than others, or asked to do more work than is reasonable. As a result the employee leaves, thus contributing to the vicious circle of managers spending an undue amount of time on recruiting staff who never become fully effective because they never stay long enough, and who create disruptions to the business by leaving suddenly. This may be a hypothetical example, but it is one which features all too often in the literature, although the costs of such labour turnover are difficult to quantify (see Johnson, 1980, 1981, 1985, 1986; Riley, 1981). It should not be forgotten that some managers seek to perpetuate high levels of turnover to control staff numbers (Simms *et al.*, 1988).

There is evidence of a gap between organizational statements of intent about the importance of 'good' rewards and what actually happens in practice at unit level in larger companies (see also the section on 'policy' in Chapter 2). The systems and control mechanisms set in place by head office do not actually allow a platform of 'good' rewards to be developed, and this 'pay policy' may serve to encourage managerial actions and behaviours that are ultimately more damaging to organizational effectiveness, as the following example illustrates.

Pay is the most directly tangible and visible aspect of labour costs and organizations commonly set wage cost : sales revenue ratios that unit managers must adhere to as a measure of operational efficiency. In reality this gives the unit manager relatively little autonomy and scope and, therefore, personal control over what can be paid to staff, because the basis on which head office has calculated the ratios in the first place is likely to be more tight than loose. The only realistic options available to the unit manager are to keep pay rates low or keep staff numbers down; there is almost no room for manoeuvre elsewhere, unless additional reward can found from other sources. It is just conceivable that managers want to do more than simply 'administrate' head office pay policy but the only way they can find to do so is through the provision of less tangible and often illicit rewards. In other words, the practical effects of tightly drawn head office pay policies may be the opposite of what is intended because they drive managers who want to demonstrate autonomy and scope into illicit practices that are ultimately more damaging to organizational efficiency.

A low pay, tight labour cost control policy may thus produce more negative than positive consequences (see also productivity and flexibility in Chapter 3). Reward management in hotels and catering can be said not to fit with Conservative administrations' aims of economic improvement (see also Nolan, 1989; Smith, 1993). Alternatively, Lucas and Bailey's (1993) findings suggested that employers' utilisation of cheaper sixth-formers in part-time employment seemed to be effective. Perhaps if a low pay policy is managed reasonably well, it can work in certain circumstances.

Employees

It is difficult to see how the pay scenario presented in this chapter satisfies Torrington and Hall's (1991) employee objectives in the contract for payment in relation to the 'felt-fairness' principle, a share in the company's success, reasonable pay relative to that of others, adequate recognition for personal contribution and a good overall package. Many workers have little choice other than to accept poorly remunerated, unskilled employment (Gabriel, 1988; Macaulay and Wood, 1992). To some, both men and women, the work may be readily available because, for example, it is close to home or the hours fit with family commitments or full-time education requirements. In one sense these workers are being exploited because employers are able to take advantage of their disadvantaged situation. They may, in spite of their low paid status, be performing well and to the best of their ability. Yet most receive no performance-related pay.

Studies have shown that low paid workers, partly because of other personal attributes, are not likely to be as effective a factor of production as better trained and rewarded workers (for example, Craig *et al.*, 1982; Prais *et al.*, 1989). Peripheral workers who have more limited access to informal rewards will feel that they have been unfairly treated and may become less effective; given the high reliance on such workers, the organization is likely to function less well.

Customers

Finally the evidence points largely to customer dissatisfaction about their role in the effort–reward bargain. Many of the studies of tipping have found that the British customer is uneasy about the practice. Disquiet about a service charge is, perhaps, even more pronounced because the customer feels that he or she has already paid the price for the product, and there is no guarantee that staff will receive any money from the charge. On the other hand a customer may feel that he or she is buying more effective service by giving a waiter an advance tip.

Summary propositions

- The contract for payment in hotels and catering does not satisfy the principles of 'good' practice.
- Organizations need to think much more carefully about pay policies and their effects on managerial and employee behaviour.

Discussion points

- What are the pros and cons of the total reward system for managers and workers?
- What are the consequences of a low pay policy for managers and workers?
- Are tips a desirable element of the reward package for managers, workers or customers?

REFERENCES

Analoui, F. and Kakabadse, A. (1989) Defiance at work. *Employee Relations*, **11** (3), 1–62.

Analoui, F. and Kakabadse, A. (1993) Industrial conflict and its expressions. *Employee Relations*, **15** (1), 46–62.

Atkinson, J. (1984) Manpower strategies for flexible organizations. *Personnel Management*, August, 28–31.

Boyd Ohlen, J. and West, J.J. (1993) An analysis of fringe benefit offerings on the turnover of hourly housekeeping workers in the hotel industry. *International Journal of Hospitality Management*, **12** (4), 323–36.

Brown, M. and Winyard, S. (1975) *Low Pay in Hotels and Catering*. London: Low Pay Unit.

Butler, S.R. and Skipper, J. (1981) Working for tips. *The Sociological Quarterly*, **22**, 15–27.

Byrne, D. (ed.) (1986) *Waiting for Change*. London: Low Pay Unit.

Commission on Industrial Relations (1971) *The Hotel and Catering Industry, Part 1. Hotels and Restaurants, Report No. 23*. London: HMSO.

Craig, C., Rubery, J., Tarling, R. and Wilkinson, F. (1982) *Labour Market Structure, Industrial Organisation and Low Pay*. Cambridge: Cambridge University Press.

Croney, P. (1988) An investigation into the management of labour in the hotel industry. MA thesis, University of Warwick.

Dann, D. and Hornsey, T. (1986) Towards a theory of interdepartmental conflict. *International Journal of Hospitality Management*, **5** (1), 23–8.

Dronfield, L. and Soto, P. (1980) *Hardship Hotel*. London: Counter Information Services.

Gabriel, Y. (1988) *Working Lives in Catering*. London: Routledge & Kegan Paul.

Hotel and Catering Economic Development Committee (1968) *Why Tipping?* London: HMSO.

Hotel and Catering Economic Development Committee (1975) *Manpower Policy in the Hotel and Catering Industry – Research Findings*. London: HMSO.

Hotel, Catering and Institutional Management Association (HCIMA) and Touche Ross Greene Belfield-Smith Division (1992a) *Salaries and Benefits in the Hotel and Catering Industry 1991 Survey. Volume 1, Survey*. London: Touche Ross Greene Belfield- Smith Division.

Hotel, Catering and Institutional Management Association (HCIMA) and Touche Ross Greene Belfield-Smith Division (1992b) *Salaries and Benefits in the Hotel and Catering Industry 1991 Survey. Volume 2, Appendices*. London: Touche Ross Greene Belfield-Smith Division.

Johnson, K. (1980) Staff turnover in hotels. *Hospitality*, February, 28–36.

Johnson, K. (1981) Towards an understanding of labour turnover? *Service Industries Review*, **1**(1), 4–17.

Johnson, K. (1982) Fringe benefits: the views of individual hotel workers. *Hospitality*, June, 2–6.

Johnson, K. (1983) Payment in hotels: the role of fringe benefits. *The Service Industries Journal*, **3** (2), 191–213.

Johnson, K. (1985) Labour turnover in hotels – revisited. *The Service Industries Journal*, **5** (2), 135–52.

Johnson, K. (1986) Labour turnover in hotels – an update. *The Service Industries Journal*, **6** (3), 362–80.

Jordan, D. (1978) *Low Pay on a Plate*. London: Low Pay Unit.

Kelliher, C. (1984) An investigation into non-strike conflict in the hotel and catering industry. MA thesis, University of Warwick.

Knight, I. (1971) *Patterns of Labour Mobility in the Hotel and Catering Industry*. London: HCITB.

Lennon, J.J. and Wood, R.C. (1989) The sociological analysis of hospitality labour and the neglect of accommodation workers. *International Journal of Hospitality Management*, **8** (3), 227–37.

Lucas, R.E. (1991) Remuneration practice in a wages council sector: some empirical observations in hotels. *Industrial Relations Journal*, **22** (4), 273–85.

Lucas, R.E. (1992) Minimum wages and the labour market – recent and contemporary issues in the British hotel industry. *Employee Relations*, **14** (1), 33–47.

Lucas, R.E. and Bailey, G. (1993) Youth pay in catering and retailing. *Personnel Review* **22**, (7), 15–29.

Lucas, R.E. and Laycock, J. (1991a) Developing an interactive personnel function for managing budget hotels: how the personnel function operates in Campanile. Paper presented to the Third International Journal of Contemporary Management Conference, October, Bournemouth.

Lucas, R.E. and Laycock, J. (1991b) An interactive personnel function for managing budget hotels. *International Journal of Contemporary Hospitality Management*, **3** (3), 33–6.

Lucas, R.E. and Laycock, J. (1992) Quality and the human resourcing function in Campanile UK. Paper presented to the Second Annual Conference on Human Resource Management in the Hospitality Industry, Quality and Human Resources, London, December.

Macaulay, I.R. and Wood, R.C. (1992) *Hard Cheese: a Study of Hotel and Catering Industry Employment in Scotland*. Glasgow: Scottish Low Pay Unit.

Marchington, M. (1992) Managing labour relations in a competitive environment. In A. Sturdy, D. Knights and H. Willmott (eds), *Skill and Consent: Contemporary Studies in the Labour Process*, pp. 149–84. London: Routledge.

Mars, G. (1973) Hotel pilferage: a case study in occupational theft. In M. Warner (ed.), *The Sociology of the Workplace*, pp. 200–10. London: Allen & Unwin.

Mars, G. (1983) *Cheats at Work: an Anthology of Workplace Crime*. London: Unwin Paperbacks.

Mars, G. and Mitchell, P. (1976) *Room for Reform?* Milton Keynes: Open University Press.

Mars, G. and Mitchell, P. (1977) *Catering for the Low Paid: Invisible Earnings. Low Pay Bulletin No. 15*. London: Low Pay Unit.

Mars, G. and Nicod, M. (1981) Hidden rewards at work: the implications from a study of British hotels. In S. Henry (ed.), *Can I Have It in Cash? A Study of Informal Institutions and Unorthodox Ways of Doing Things*, pp. 58–72. London: Astragal Books.

Mars, G. and Nicod, M. (1984) *The World of Waiters*. London: Allen and Unwin.

Mars, G., Bryant, D. and Mitchell, P. (1979) *Manpower Problems in the Hotel and Catering Industry*. Farnborough: Gower.

Marshall, G. (1986) The workplace culture of a licensed restaurant. *Theory, Culture and Society*, **3** (1), 33–47.

Nailon, P. (1978) Tipping: a behavioural review. *HCIMA Review*, **2**, 231–41.

Nolan, P. (1989) Walking on water? Performance and industrial relations under Thatcher. *Industrial Relations Journal*, **20** (2), 81–92.

Peters, R. (1992) *Essential Law for Hotel and Catering Students*. London: Hodder and Stoughton.

Prais, S.J., Jarvis, V. and Wagner, K. (1989) Productivity and vocational skills in services in Britain and Germany: hotels. *National Institute Economic Review*, November, 52–74.

Ralston, R. (1989) The changing nature of personnel management in the hotel and catering industry. MSc thesis, University of Manchester.

Riley, M. (1981) Labour turnover and recruitment costs. *Hospitality*, September, 27–9.

Robinson, O. and Wallace, J. (1984) Earnings in the hotel and catering industry in Great Britain. *The Service Industries Journal*, **4** (2), 143–60.

Saunders, K.C. and Pullen, R.A. (1987) *An Occupational Study of Room-maids in Hotels*. London: Middlesex Polytechnic Research Monographs.

Shamir, B. (1981) The workplace as a community: the case of British hotels. *Industrial Relations Journal*, **12** (6), 45–56.

Shamir, B. (1983) A note on tipping and employees' perceptions. *Journal of Occupational Psychology*, **56**, 255–9.

Simms, J., Hales, C. and Riley, M. (1988) Examination of the concept of internal labour markets in UK hotels. *Tourism Management*, **9** (1), 3–12.

Smith, I.G. (1993) Reward management: a retrospective assessment. *Employee Relations*, **15** (2), 45–59.

Snow, G. (1981) Industrial relations in hotels. MSc thesis, University of Bath.

Torrington, D. and Hall, L. (1991) *Personnel Management: a New Approach*, 2nd edition. London: Prentice Hall.

Wood, R.C. (1992a) *Working in Hotels and Catering*. London: Routledge.

Wood, R.C. (1992b) Hospitality industry labour trends: British and international experience. *Tourism Management*, **13** (3), 297–301.

SIX

Decision-Making: Methods and Processes

The purposes of this chapter are to:

- consider the main forms of decision-making that can be used to determine remuneration and other terms of employment;
- outline the main types of process that are used to make decisions in the employment relationship;
- assess the relevance and effectiveness of these mechanisms in hotels and catering.

6.1 INTRODUCTION

The first main topic of this chapter is the means and methods by which pay and other related issues are determined and regulated. The WIRS3 found that formal and systematic methods of pay regulation were largely absent from hotels and catering at company and workplace level. As noted in Chapter 4 (see Table 4.1), very few workplaces bargain with recognized trade unions. The only formal method of pay regulation at industry level used to be the wages councils; they were abolished in 1993. Even though three wages councils determined the pay of approaching one million workers in the industry, their effectiveness has been questioned because pay remained low and voluntary collective bargaining arrangements did not develop (Lucas, 1989, 1991a).

Therefore the emphasis falls on more individualistic methods of decision-making, mainly unilateral management decision-making. Methods of decision-making in hotels and catering are more likely to be informal than formal and, as noted in Chapter 5, they sometimes incorporate illicit behaviours and actions and an input from the customer. While formal collective procedural arrangements may be said to have been associated with inadequate substantive arrangements, there is no reason to suggest that less formal individual arrangements will prove to be more adequate.

Gospel and Palmer's (1993) spectrum of methods of decision-making used to establish employment rules informs much of the discussion in the first part of the chapter. Although the emphasis falls on the manager–worker relationship, it is necessary to address the position of the customer at appropriate points.

The interpersonal encounter, on either an individual or a collective basis, underpins the process of decision-making. The most common form of individual encounter is the interview, a process which is used formally and informally to solve problems and reach decisions, including the selection of staff, grievance resolution, determining disciplinary action and appraising performance. Negotiation, in an employee relations context, is usually a collective activity, although bargaining in order to reach an agreement can still take place between individuals (manager and worker or manager/worker and customer). While good procedures are a necessary prerequisite to ordered relations (see Chapter 7), it is what actually goes on in the process of the personal interface between individuals that is the crux of resolving the matter satisfactorily (see Torrington, 1991; Torrington and Hall, 1991, pp. 149–62). Interviewing and negotiating are highly skilled and ritualized processes that justify some analysis and evaluation at a later point in this chapter. Success in the interpersonal encounter also depends on communicating effectively. Although some aspects of communicating are mentioned in this chapter, communications and the employee–guest encounter are addressed as separate topics in more detail in the context of managing involvement and commitment in Chapter 8.

6.2 METHODS OF DETERMINING PAY AND TERMS OF EMPLOYMENT

The means by which the range of terms of employment are determined in an organization may vary. For example, although wages used to be underpinned by statutory arrangement, this did not prevent managers from dealing with pay on a joint consultative basis or on the basis of individual negotiations between a manager and subordinate employee. Work rosters may be determined on the basis of collective discussions with the employees of a department but the company sick-pay scheme may be fixed unilaterally by senior management. Customers are also decision-makers in the sense that they may provide a significant input into an employee's remuneration.

Gospel and Palmer (1993, p. 196) have identified

> a spectrum of methods of decision making and ways of generating rules which control the employment relationship, the organization of work, and relations between employers and their employees ... All of the processes involve elements of negotiation between individuals and between groups; within groups there will also be bargaining and it is important to recognise this is an important aspect of industrial relations.

These five methods, unilateral decision-making, individual bargaining, joint consultation, collective bargaining and unilateral employee regulation, plus a sixth, statutory regulation, are discussed first. However, it should be noted that there is an inherent weakness in Gospel and Palmer's model: it is two-dimensional and thus ignores the complexities generated by a three-dimensional link that underpins some hotel and catering employment relationships. The next section considers some issues relating to power and their implications for decision-making.

Unilateral decision-making

Unilateral decision-making is one extreme of the spectrum, which Gospel and Palmer (1993, p. 175) allege is 'the predominant form in firms which do not have formal joint consultative arrangements or collective bargaining arrangements', although they maintain that unilateral control is never total because it always involves some element of negotiation and bargaining. This observation is certainly supported by the findings of the WIRS3 which are summarized in Table 6.1. There is a substantially higher degree of unilateral decision-making in relation to pay in hotels and catering in comparison with elsewhere, although this tendency decreases as establishment size increases.

Table 6.1 *Method of decision-making used to determine pay where no recognized trade unions.*

			PSS (by no. of employees)				HCI (by no. of employees)			
	AIS	SI	All	25–49	50–99	100+	All	25–49	50–99	100+
Manual workers										
Unweighted base	556	364	319	89	76	154	48	17	16	15
Weighted base	789	544	463	285	110	68	82	49	22	11
Management alone (%)	74	75	81	83	82	71	92	97	87	75
Discuss/consult with employees/ representatives (%)	23	21	17	15	15	28	7	3	6	5
Non-manual workers										
Unweighted base	874	471	456	134	104	218	52	18	17	17
Weighted base	1111	696	660	404	154	103	90	52	26	12
Management alone (%)	78	78	81	78	89	80	92	87	100	95
Discuss/consult with employees/ representatives (%)	19	19	17	20	8	18	8	13	-	5

Table does not include percentage figures for missing cases.
Source: WIRS3 (1990).

Other empirical evidence from a survey among hotels (Lucas 1991a, b, 1992) has also confirmed the predominance of unilateral decision-making, particularly among hotel groups, in regard to remuneration and other employment terms, including sick pay and pension schemes. Ralston (1989, p. 289) identified a significant degree of managerial decision-making over pay for operatives, administrative staff and managers. In the smaller, generally individually owned or managed hotels, unilateralism was the method used by just over half of the responding hotels (Lucas, 1991b, p. 282). Ralston (1989, p. 288) found that although pay was underpinned by the wages council, management took the final decision on actual rates.

While unilateralism enables managers to preserve managerial prerogative and to maintain flexibility and control, Gospel and Palmer (1993) assert that it may lead to an alienated workforce that is difficult to manage and unproductive, a scenario that may be all too familiar in hotels and catering (for example, Dronfield and Soto, 1980; Byrne, 1986; Wood, 1992). Although there is scepticism about the value of unilateralism in relation to the employment relationship, some managerial decisions, including planning and investment, are nearly always unilateral decisions.

Individual bargaining

Individual bargaining is an all-pervasive process within organizations. Although Gospel and Palmer maintain that individual bargaining is limited in scope because of a power imbalance between employer and employee, in practice it is never as simple as 'take it or leave it'. For example, individual bargaining takes place in order to arrive at an employment decision and to fix the initial terms of employment on commencing work. However, the employer's power is circumscribed to the extent that it is the candidate who exercises the right to accept or refuse a job offer (see Torrington and Hall, 1991, pp. 283–4). On the other hand, if this method is associated with illicit deals in the total reward system (Mars *et al.*, 1979), then there is something of a mutual dependency between manager and subordinate.

Johnson (1983) found that core workers were able to obtain additions to basic pay by individual bargaining but these were 'technological' (that is, special to hotels) rather than 'institutionalized' (for example, sick pay and pensions). He concluded that because trade unions were concerned with institutionalized benefits, individual and collective bargaining were not mutually exclusive. Lucas (1991b) also found evidence of individual bargaining in hotels in relation to basic pay 'We have individual negotiations using the wages council rate as a start' (p. 282). The benefits and drawbacks of this method are similar to those outlined in relation to unilateralism, with the additional drawback that it can be time-consuming and create anomalies that generate discontent.

There is an element of individual bargaining between worker and customer in that the worker is selling an 'attitude' that constitutes good service. A customer complaint that the worker has shown the wrong attitude may be sufficient grounds to persuade the worker's manager to dispense with his or her services, so customer influence extends to more than just tipping.

Joint consultation

Gospel and Palmer (1993) define joint consultation to mean the existence of formal representation and formal mechanisms designed to give employees a voice in decision-making, although they do acknowledge that it encompasses a wide spread of arrangements from very informal daily briefing sessions to formal joint consultative committees and works councils, which are required by law in some European countries (see also Chapter 8). Marchington (1989) views joint consultation from a broader perspective, seeing it as taking 'any one of a number of different forms, depending upon the objectives of the parties involved and the climate within which it takes place' (p. 378), and this is discussed more fully in Chapter 8.

As shown in Table 6.1, the WIRS3 found that in the absence of collective bargaining with trade unions, discussions or consultations about pay with employees or their representatives were a marginal practice (7 per cent of respondents) in hotels and catering, although this method was more prominent elsewhere (at or around 20 per cent of respondents). Lucas (1991b) found that a sizeable minority of smaller hotels used participative means to determine remuneration, although this practice was much more limited in the major hotel companies (these were more likely to feature in the WIRS3 than the Lucas study). Lucas noted that 'Closer proximity to smaller numbers of staff would appear to be related to a more participative approach in a significant proportion of small units, seemingly dispelling some of the perceptions that most hotel managers are autocrats' (Lucas, 1991b, p. 281).

This method is also all-pervasive because there is always some information flow between employer and employee and some discussion. While management retains ultimate discretion and control, employees are given a voice, which may increase their commitment. Joint consultation, employee involvement and other related issues form the main subject matter for Chapter 8.

Collective bargaining

For collective bargaining to take place, employees must associate into a workgroup, trade union or professional association (employers are not required to associate, although some may choose to do so) and each side must have something the other side wants. Each side is able to apply a certain amount of pressure and a process of give and take leads to a negotiated agreement (the negotiating process is outlined later in this chapter).

Although Gospel and Palmer (1993) note that collective bargaining has a wide impact on the economy, this is not the case in hotels and catering because employees have not associated in any way. There have been very few formal voluntary collective bargaining arrangements in the industry (CIR, 1971, 1972, 1973; ACAS, 1980; Lucas, 1991a, b, 1992) and this situation still exists, as confirmed by the WIRS3 (see Table 3.1).

The absence of adequate collective bargaining arrangements and the presence of unbridled individualism that led to undue worker exploitation were the reasons why wages councils were established in certain employment sectors, including hotels and catering. This state-sponsored form of collective bargaining was removed when the wages councils were abolished in 1993. (For further discussion on collective bargaining see Gospel and Palmer, 1993, pp. 177–78, 198–232; Kessler and Bayliss, 1992, pp. 176–90.)

Unilateral employee regulation

Unilateral employee regulation centres on issues of 'custom and practice'; that is, practices that have evolved over time that become part of the organization's 'unofficial' rules. They are almost never written down, and may include starting times, tea breaks, transfers, work allocation and pilferage. These can develop into irksome restrictive practices, the removal of which may generate conflict. This could include a tolerated level of pilferage that becomes identifiable, and no longer tolerable, as a result of newly instigated head office accounting procedures.

Butler and Snizek (1976) found that service workers practise promotional manipulation in an attempt to control the work situation, thereby maximizing the tab and reward by way of eventual tips. Customer 'manipulation' also features largely in work by Mars and others (including Mars and Mitchell, 1976; Mars *et al.*, 1979; Kelliher, 1984; Analoui and Kakabadse, 1989). In relation to a customer care programme designed to empower employees in a luxury hotel, Wynne (1993) identified a situation whereby employees are able to manipulate the rules by way of a coping strategy. Employees at the front desk are trained to initiate and maintain eye contact with customers in order to acknowledge their presence. However, this may often initiate verbal encounters. If a visual encounter is avoided, most guests will wait quietly to be served and the receptionist can slow down work to a more controllable level. A similar observation has been made by Marchington (1992a), who found that although customer care programmes may enable employees to identify their own contribution to the organization, they are no more than 'getting by' by making the work more tolerable.

Statutory regulation

Statutory regulation in the employment relationship refers to the mass of individual and collective laws that regulate a wide range of employment issues, including health and safety, discrimination, discipline, dismissal, maternity rights and trade union activities. Most of the individual rights can be invoked by workers through the industrial tribunal system, although much of collective law and health and safety law is enforceable through the courts. Britain is becoming increasingly subject to legal decisions reached at the European Court of Justice. Employment law is discussed in more depth in Chapters 9, 10 and 11.

A one-sided state of affairs?

Without more systematic research among small establishments (fewer than 25 employees), it is only possible to speculate about the extent to which these various forms of decision-making typify the way in which decisions about the employment relationship are made in hotels and catering. Clearly, collective bargaining, which may be 'an institution of major importance which dominates the determination of pay and conditions of employment for British employees' (Gospel and Palmer, 1993, p. 198), is an institution of virtually no importance in hotels and catering. It is also probable that joint consultation about pay, whether formal or informal, is at best a minority practice, although this method may be used to determine other matters (Croney, 1988; Marchington *et al.*, 1992). As Marchington (1992a) notes, the competitive environment in which hotel and catering companies operate 'is hardly the kind of climate in which employee involvement and consent can be easily developed' (p. 171). Nevertheless, some form of participation does exist in smaller units, and this may be more marked than among larger establishments, but it is likely to be informal (Lucas, 1991b). Individual bargaining and unilateral employee regulation are also likely to be found in varying degrees across hotels and catering.

Unilateralism over pay does emerge as the predominant practice, and this should not come as a complete surprise in terms of 'managing in hotels and catering', discussed in Chapter 2. However, the extent to which managers *actually* manage in this way is cast into doubt by more 'hidden' factors associated with methods such as informal consultation and unilateral employee regulation. Managers may like to think that they are in charge but fail to recognize circumstances or identify factors that in reality serve to undermine their managerial prerogative.

At a more pragmatic level it is probable that combinations of the various types of decision-making identified by Gospel and Palmer (1993) are found in many organizations and that, overall, different methods are used to determine different kinds of rules which extend to issues of substance other than pay. It is probably true to state that some form of bargaining takes place in most employment decisions, although this may be more covert than overt in many cases. However, the main weakness of Gospel and Palmer's model is that is does not allow for a triadic relationship that involves manager, worker and customer, and therefore fails to address the complexities, tensions and role conflict that arise from this situation. The bargaining process encompasses issues about power, to which this chapter now turns.

Summary propositions

- Unilateral decision-making, individual bargaining and joint consultation are the main overt procedural means that are used to determine the substance of the effort–reward bargain. In practice, all involve an element of bargaining.
- Different methods may be used to determine different issues.
- Managers may not be fully aware that unilateral employee regulation does occur. Because it represents a more covert means of regulating the employment relationship, it may be more difficult to identify and manage.

Discussion points

- What are the pros and cons of unilateral decision-making, individual bargaining, joint consultation, collective bargaining, unilateral employee regulation and statutory regulation from a managerial perspective?
- What are the pros and cons of unilateral decision-making, individual bargaining, joint consultation, collective bargaining, unilateral employee regulation and statutory regulation from a worker's perspective?
- As a practising manager, which method(s) would you choose to determine the pay and employment conditions of your workforce?
- As an employee, which method(s) would you prefer were used to determine your pay and employment conditions?
- Evaluate the role of the customer in the decision-making process surrounding pay and employment conditions.

6.3 POWER

One of the difficulties that would still exist even if more research in the field of decision-making methods were carried out in hotels and catering concerns the complex subject of power (for example, Marchington, 1982, pp. 94–108; Mintzberg, 1983;

Kirkbride, 1992; Gospel and Palmer, 1993, pp. 189–96; Wynne, 1993; Wood, 1994, pp. 173–82). As Wynne (1993, p. 42) notes,

> A precise definition of power is difficult to locate since, although at a very general level power is essentially the ability of an individual or group to gain some kind of advantage, the ability is never held absolutely or in isolation. It is always held in terms of a relationship with others and is always contingent on the particular situation.

Marchington (1982, p. 96) notes that a 'satisfactory analysis of power must incorporate both decision-making and *non-decision-making*, to examine which issues are *excluded* from ... discussions and to assess their effect on employees.' He goes on to argue that where a manager can get a worker not even to desire involvement in a specific issue, this is a particularly effective use of managerial power. By implication, the authoritarian manager who quashes any staff dissent and prides himself or herself on the absence of grievances constituting good employee relations is powerful, but employee relations may not be good if defiant behaviours are being manifested (see Analoui and Kakabadse, 1989). On the other hand, power does not have to be used to be effective. Reputation or the threat of sanctions is often sufficient to get demands met by the other party.

Managers may believe that they are acting unilaterally in fixing pay, whereas in reality employees are able to exert rather more influence, not necessarily on the level of pay itself, but in relation to the amount of effort they are prepared to expend in return for that element of reward. This has parallels with Wynne's (1993) study, whereby the employer determines the training policy in relation to standards of performance, but the employee is able to manipulate this to alter the standards of performance to slow down the work rate. In this scenario the customer enters, and is integral to, the power relationship.

Gospel and Palmer (1993, pp. 189–90) suggest that 'One way to view power is as the probability that one actor within a relationship will be in a position to carry out his or her will despite resistance.' The movement of a top chef from a high-class restaurant to another establishment elsewhere, which could cause the first restaurant to close, exemplifies this approach. Gospel and Palmer go on to note that employer and employee power bases derive from strategic importance and non-substitutability, so that the chef in the example given can be said to be a powerful employee. Most workers do not come anywhere close to that hallowed position, illustrated by Macaulay and Wood's (1992) finding that most people took hotel and catering jobs 'because they needed the money as a result of being unemployed and/or needing a job to bring extra money into the household' (p. 81). This weak employee position serves to increase the power of the employer. Similarly, an influential customer may exercise considerable power by complaining about staff and getting them sacked.

'Management still possesses significantly more power and has greater resources at its disposal' (Marchington, 1982, p. 94) than workers in the workplace. Ownership of the means of production (the majority of hotel and catering businesses are owner-managed) tilts the balance of power much more strongly in favour of the employer (Goss, 1988, p. 119). Managerial strategies that are designed to reduce the strategic importance and increase the substitutability of employees at work, which lie at the heart of the labour process approach, are all too readily evident in much of hotels and catering. Managerial power can be enhanced by reducing dependency on particular groups (the use of cook–chill products in food preparation, contracting out services), inhibiting the development of solidarity (perpetuating high labour turnover, heavy reliance on part-time and casual workers), making work routine (deskilling in fast-food) and the threat of dismissal (arbitrary practices) (see also Marchington, 1982, p. 108).

Thus in both a real and a theoretical sense employees are weak, and this position is reinforced by difficult labour market conditions, a highly competitive product market based on, in many cases, a relatively low-value intangible product, and work that is generally unskilled. Conversely the relative ease with which an employee can take his or her services elsewhere and can 'manipulate' the effort–reward exchange perhaps undermines employers' power more effectively than has so far been recognized, with the consequence that productivity, efficiency and even survival of the organization are placed in jeopardy.

What follows from the consideration of Gospel and Palmer's categorization of methods of decision-making is that the issue of legitimate power can be perceived differently in hotels and catering from elsewhere in the sense that it does not derive from employer or, more particularly, employee collectivism.

Summary proposition

- Power in a relationship is a relative concept, is not always visible or tangible, and may be more perceived than real.

Discussion points

- What is legitimate managerial power and how can it be exercised?
- What is illegitimate managerial power and how can it be exercised?
- What is legitimate workforce power and how can it be exercised?
- What is illegitimate workforce power and how can it be exercised?
- What is legitimate customer power and how can it be exercised?
- What is illegitimate customer power and how can it be exercised?

6.4 MANAGING INDIVIDUAL AND COLLECTIVE ENCOUNTERS

From what has been noted thus far, it is possible to suggest that the emphasis in decision-making will fall on the individual encounter, which may have a three-dimensional implication. Here the main vehicle is the interview, which may involve, to a greater or lesser extent, formality, informality, an element of bargaining and problem-solving. Although very few hotel and catering managers will be involved in formal collective bargaining with trade unions, virtually all managers will negotiate on varying bases as part of managerial work.

Gospel and Palmer define bargaining as 'an all-pervasive social process which involves the interaction between two or more individuals or groups which are attempting to define or redefine the terms of their relationship' (Gospel and Palmer,

1993, p. 180). Negotiation may take place at an individual level between a manager and a subordinate, between two managers, between two employees, between a manager and a customer, and between a worker and a customer. It also occurs at a collective level between a number of managers and employees, among groups of managers, and among groups of employees. While some distinguish between bargaining between managers – termed 'social exchange' by Fox (1974) – and collective bargaining, for the purposes of this book the main principles of interpersonal skills and negotiating behaviours and practice that underpin both situations will be outlined.

Interpersonal communication

Discussion on interpersonal communication is a major feature of managerial life, featured particularly in the work of Stewart (1970) and Mintzberg (1973) and in the studies of managerial work in hotels and catering discussed in Chapter 2. Brownell's (1992) study offers a useful insight into the communication practices of hospitality managers, not only up, down and across the organization, but also with customers. Managers communicate directly not only on a one-to-one basis but also as members of working groups and teams. Resulting from an emphasis on customer care programmes in the 1980s, more organizations have placed pressure on the way that employees behave towards customers; that is, how and what they communicate (Sparks and Callan, 1992; Wynne, 1993, 1994). Torrington (1991) and Torrington and Hall (1991, pp. 149–62) offer commentaries on interpersonal communication, which has been useful in informing some of the subsequent analysis (see also Marchington, 1992b, pp. 57–102).

Torrington and Hall (1991, p. 149) state that 'Interpersonal interactions provide conundrums for those who participate, because all participants are performing rather than being themselves'. Communication, verbally and in writing, requires interpretation. People speaking use certain behaviours, make gestures and may adopt a particular type of posture that can convey additional meaning to the actual words used (Torrington, 1991, pp. 17–39). Individuals may signify the opposite of what they intend; for example, the phrase 'with respect' usually means that someone does not respect the other's point of view at all, and wishes to propose an alternative point. Everyone judges others by 'appearance', but this is subjective and may lead to false conclusions. Much can also be read into the nature of the language used in memoranda or written correspondence.

As Torrington (1991, p. 6) notes, there are inhibitors to effective interaction primarily because few people believe they are bad communicators. Interactive behaviours are deeply ingrained and well practised, such that they cannot easily be altered. Additionally, many people simply hear what they expect to hear or stereotype people according to preconceived ideas. At least being aware of such dangers and trying to exercise self-discipline can reduce the degree to which misunderstandings occur. Mary may well seem to be a regular moaner, so some skill is needed not to miss the occasion when she really has a genuine grievance. Some managers and individuals may simply work from competing frames of reference and be completely unable to recognize the other's point.

Communication skills have been identified as an important facet of hospitality management competence (Hales and Nightingale, 1986; Worsfold, 1989). Brownell

(1992) argued that little was known about the specific communication activities that hospitality managers perform on the job, although they are essential components of leadership and managing. Even among managers themselves 'the very nature of social skills is generally insufficiently understood . . . where there is some understanding there is a genuine problem in articulating requirements' (Clark, 1993, p. 59).

Drawing on the approach used by Hales and Nightingale (1986) – with whom do managers work and what do managers do? (see also Chapter 2) – Brownell (1992) examined the communication activities of managers. Although she found a complex picture, general managers were more likely to communicate frequently with guests than middle managers, while middle managers were more likely to communicate frequently with hourly paid staff than general managers (p. 117). Both general managers and middle managers identified hourly paid employee communication with middle managers as most troublesome, while internal management communication was perceived as least problematic (p. 118). Perhaps not surprisingly, middle managers were found to operate in a largely verbal environment with subordinates and colleagues. It would seem that technological developments such as computer-generated messages and electronic mail are likely to have relatively little effect on this practice. The increasing use of mobile telephones might conceivably facilitate verbal communications within the workplace, although this would involve a lesser extent of personal interface.

Effective service is often synonymous with effective communication (Brownell, 1992, p. 111). 'Success of a service business depends quite considerably upon the effectiveness of staff managing their interpersonal relations with customers' (Sparks and Callan, 1992, p. 215). Acknowledging the difficulty of establishing rules that apply in every situation to reduce product failure, Sparks and Callan (1992) propose the use of a convergent style of communication with customers. By altering behaviour to become more similar to the customer, staff may be better able to develop a rapport, although there is the risk that a choice of incorrect behaviour could be evaluated negatively by the customer.

In the more formal management–worker individual encounter, such as the grievance, discipline, selection, counselling or appraisal interview, certain basic ground rules can be established, although the way in which these are used will vary according to the context within which they apply. Torrington (1991, p. 64) categorizes interactive situations into four types. These are *finding out* (selection or exit interviews), *putting it over* (selling a service or product), *joint problem-solving* (counselling, appraisal, grievance and discipline interviews) and *conflict resolution* (negotiating). There are power issues in the individual encounter. An individual will normally meet a manager on his or her own in a selection and appraisal interview. In a grievance or disciplinary hearing, an individual has the right to be accompanied by a fellow worker or, if appropriate, a trade union representative.

The individual encounter can be divided into three main phases, labelled preparation, encounter and follow-up, to use Torrington's (1991) terminology.

Preparation

Successful outcomes start from having clear aims and objectives and from gathering relevant information from appropriate sources, including talking to others and/or reading documentation. Outline questions or points at issue may need to be prepared. A suitable time and location for the encounter to take place need to be organized.

Encounter

Torrington (1991, p. 41) identifies three dimensions in face-to-face situations: speaking, listening and questioning. The first task for the manager is to *open* the encounter by establishing rapport with the interviewee using a friendly easy manner, by being attentive, using an easy and non-controversial opening topic, maintaining calm and initiating eye contact (Torrington and Hall, 1991, p. 154).

The interaction then has to be *sustained*. Managers can display interest by signifying agreement wherever possible, using positive eye behaviour (look at the interviewee, not out of the window) and encouraging noises (words will interrupt the flow) and gestures (nodding). In certain encounters, mainly grievance and discipline, it may be necessary to bring latent, suppressed feelings into the open, in order to clear the air for a more constructive outcome.

The choice and use of types of *questions* and the deployment of effective *questioning technique* are vital to eliciting facts that will lead to a resolution of the issue. Closed questions are designed to elicit a precise answer, including yes and no. Open questions allow the interviewee to open up on a subject. Follow-up questions allow the manager to develop points raised from open questions. Direct questions are designed to evoke a precise reply. Some common behaviours may produce unintended effects, including leading questions, unreasonable exhortation and multiple questions.

Feedback to the interviewee by a periodic summarizing of points to gain the interviewee's confirmation enables the encounter to move forward. Probing may be necessary in order to get information that the interviewee is unwilling to divulge. Braking may be needed to slow down the response or for the manager to regain control of the encounter. The ability to *close* the encounter by disengaging through verbal prompts and the gathering of papers will indicate that the interview is about to finish. Recapitulation should take place so that the outcome is clear to both parties.

Follow-up

Further action, which could involve the issue of a written warning, pursuing a grievance to a higher level of management, or making arrangement for further training, should be expedited quickly.

Negotiating and bargaining

Negotiating may take place between individuals and between groups. In hotels and catering, even though most managers think they act unilaterally, there is still an element of bargaining even in the least democratic forms of management decision-making (Gospel and Palmer, 1993), as noted earlier in the chapter. The WIRS3 (discussed in Chapter 8 – see Tables 8.3 and 8.4) also shows that joint committees have a negotiating role, although this is unlikely to involve trade unions.

Bargaining is about striking a deal that satisfies both parties, and for the deal to be effective and lasting both will have to give and take. It may be ritualistic and time-consuming, but as Torrington and Hall (1991, pp. 534-5) note, attempts by Lemuel Boulware to side step all this by adopting the one 'first and final' offer approach soon failed because both sides needed to articulate, ritualize and so on.

Negotiation centres on commonality and conflict. Manager and subordinate may strive to arrive at a fair deal about work allocation but may conflict over what that represents and how it can be achieved. Restaurant staff may need to negotiate a settlement with a customer dissatisfied with a meal or the level of service provided. The principle of negotiation presupposes that conflict is inevitable and, therefore, is integral to the pluralistic approach that conflicts of interest and of right are best managed through the process of bargaining. The pros of bargaining include allowing the air to be cleared and increasing understanding through articulation, arguing and debating. The cons of bargaining include that it is time-consuming and may involve emotional stress for the participants and the organization (see also HCITB, 1981).

'Negotiations only begin because each party has some power over the other' (Torrington, 1991, p. 194). Power is not equal but is perceived to be so, as Torrington and Hall (1991, p. 519) note thus: 'when both sides set out to reach an agreement that is satisfactory to themselves and acceptable to the other, then their power is equalized by that desire'. The most widely accepted effective bargaining strategy is what Torrington and Hall (1991) call 'confrontation'. This produces win-win outcomes for both parties, who confront the situation and find a solution that accommodates their differences.

The negotiating process

Negotiating can be seen as cyclical and can be broken into the preparation, encounter and follow-up modes (see also HCITB, 1981; Fowler, 1986, 1990; Torrington and Hall, 1991).

Preparation

If there is no agreed *agenda*, the encounter may degenerate into 'talks about talks' rather than dealing with the real issues. The agenda helps to structure the strategy. Both sides may need to exchange *information* in advance to clarify understanding. Managers might wish to give details of the poor financial state of the company to support the proposal not to increase wages on the usual pay anniversary.

Most important is the preparation of a *strategy*. What is the objective and what outcome do you want to secure from this negotiation? What is the resistance point (the point beyond which you cannot go), the ideal settlement point (where you would like to go) and the realistic target point (where you are prepared to go)? Much skill lies not only in preparing your strategy but also in anticipating the other side's strategy and predicting counter claims. In group negotiations, roles may need to be allocated, including those of chair and specialist adviser (from personnel).

Encounter

Both sides may commence proceedings with firm statements of intent but statements are likely to become more tentative as matters proceed in pursuit of finding an accommodation. Adjournments may be required at particular points to enable either party to consider an unforeseen issue or to cool things down. Skilled and successful negotiators adopt certain behaviours (HCITB, 1981, p. 10). These include behaviour flagging (introducing remarks by 'my proposal is . . .'), testing understanding and summarizing, 'building' behaviours (building on proposals already made) and open behaviour (admitting to mistakes). Bad behaviours, including linking too many issues in one statement, attacking people on the other side and too many counter proposals, should be avoided.

Follow-up

Any agreement should be produced in writing and ratified by both parties.

Torrington and Hall (1991, p. 532) argue that collective bargaining is different from individual bargaining because in the former the parties must reach an agreement or else the matter becomes a dispute. In individual bargaining either party can opt out at any stage, but this is not so in grievance handling, where there is undoubtedly an element of bargaining.

Summary propositions

- More effective management of individual and collective encounters is likely to be based on an understanding of the principles of interpersonal communications and negotiating.
- Experience and skills development will only come about through continuous practice.

Discussion points

In the context of a triadic relationship – managers, workers and customers – consider the following.

- Identify key interpersonal managerial communications skills.
- How can these skills be developed?
- Identify key interpersonal workforce communications skills?
- How can these be developed?
- To what extent is the individual encounter a more appropriate means of resolving problems and reaching decisions than the collective encounter?

REFERENCES

Advisory, Conciliation and Arbitration Service (1980) *Licensed Residential Establishment and Licensed Restaurant Wages Council Report No. 18*. London: HMSO.

Analoui, F. and Kakabadse, A. (1989) Defiance at work. *Employee Relations*, **11** (3), 1–62.

Brownell, J. (1992) Hospitality managers' communication practices. *International Journal of Hospitality Management*, **11** (2), 111-28.

Butler, S.R. and Snizek, W.E. (1976) The waitress-diner relationship. *Sociology of Work and Occupations*, **3** (2), 209–22.

Byrne, D. (ed.) (1986) *Waiting for Change*. London: Low Pay Unit.

Clark, M. (1993) Communication and social skills: perceptions of hospitality managers. *Employee Relations*, **15** (2), 51–60.

Commission on Industrial Relations (1971) *The Hotel and Catering Industry, Part 1. Hotels and Restaurants, Report No. 23*. London: HMSO.

Commission on Industrial Relations (1972) *The Hotel and Catering Industry, Part II. Industrial Catering, Report No. 27*. London: HMSO.

Commission on Industrial Relations (1973) *The Hotel and Catering Industry, Part III. Public Houses, Clubs and Other Sectors, Report No. 36*. London: HMSO.

Croney, P. (1988) An investigation into the management of labour in the hotel industry. MA thesis, University of Warwick.

Dronfield, L. and Soto, P. (1980) *Hardship Hotel*. London: Counter Information Services.

Fowler, A. (1986) *Effective Negotiation*. London: Institute of Personnel Management.

Fowler, A. (1990) *Negotiation Skills and Strategies*. London: Institute of Personnel Management.

Fox, A. (1974) *Work, Power and Trust Relations*. London: Faber and Faber.

Gospel, H.F. and Palmer, G. (1993) *British Industrial Relations*, 2nd edition. London: Routledge.

Goss, D. (1988) Social harmony and the small firm: a reappraisal. *Sociological Review*, **36**, 114–32.

Hales, C. and Nightingale, M. (1986) What are unit managers supposed to do? A contingent methodology for investigating managerial role requirements. *International Journal of Hospitality Management*, **5** (1), 3–11.

Hotel and Catering Industry Training Board (1981) *Getting a Good Deal: a Negotiating Skills Training Aid*. Wembley: HCITB.

Johnson, K. (1983) Trade unions and total rewards. *International Journal of Hospitality Management*, **2** (1), 31–5.

Kelliher, C. (1984) An investigation into non-strike conflict in the hotel and catering industry. MA thesis, University of Warwick.

Kessler, S. and Bayliss, F. (1992) *Contemporary British Industrial Relations*. London: Macmillan.

Kirkbride, P. (1992) Power. In J.F. Hartley and G.M. Stephenson (eds), *Employment Relations: the Psychology of Influence and Control at Work*, pp. 67–88. Oxford: Blackwell.

Lucas, R.E. (1989) Minimum wages – straitjacket or framework for the hospitality industry into the 1990s? *International Journal of Hospitality Management*, **8** (3), 197–214.

Lucas, R.E. (1991a) Promoting collective bargaining: wages councils and the hotel industry. *Employee Relations*, **13** (5), 3–11.

Lucas, R.E. (1991b) Remuneration practice in a wages council sector: some empirical observations in hotels. *Industrial Relations Journal*, **22** (4), 273–85.

Lucas, R.E. (1992) Minimum wages and the labour market – recent and contemporary issues in the British hotel industry. *Employee Relations*, **14** (1), 33–47.

Macaulay, I.R. and Wood, R.C. (1992) *Hard Cheese: a Study of Hotel and Catering Industry Employment in Scotland*. Glasgow: Scottish Low Pay Unit.

Marchington, M. (1982) *Managing Industrial Relations*. Maidenhead: McGraw-Hill.

Marchington, M. (1989) Joint consultation in practice. In K. Sisson (ed.), *Personnel Management in Britain*, pp. 378–402. Oxford: Blackwell.

Marchington, M. (1992a) Managing labour relations in a competitive environment. In A. Sturdy, D. Knights and H. Willmott (eds), *Skill and Consent: Contemporary Studies in the Labour Process*, pp. 149–84. London: Routledge.

Marchington, M. (1992b) *Managing the Team: a Guide to Successful Employee Involvement*. Oxford: Blackwell.

Marchington, M., Goodman, J., Wilkinson, A. and Ackers, P. (1992) *New Developments in Employee Involvement. Employment Department Research Series No. 2*. London: HMSO.

Mars, G. and Mitchell, P. (1976) *Room for Reform?* Milton Keynes: Open University Press.

Mars, G., Bryant, D. and Mitchell, P. (1979) *Manpower Problems in the Hotel and Catering Industry*. Farnborough: Gower.

Mintzberg, H. (1973) *The Nature of Managerial Work*. New York: Harper and Row.

Mintzberg, H. (1983) *Power in and around Organizations*. Englewood Cliffs, New Jersey: Prentice Hall.

Ralston, R. (1989) The changing nature of personnel management in the hotel and catering industry. MSc thesis, University of Manchester.

Sparks, B. and Callan, V.J. (1992) Communication and the service encounter. *International Journal of Hospitality Management*, **11** (3), 213–24.

Stewart, R. (1970) *Managers and Their Jobs*. London: Pan Piper.

Torrington, D. (1991) *Management Face-to-face*. London: Prentice Hall.

Torrington, D. and Hall, L. (1991) *Personnel Management: a New Approach*, 2nd edition. London: Prentice Hall.

Wood, R.C. (1992) *Working in Hotels and Catering*. London: Routledge.

Wood, R. C. (1994) *Organizational Behaviour for Hospitality Management*. Oxford: Butterworth Heinemann.

Worsfold, P. (1989) A personality profile of the hotel manager. *International Journal of Hospitality Management*, **8** (1), 51–62.

Wynne, J. (1993) Power relationships and empowerment in hotels. *Employee Relations*, **15** (2), 42–50.

Wynne, J. (1994) Luxury hotels, guest–employee interactions, accusations and justifications. Paper presented to the Third Annual CHME Research Conference, Napier University, April.

SEVEN

Procedures in the Employment Relationship

The purposes of this chapter are to:

- consider the main types of procedure and process that are used to make continual adjustments in, and to, the employment relationship;
- assess the relevance and effectiveness of these mechanisms in hotels and catering.

7.1 INTRODUCTION

Many rules, other than those connected with payment, surround the effort–reward bargain. These cover the whole range of human resourcing issues, such as what work is done, how jobs are constituted, training and promotion, hours of work, health and safety, and standards of behaviour and performance. But rules are not set in tablets of stone. The employment relationship is a dynamic nexus which is subject to constant adjustment by both parties in order to resolve matters of dispute or conflict. Thus if an employer is dissatisfied with a worker's effort or performance, the disciplinary process is likely to be invoked. Conversely a worker dissatisfied with his or her pay may take up the matter with management as a grievance, rather than by resorting to 'fiddles'.

Procedures provide the framework to guide the process of managerial behaviours and actions. Torrington and Hall (1991, p. 537) assert that the objective of the grievance/discipline process is to work towards getting as 'good a fit' in the employment relationship as possible, although this concept of 'good fit' may represent more of an idealistic long-term goal than a readily achievable practical reality. Rollinson (1992) rejects this 'complementary facet' notion, arguing that discipline is tantamount to a system of control whereas grievances are 'part of a system of rule making' (pp.

48–9). The employment relationship is inevitably more management driven than employee driven and, therefore, discipline is the major issue.

7.2 THE CASE FOR PROCEDURES

The roots of the movement towards an extension of formal employment procedural arrangements in the workplace can be traced to the Donovan Report (1968) (see Kessler and Bayliss, 1992, pp. 4–39, 176–90). Although Donovan envisaged that such procedures would be established through the method of collective bargaining, formal procedural arrangements are more likely to have been management determined than jointly determined in industries such as hotels and catering. Although in some cases formality may equal complexity, formal employment procedures need not be over-complicated. Some kind of articulated order is seen to be the route to organizational effectiveness, profitability, growth and job creation, even in small firms (ACAS, 1987b).

Torrington and Hall argue that procedures are an essential part of the framework of organizational justice surrounding 'the everyday employment relationship so that managers and supervisors, as well as other employees, know where they stand when dissatisfaction develops' (Torrington and Hall, 1991, p. 542). Procedures, which are essentially for the purpose of resolving 'disputes' about a range of issues, 'play a key role in bringing fairness and consistency in employment relationships' (Millward *et al.*, 1992, p. 185); here the emphasis is on conflict resolution. Conversely, procedures may also induce inflexibility, be time consuming and be restrictive, more so if the procedure is in writing.

The WIRS series (1980, 1984, 1990) has shown that formal disciplinary and dismissal procedures are now 'almost universal in all but the smallest workplaces' (Millward *et al.*, 1992, p. 212); these procedures exclude redundancy (dismissals and redundancy are discussed more fully in Chapter 9). The relative absence of formality in smaller firms is supported by evidence from another survey, which found that only one-third of firms employing fewer than 20 employees had a written disciplinary procedure (Evans *et al.*, 1985, p. 30). The Employment Act 1989 has since exempted employers with 20 or fewer employees from providing discipline procedures (such firms are not included in the WIRS series as yet). The WIRS series also shows that the extent of grievance and health and safety procedures has increased.

Croney's (1988) study of four hotel groups showed that all had disciplinary procedures and that they were similar to those operating more generally, although Price's (1993, 1994) study of 241 hotels and restaurants, mentioned below, found significantly less comprehensive disciplinary procedural arrangements in place. Ralston's (1989) study also found less comprehensive procedural coverage; the most common types of procedure in operation were for discipline and grievances, but this was more likely to be the case for operatives (76 per cent) than for managerial (60 per cent) and administrative (53 per cent) staff (p. 279).

Evidence from the WIRS3, shown in Table 7.1, does indicate a lower incidence of formal procedures in hotels and catering than elsewhere, most marked in relation to disputes about pay and conditions (see below). The incidence of formal procedures decreases with workplace size. In other areas, a greater extent of formal procedures is likely to result from legislative pressures. The requirement for grievance and discipline procedures is underpinned by legislation in relation to the written statement of terms and conditions of employment (see Chapter 9) and, therefore, their extent is likely to be greater than for procedures relating to pay and conditions, where agreements remain voluntary and not legally binding. The legal requirement for firms to have formal safety policies is also likely to have been an influence on the relatively great extent of formal health and safety procedures.

Table 7.1 *Extent of formal procedures.*

			PSS (by no. of employees)				HCI (by no. of employees)			
	AIS	SI	All	25–49	50–99	100+	All	25–49	50–99	100+
Unweighted base	2061	1299	702	174	153	375	61	21	18	22
Weighted base	2000	1462	886	518	218	151	99	59	26	14
Disputes on pay and conditions (%)	65	67	60	57	64	64	51	42	59	77
Discipline and dismissals (%)	91	92	88	84	92	96	84	79	89	98
Individual grievances (%)	87	90	86	81	92	96	82	81	80	91
Health and safety (%)	87	89	84	79	91	92	88	89	80	98
No dismissal or grievance procedure (%)	7	6	9	13	4	2	14	19	11	[a]

[a]Fewer than 0.5%.
Any slight discrepancies between figures are due to the rounding of decimal points.
Source: WIRS3 (1990).

Disputes about pay and conditions

A lower incidence of formality could be expected in relation to disputes over pay and conditions because, apart from the fact that they are not legally binding, these substantive rules are determined by less formal institutionalized means in hotels and catering than elsewhere (see Table 3.1), with decisions by management alone constituting the most significant method of decision-making (see Table 5.1). As Table 7.2 shows, disputes procedures relating to pay and conditions are less likely to be contained in a written document in hotels and catering than elsewhere. Surprisingly, larger hotel and catering workplaces are less likely to have such procedures in writing than smaller workplaces. As Table 7.2 also shows, very few such procedures are agreed between management and unions in hotels and catering compared to elsewhere, largely because there is no union representation in the vast majority of relevant hotel and catering workplaces.

A feature of many pay disputes procedures is the provision for third party intervention, often as a final stage. In terms of the WIRS3, third party intervention or reference to an outside body includes management outside the establishment but inside the organization. The vast majority of disputes are resolved within the organization, although provision for third party intervention is less common in hotels and catering than in the economy as a whole. The pattern in hotels and catering also

Table 7.2 *Status of written procedures.*[a]

	AIS	SI	PSS (by no. of employees) All	25–49	50–99	100+	HCI (by no. of employees) All	25–49	50–99	100+
In writing										
Pay and conditions (%)	91	94	91	92	90	90	81	86	84	65
Discipline and dismissals (%)	93	95	92	91	93	97	94	93	94	100
Agreed with unions										
Pay and conditions (%)	65	67	48	47	46	56	9	–	16	17
Discipline and dismissals (%)	55	58	37	36	36	42	3	1	[b]	13

[a]The base for pay and conditions is all establishments with provision for third-party intervention. The base for discipline and dismissals is all workplaces with the appropriate procedures. As numbers in each base differ, they are excluded to simplify presentation.
[b]Fewer than 0.5 per cent.
Table excludes missing cases.
Source: WIRS3 (1990).

shows a marked difference from elsewhere because, in the vast majority of cases (91 per cent), the provision for such references was 'in-house', to management outside the establishment but inside the organization. The only other significant body for hotels and catering was ACAS, mentioned in 20 per cent of cases. Although this 'in-house' provision was still the most common type elsewhere, reference to ACAS was much more marked among sectors other than hotels and catering, particularly among private sector services (45 per cent of cases). As a general rule, 'in-house' pay disputes references and references to ACAS were more likely to be found in smaller firms. References outside involving trade unions, often in a joint capacity with some level of management, were significant alternatives in all industries and services and service industries. Provision for arbitration (other than ACAS) was virtually absent from hotels and catering although it was a minority (typically 10 per cent) practice elsewhere (see Table 7.3).

Discipline and dismissals procedures

The WIRS3 established that a high proportion of discipline and dismissals procedures were set out in written documents across all comparator groups, at or around 95 per cent of cases (see Table 7.2) and this practice is no different in hotels and catering compared to elsewhere. Table 7.2 also shows that a smaller proportion of discipline and dismissals procedures than disputes procedures about pay and conditions are agreed with trade unions across all sectors of the economy, with hotels and catering having very few such arrangements.

The WIRS3 data show a more formalized scenario among workplaces with 25 employees or more than that found in Price's survey of 'relatively large and relatively high quality establishments' (Price, 1993, p. 23) in the hotel and restaurant sector. Price found a strong degree of correlation between establishment size and the degree

Table 7.3 *Outside body to which issues are referred (percentages).*[a]

	AIS	SI	PSS (by no. of employees) All	PSS 25–49	PSS 50–99	PSS 100+	HCI (by no. of employees) All	HCI 25–49	HCI 50–99	HCI 100+
Management outside establishment but inside organization										
Pay and conditions	48	57	66	71	62	53	91	94	99	78
Discipline and dismissals	40	48	38	38	41	37	60	51	76	62
Not referred										
Pay and conditions	–	–	–	–	–	–	–	–	–	–
Discipline and dismissals	36	31	46	50	39	43	33	40	24	23
Arbitration (not ACAS)										
Pay and conditions	10	10	11	16	4	6	1	–	–	3
Discipline and dismissals	3	3	2	2	1	1	4	7	–	2
ACAS										
Pay and conditions	31	29	45	48	41	43	20	42	1	14
Discipline and dismissals	9	8	11	10	11	13	3	1	[b]	13
Some union involvement										
Pay and conditions	43	43	17	12	19	30	5	–	–	13
Discipline and dismissals	19	19	7	4	9	15	1	1	–	8
Other										
Pay and conditions	10	6	3	3	4	2	1	–	–	3
Discipline and dismissals	7	5	4	4	3	4	4	7	–	2

[a]The base is all workplaces with the appropriate procedures. As numbers differ for each procedure, they are excluded to simplify presentation.
[b]Fewer than 0.5 per cent.
Table excludes missing cases.
Source: WIRS3 (1990).

of formality; managers in smaller firms 'saw little need for formality, given their caring approach to staff' (Price, 1993, p. 21). Because Price's sample contains a significant number of smaller establishments employing fewer than 25 employees (excluded from the WIRS3 sample), overall only 58 per cent of establishments had a written procedure compared to the WIRS3 results of 94 per cent of hotel and catering workplaces.

Table 7.3 shows a different pattern of outside bodies to which discipline and dismissals issues are referred in comparison with pay and conditions issues. More workplaces made provision for reference outside in relation to discipline and dismissals than for pay disputes. However, a sizeable percentage of workplaces made no provision for reference to any body. In the service sector (31 per cent) and in hotels and catering (33 per cent), a lower proportion of establishments made no provision for reference outside than was the case in the economy as a whole. The most significant method overall was 'in-house' to management outside the establishment but inside the organization. Unions were less likely to be involved in disciplinary issues than pay issues, except in hotels and catering, but in both cases the extent of involvement is extremely small.

Summary propositions

- Some kind of articulated order in relation to issues such as discipline, grievance and health and safety is more likely to have positive outcomes than negative consequences for the organization.
- The extent of formal procedural arrangements in hotels and catering is quite high in workplaces employing 25 employees or more but is likely to be much more limited in smaller workplaces.

Discussion points

- What are the pros and cons of formal procedures?
- In which areas of employee relations are formal procedures most needed and why?
- In the absence of formal procedures how can organizations ensure fairness and consistency in the employment relationship?

7.3 HANDLING GRIEVANCES

What is a grievance?

Complaints and grievances are a pervasive fact of workplace life and, given their potential for producing constructive or destructive outcomes, there is a surprisingly limited amount of literature on them (Salipante and Bouwen, 1990, p. 17). Most literature has concentrated on complaints and grievance procedures, rather than their causes and effects and the role of the organization. 'Grievances are usually complex but are typically reformulated and misrepresented to fit acceptably simplistic classifications' (Salipante and Bouwen, 1990, p. 17).

Manifestations of employee disquiet have been broken into three areas by Pigors and Myers (1977, p. 229), which broadly seek to differentiate the formal from the informal (see also Torrington and Hall, 1991, pp. 540–2). A *dissatisfaction* represents something that bothers an individual, although this is not always expressed out loud. An individual might feel upset because he or she has been wrongly reprimanded for someone else's mistake or because he or she has been given insufficient credit for dealing well with a difficult situation. Often a dissatisfaction may arise because the aggrieved has got the 'wrong end of the stick' or is not in possession of the full facts. On their own, such incidents are likely to have a short-term negative effect, but tend to be fairly quickly forgotten. If they recur fairly often, the consequences are likely to be more serious, including a more marked lowering of performance, negative attitudes that lead to pilfering and sabotage, and high staff turnover. Such disquiet is more problematic for managers because it is often more difficult to identify and, therefore, to resolve. This manifestation of disquiet may well be the most typical employee response in hotel and catering employment.

A dissatisfaction becomes a *complaint* when it is articulated and brought to the attention of someone, usually a more senior employee. Managers are better able to deal with the matter because it has been brought into the open and, as a result, it is more tangible. Quite often the problem is resolved by further explanation rather than by righting a wrong.

A *grievance* is essentially a more formal complaint that is presented to management and, as such, represents a direct challenge to managerial authority. Grievances are relatively rare because of this adversarial point, and because many employees genuinely believe that nothing will be done anyway. Salipante and Bouwen (1990) use a similar approach to that used by Pigors and Myers in that a grievance is 'a complaint about an organisational policy, procedure or managerial practice that is acted upon in any way by the subordinate' (p. 18).

Resolving the problem

Salipante and Bouwen (1990) seek to differentiate sources of conflict in three ways, although they acknowledge that grievances may come from more than one source and can be extremely complex. *Environmental conflict* arises from working conditions and the nature of work itself. *Social-substantive conflict* stems from perceived inequities in treatment or from disagreements over goals or means. *Social-relational conflict* derives from relationships between individuals, groups and organizations, personality conflicts and racism and sexism. Any analysis of grievance records would not capture the complexities involved because grievances are recorded in a narrow way.

Salipante and Bouwen (1990) further argue that grievances cannot be viewed as discrete events to be handled properly in order to resume normal organizational life: 'Grievances may be seen as continuous, without clear beginnings or endings' (p. 21), that are endemic in organizations. Grievances are inextricably linked to the ongoing relationship between organizational members; thus a grievance is 'not only an episode in a continuing relationship, but is episodic in itself' (Bouwen and Salipante, 1990, p. 27). Behavioural studies of grievances thus make a useful contribution to role conflict and conflict resolution in organizations.

Some problems are easier to articulate and resolve than others. The most straightforward problems are tangible (the meat slicer keeps breaking down), followed by those that arise from an employee's response to an unpleasant situation (it is too hot in the kitchens). Such problems are examples of Salipante and Bouwen's environmental conflict, which tends to be the least frequent source of conflict. Less complex grievances have a better chance of achieving a successful outcome.

Problems which centre on a managerial decision are examples of social-substantive conflict (Why has Joan been promoted to bar supervisor when I am better qualified and have been working here longer?). These are more difficult to handle because managers may take the matter too personally and employees are generally reluctant to challenge management on a one-to-one basis, at least openly in this way (see also Torrington and Hall, 1991, pp. 541–2). As noted earlier, many grievances in hotels and catering are expressed through covert means, such as pilfering. Salipante and Bouwen (1990) assert that there is almost always more to a grievance or complaint than the grievant ever states to his or her superiors and, as a result, 'many grievances result in unresolved conflict that can have a lasting effect ... as individuals seek to achieve some measure of satisfaction of their frustrations' (Bouwen and Salipante, 1990, p. 31).

Evidence from the WIRS3 identified 23 different types of grievance, and although no one area predominates, grievances about pay, disciplinary matters and relations with employees were the three main sources of complaint in hotels and catering. The pattern was more varied in other sectors, although pay and discipline were the two most frequently mentioned issues everywhere. Grievances were more likely to have been raised in hotels and catering than elsewhere, particularly in the smaller workplaces, shown in Table 7.4, and employers were more confident that the procedure could effectively handle most issues. Kelliher (1984, p. 35) noted that little attention had been paid to how workers in unorganized sectors express their grievances, although Analoui and Kakabadse (1989, 1993) in particular, Marchington (1992) and Wynne (1993) shed some light on this issue.

Procedural steps

Although Torrington and Hall (1991, pp. 547–9) limit the number of steps to three, ACAS recommends no more than two stages in small firms (ACAS, 1987b, p. 41). As a first stage the individual should raise the grievance with his or her supervisor informally. If unresolved, the matter should then be raised with the employer, ideally within five working days, and the individual should have the right to be accompanied by a colleague if desired. This second stage may allow the matter to be considered more dispassionately and broadly, particularly if relations between grievant and supervisor are not good. Additionally, the influx of fresh thought may serve to clarify issues that have become entrenched and confused. Torrington and Hall's third stage is a further level of appeal to a more senior manager. Time limits between stages will enable the issue to be expedited quickly and not left to drag on.

Bouwen and Salipante (1990) acknowledge that formal approaches do aid grievants in achieving a measure of success. Proper resolution which focuses on the sources of conflict that led to the grievance will typically involve third party intervention.

Table 7.4 *Grievances.*

			PSS (by no. of employees)				HCI (by no. of employees)			
	AIS	SI	All	25–49	50–99	100+	All	25–49	50–99	100+
Unweighted base	2061	1299	702	174	153	375	61	21	18	22
Weighted base	2000	1462	886	518	218	151	99	59	26	14
Most common issues										
Pay[a] (%)	8	6	7	6	9	10	11	15	5	5
Discipline (%)	6	6	9	8	9	12	9	6	15	11
Relations between employees (%)	1	1	2	2	1	1	5	6	5	5
No issues raised (%)	63	67	62	60	67	61	56	49	65	65
Procedure ineffective (%)	12	12	8	8	5	10	–	–	–	–

[a]Pay and conditions and pay and allowances.
Any slight discrepancies between figures are due to the rounding of decimal points.
Source: WIRS3 (1990).

The interview

In any interview context, the ability to 'read between the lines' in order to interpret the meaning of certain words, phrases, gestures and silences is likely to prove to be an invaluable managerial skill. This is perhaps more important in grievance handling than in some other situations because of the problems of articulation noted above. The pros and cons of the interpersonal encounter, put forward by Torrington and Hall (1991, pp. 149–62) and Torrington (1991), have been outlined in Chapter 6.

Salipante and Bouwen (1990) note that the articulation of grievances by employees transforms the grievance. They cite the example of one individual who, having planned for several months to make a complaint about inequitably low wages, made personal need for income the basis of the complaint. Following a negative response from the superior, the employee quickly changed argument, alleging that the wage inequity stemmed from discriminatory stereotyping. The implication that an appeal for redress could be made to an outside body led to a more satisfactory response.

Reformulation of grievances leads to their refinement, and 'Persistence in interacting with others is a critical factor for grievants eventually achieving some degree of success' (Bouwen and Salipante, 1990, p. 29). Successful outcome is also dependent on a multivariant approach to grievance presentation and a variety of sets of actions. Persistence by the grievant is often enough to convince the superior that the grievance is founded on problems that need to be addressed.

Although grievances may seem to be a reasonably straightforward matter in comparison to discipline, Rollinson (1992) disagrees by asserting that 'Grievance, however, which is concerned with resolving what the rules of behaviour should be, and political actions such as trade-offs, alliances and bargaining are likely to be involved ... its complexity and dynamics could turn out to be far more problematic' (p. 49). Unfortunately, he does not develop this point in more detail.

Summary propositions

- Grievances are an endemic, ongoing feature of organizational life that can rarely be classified simply or resolved in a neat and tidy fashion.
- Many individual employment problems may never be formally and openly articulated by employees, ostensibly creating a range of problems, such as high turnover and low morale, that are more difficult for managers to deal with.
- The extent of formal grievances is higher in hotels and catering than elsewhere.
- Grievance procedures are better kept simple, with the provision that matters can be expedited quickly.

Discussion points

- In what form do employee grievances manifest themselves in hotels and catering? How can managers 'manage' these situations more effectively?
- What problems do workers encounter in attempting to air grievances and find solutions to their problems? How can these be overcome?
- Why are there more grievances in hotels and catering than elsewhere?

7.4 MANAGING DISCIPLINE

The subject of discipline is, perhaps, conceptually more difficult for managers to deal with than many issues in the employment relationship because of the range of meanings attached to it (it has already been noted above that Rollinson (1992) holds the opposite view, that grievance is more problematic than discipline). The purpose of discipline is to establish standards that are acceptable to management. Discipline is as much about punishment as it is about developing team autonomy and responsibility to produce high standards of performance. Some examples of different approaches to discipline are now identified.

What is discipline?

Edwards (1989) sees discipline in terms of correction for breaking rules. He proposes three faces of discipline: punishment for breach of the rules, the formulation of the rules themselves and the creation in practice of the expectations, norms and understandings that govern behaviour (Edwards, 1989, p. 296). Others see discipline in terms of managerial control (see, for example, Edwards, 1979; Fox, 1985; Fenley, 1986). Edwards's (1979) 'bureaucratic' control is exercised through 'ensuring adherence to myriads of rules and regulations which govern the conduct, the appearance and rights of all those employed in the company' (Gabriel, 1988, p. 115). Alternatively, Torrington and Hall (1991, pp. 538–9) see discipline in terms of producing a controlled human performance, whether this is management led or employee led; that is, a performance based on the exercise of a high degree of self-discipline or based on teamwork (see also Osigweh and Hutchison, 1990). (Self-discipline and teamwork are two elements of an HRM approach that are considered in the section relating to employee involvement in Chapter 8.) ACAS takes a more pragmatic two-pronged approach to discipline: 'Disciplinary procedures should not be viewed primarily as a means of imposing sanctions. They should also be designed to emphasise and encourage improvements in individual conduct' (ACAS, 1977, p. 3).

Because the requirements for 'good' disciplinary practice and procedure are a major factor that underpin a 'fair' dismissal (see Chapter 9), there is, perhaps, a tendency to pigeonhole discipline in terms of employers avoiding the embarrassment of losing an unfair dismissal case at an industrial tribunal. Disciplinary issues are, however, much broader than those which focus on conduct and capability. Even so, the wisdom of ACAS in this regard, provided in the ACAS Code *Disciplinary Practice and Procedures in Employment* (ACAS, 1977) and the ACAS Advisory Handbook *Discipline at Work* (ACAS, 1987a), which also incorporates the ACAS Code, does not need to be viewed too narrowly. ACAS states that 'Proper procedures are an aid to good management' (ACAS, 1987a, p. 4).

The ACAS guidance is influential in informing the subsequent analysis. Although the guidance tends to verge on the prescriptive it is invaluable. The Handbook also covers other issues which may not always be disciplinary issues. These include dealing with absences, including long-term sickness, and sub-standard work. Failure to do the job correctly may result from inadequate or no training, and is not necessarily a disciplinary matter. Copies of the Handbook, for which there is a small charge, are available from ACAS offices.

Rules

Much of what has already been mentioned in this book has suggested that the process of rule making is complex. While in broad terms it is true to state that rules derive from statute (no smoking in the kitchens) and from organizational standards (persistent absenteeism will not be tolerated), their formulation and application are considerably more problematic. Some rules are pretty absolute, while others are considerably more permissive. Rules may be formal and informal. Mock rules are ignored, representative rules enjoy a general level of support by everyone and punishment-centred rules generate an adverse reaction from workers (Edwards, 1989, pp. 308–10, 315–16). Edwards also notes that there is no reason to presuppose that, because small firms work on the basis of informal methods, strict discipline does not exist.

Analoui and Kakabadse (1989) note how an in-built flexibility in an hotel and catering organization's formal rules and procedures resulted in acts of defiance that were intended to facilitate the achievement of specific objectives. To save time and facilitate service, staff poured drinks from bottles rather than using measures (gills) or helped themselves to alcoholic drinks in order to cope with the pressures of the job.

In terms of ACAS's wisdom, the most difficult task for managers is to establish rules that are clear and unambiguous, a point that is also addressed by Edwards (1989). Changes in standards do occur and, therefore, ACAS recommends giving warning where tightening occurs of a rule that is not strictly enforced (in cases where this has led to incidences of unilateral employee regulation on matters such as length of breaks and acceptable pilferage, the problem may be more difficult to manage). Rules should be available in written form, and in the languages of the staff employed.

ACAS also identifies a simpler form of rules for small companies (ACAS, 1987a, p. 58).

Offences

Different offences may be dealt with by different degrees of severity. Thus ACAS recommends that managers indicate the type of conduct leading to dismissal as a first offence. If rules are ambiguous as to the seriousness of a particular matter, but an employer dismisses for the first breach (without warning), a tribunal may well consider that unfair, as in *Trusthouse Forte (Catering)* v *Adonis* [1984] IRLR 382 EAT. Unequivocally unacceptable conduct, usually termed 'gross misconduct', will justify summary dismissal and could include theft, fraud, safety offences and physical violence. Conduct (or a level of incapability) which is still unacceptable, but on a lesser scale, will normally justify a less severe approach that may be managed through a series of warnings that could culminate in dismissal. Such less serious conduct includes poor timekeeping and persistent absence, and these types of misconduct have been found to be the most common disciplinary issues in small firms (Evans *et al.*, 1985). As noted above, incapability may not always be a disciplinary issue.

Criminal offences are not automatic reasons for disciplinary action. Any employer must have reasonable grounds for believing an employee has stolen rather than the absolute proof which is necessary to establish guilt in criminal proceedings, as decided in *Trust Houses Forte Leisure Ltd* v *Aquilar* [1976] IRLR 251 EAT.

Penalties

The range of sanctions available to managers is quite diverse in practice but, in terms of 'defendable practice', is likely to centre on the 'rule of three warnings': verbal warning, written warning, final written warning and, ultimately, dismissal. Evidence from the WIRS3 shown in Table 7.5 confirms a more marked use of formal disciplinary sanctions in hotels and catering than elsewhere, particularly in the larger workplaces. Dismissals per thousand employees are more than three times greater in hotels and catering than in service industries and over twice as high as in private sector services.

It is probably the case that one dimension of managerial effectiveness derives from the close and carefully considered observation of staff behaviour. Knowledge about individual employees and their behaviour is a central issue in staff motivation. Furthermore, problems that are 'nipped in the bud' are, perhaps, more easy to resolve than those which have become chronic and institutionalized. The point here is that managers, as an ongoing part of managerial work, have to indicate that 'something is not quite right or could be done better', but this need not be an automatic trigger of the disciplinary process or even be tantamount to it. Perhaps the 'quiet but firm word in the ear' is an approach that could be utilized more often.

What is paramount in terms of the disciplinary process is that managers have to distinguish between formal and informal instances of 'words in the ear', particularly in terms of the consequences of continued misdemeanour. Failure to do so could jeopardize the successful defence of a claim for unfair dismissal. Here it is important to differentiate between an informal caution and a formal verbal warning, the latter of which will be recorded on the employee's file because it constitutes a first formal step in procedural terms.

Table 7.5 *Types of disciplinary sanctions used in last year.*

			PSS (by no. of employees)				HCI (by no. of employees)			
	AIS	SI	All	25–49	50–99	100+	All	25–49	50–99	100+
Unweighted base	2061	1299	702	174	153	375	61	21	18	22
Weighted base	2000	1462	886	518	218	151	99	59	26	14
Formal written warning (%)	58	52	62	55	63	84	72	61	87	90
Suspension (%)										
Full pay	15	16	15	9	15	32	11	–	21	41
Reduced pay	1	1	[a]	–	1	2	–	–	–	–
No pay	8	5	5	4	6	8	6	–	20	5
Deduction from pay (%)	6	7	7	9	5	5	7	11	–	–
Dismissals per 1000 employees[b]	15	13	20	28	28	12	45	42	75	25

[a]Fewer than 0.5 per cent.
[b]Where number of dismissals known.
Any slight discrepancies between figures are due to the rounding of decimal points.
Source: WIRS3 (1990).

Once formal procedure has been invoked (for example, a final written warning has been issued) ACAS (1977) recommends that the consequences of further misdemeanour should be spelt out. For example, continued failure to arrive for work at the due time will result in dismissal.

Other penalties include suspension (with, without or on reduced pay), fines and disciplinary transfer (to another place of work; this could also embrace demotion). There are some practical and legal difficulties associated with these alternatives in terms of the Wages Act 1986 (this specifies lawful deductions from pay, which must generally be part of the contract) and constructive dismissal (see Chapter 9). Employers can deduct up to '10% of the gross amount of wages payable on any day (including any deduction due to alleged dishonesty' (ACAS, 1987a, p. 50) in order to make good cash shortages or stock deficiencies. However, as Table 7.5 shows, suspension is quite often used, but as this is mainly with full pay, presumably pending investigation of an alleged offence, the incidence of cases of suspension with legal implications is relatively small.

A final point in relation to penalties relates to the 'principle of redemption'. Improved performance or behaviour should, according to ACAS, be rewarded by the deletion of any formal warnings from an employee's file after an appropriate period of 'good behaviour'. If discipline is to be positive and to reflect improved conduct and performance, then an individual should not be stigmatized by a 'blot' from the past, and such an occurrence should not be usable in evidence against the individual at an appropriate future juncture.

Procedural steps

ACAS recommends that the procedure, which should be in writing, will incorporate a number of steps that specify the authority and roles of the managers, and should allow for matters to be dealt with quickly. The power to dismiss should lie in the hands of a more senior manager than an employee's immediate supervisor. Individuals should be 'informed of the complaints against them and ... be given the opportunity to state their case before decisions are reached' (ACAS, 1977, p. 3). A full investigation must take place before any action is taken, which includes the disciplinary interview. As Torrington and Hall (1991, p. 547) note, 'Many disciplinary problems disappear under analysis, and it is sensible to carry out the analysis before making a possibly unjustified allegation of discipline.'

The nature of the penalty must be explained; a final warning is *definitely* the last chance. There should also be the facility for individuals to appeal; the grievance procedure could be used and is, as noted above from the WIRS3 (see Table 7.4). Although ACAS recommends that a separate procedure be adopted, discipline is the second most frequent issue that is pursued through the grievance procedure.

The interview

'Counselling may often be a more satisfactory method of resolving problems than a disciplinary interview' (ACAS, 1987a, p. 18); this is an extended version of the informal warning. Although interviews rarely proceed in neat, orderly stages, there are useful ground rules to be followed (see ACAS, 1987a, pp. 20–5; Torrington, 1991, pp. 163–76). ACAS recommends that employees should be given time to prepare for a disciplinary hearing and have the right to be accompanied by a representative. Management should state the complaint and attempt to discover the facts through sensitive open-ended questioning and listening to the statements of others, if appropriate. If there is no evidence to support the allegation, proceedings should be terminated. If further facts need to be established, proceedings should be adjourned.

Fair and effective in hotels and catering?

There would appear to be some paradoxes in relation to the state of discipline in hotels and catering. While the WIRS3 shows a great extent of formal written procedures, at least in the larger establishments (also found by Croney, 1988), only 24 per cent of establishments in Price's survey contained all the required elements specified by the ACAS Code of Practice (Price, 1993, pp. 19–20). In spite of this 'veneer of formality', much managerial behaviour, as described in Chapter 2, is inconsistent with the principles of good practice embodied in the ACAS Code. Many managers operate on a 'low trust' basis (Fox, 1985), which is characterized by methods of overt coercion and control. Although Croney (1988) found that hotel procedures were broadly similar to those found in most organizations in the economy, the rules were exceptionally strict. In one company, offences constituting gross misconduct included conduct prejudicial to the company and misconduct outside working hours (unrelated to the company), and managers used summary dismissals with a high degree of frequency (Croney, 1988, p. 32). Thus managerial behaviour was seen to conform to Fox's 'low trust' approach.

The WIRS3 confirms that the use of formal discipline is a major issue in hotels and catering and it may be becoming increasingly important. Ralston (1989) found that a quarter of personnel specialists and one-third of line managers spent increased amounts of time on discipline and dismissals in the preceding three years. 'There was a high level of collaboration with line management for all stages of the disciplinary process' (Ralston, 1989, p. 232). In cases of disciplinary appeals, the personnel function most often gave advice and attended meetings.

Summary propositions

- Discipline is a highly complex issue that is primarily management driven.
- Discipline is as much about motivating staff and improving performance as it is about punishing misdemeanours.
- Although there are formal procedural arrangements in hotels and catering they may not conform to the requirements of good practice.
- There is a more marked use of the disciplinary procedure in hotels and catering than elsewhere.
- The ACAS guidance probably offers the most comprehensive practical information to guide managerial actions.

Discussion points

- As a manager, how would you attempt to achieve fair and effective discipline in the workplace that conforms to good practice?
- Design a disciplinary procedure for a workplace of your choice.
- What constitutes fair and effective discipline from a worker's perspective?
- Why is use of the disciplinary procedure more marked in hotels and catering? What is the effect of this on managers, workers and the organization?

7.5 MANAGING PERFORMANCE

Performance management is a topic that has received considerable attention in the recent past and is 'intimately linked with the growth of HRM in general and with the individualization of the employment relationship in particular' (Storey and Sisson, 1993, p. 131). Performance appraisal, as Randell (1989, p. 149) notes, is a formal activity, primarily concerned to improve individual job holder performance. However, as Torrington and Hall (1991, pp. 480–98) imply, schemes that are overburdened by too many procedures and too much form filling are likely to be dismissed as a waste of time by line managers. 'Appraisal done badly is a waste of time for everyone involved; appraisal done well can improve motivation and performance for both appraiser and appraisee, and can greatly assist both individual development and organizational planning' (Torrington and Hall, 1991, p. 480).

As a general rule, performance management is discussed in the context of HRM, employee resourcing and employee development, which are largely outside the scope of this book. Because performance management is also inextricably linked to reward management (see, for example, Smith, 1993), which is generally regarded as being part of employee relations, the topic cannot be ignored completely. As noted in Chapter 5, evidence from the WIRS3 shows that individual performance is a significant factor influencing pay levels in hotels and catering. The final section of this chapter briefly outlines the main issues in performance management and presents the WIRS3 findings on performance appraisal.

Storey and Sisson (1993, p. 132) differentiate performance management in two main ways, one as a 'catch-all conceptualization' (just about everything in HRM, including planning, customer care, quality and reward) and the other as being synonymous with individual performance-related pay. Given the nature of the WIRS3 findings discussed below, and the fact that customer care programmes, quality and reward are issues that feature, to a greater or lesser extent, in this text, it would seem that a relatively permissive approach to performance management is more appropriate in terms of hotels and catering. Furthermore, there would seem to be considerable scope for a performance management approach in the industry.

Performance management, among other things, is about setting clear objectives, involves formal monitoring and review systems and is concerned with building a shared vision, all of which imply some kind of strategic approach. Given the observations made in Chapter 2 about the general lack of strategy in hotels and catering, there is a case for some scepticism about the extent to which evidence of performance management in hotels and catering really marks an issue of substance, or is simply a 'flavour of the month' approach to employee relations management. As Storey and Sisson (1993) note, in spite of the current interest in performance management and performance-related pay, 'There is a dearth of empirical evidence to show that either system produces the kind of results which managers want from them' (p. 153).

There is no known hard evidence in the literature to show whether any hotel and catering companies are utilizing measures such as planning, quality and reward measures as part of a concerted performance management strategy. The available evidence from hotels and catering centres on performance appraisal, and this is by no means conclusive, because appraisal 'means different things to different people' (Randell, 1989, p. 156). More formal methods could be expected to be found in the larger organizations, yet Croney (1988) found that no formal performance appraisal methods were used in four of the larger hotel groups that he studied.

Ralston's (1989) study sheds some light on the nature and scope of performance appraisal in hotel and catering employment. Performance appraisal was perceived as the least important activity (of 14) in terms of contributing to organizational effectiveness, perhaps because fewer than 50 per cent of establishments carried out regular performance appraisals. Yet the personnel function had considerable discretion in this area, although line managers were becoming more involved, and were most likely to conduct the appraisal. The activity was not particularly time consuming, because of the generally informal nature of the systems operating. Appraisal was used mainly to improve current performance, and to assess training and development needs. Perhaps surprisingly, it was rarely used as basis for salary review (for more details, see Ralston, 1989, pp. 231–42, 270–3).

Performance appraisal is not listed as a separate area in terms of the WIRS3 respondents' work activities (see Tables 2.1 and 2.2), so no judgement can be made about its relative importance in relation to managerial work. However, as Table 7.6 shows, performance appraisal is a more widely used system in hotels and catering than elsewhere. Its relative prominence in private sector services provides further affirmation of a link between performance appraisal and areas which are more likely to be typified by individualization in the employment relationship. The likelihood of there being a system of performance appraisal increases with workplace size.

As noted above, an appraisal can be used for different things, including 'to improve current performance, provide feedback, increase motivation, identify training needs,

Table 7.6 *Extent and uses of performance appraisal.*

			PSS (by no. of employees)				HCI (by no. of employees)			
	AIS	SI	All	25–49	50–99	100+	All	25–49	50–99	100+
Unweighted base	1644	932	482	83	109	290	50	12	17	21
Weighted base	1355	890	529	263	155	111	70	33	24	13
Yes, there is a system (%)	28	34	42	32	46	62	46	26	68	54
Used to assess (%)										
Transfers up and down	78	77	86	76	94	89	85	61	91	99
Pay increase/ decrease	53	50	58	62	56	56	71	61	74	78
Training needs	93	95	97	96	100	95	99	100	100	95
Feedback to individuals	13	14	14	10	19	14	19	–	24	30
Personal or career development	12	13	14	14	12	14	1	–	–	6
Objective goal or target setting	4	4	4	5	–	7	4	–	–	16

Table excludes missing cases and least important responses for 'used to assess'.
Source: WIRS3 (1990).

identify potential, let individuals know what is expected of them, focus on career development, award salary increases ... solve job problems ... set out objectives ... as a reward or punishment in itself (Torrington and Hall, 1991, p. 481). In the WIRS3 the predominant use of the appraisal was to assess training needs. Assessing suitability for transfers up and down was the second most important use of the appraisal, followed by assessing pay increases or decreases within the same grade. These three uses hold the same level of priority for all groups in the economy, but in hotels and catering appraisals are used more often for these purposes, particularly in relation to pay increases. The use of appraisal to assess personal or career development needs is a negligible practice in hotels and catering, although it is found, to some extent, in other sectors. Objective, goal or target setting is, perhaps surprisingly, insignificant. All employees are subject to individual appraisal in a smaller proportion of workplaces in hotels and catering (64 per cent) than in all industries and services (69 per cent), service industries (74 per cent) and private sector services (77 per cent). Conversely, 'just some' employees are more likely to be subject to appraisals in hotels and catering compared to elsewhere.

Summary proposition

- Performance appraisal is a more common practice in hotels and catering than elsewhere, particularly in relation to determining pay.

Discussion points

- Can performance be managed? If so, in what ways? If not, what are the barriers to performance management?
- What are the benefits and drawbacks of performance appraisal in relation to different occupational groups in hotels and catering?

REFERENCES

Advisory, Conciliation and Arbitration Service (1977) *Code of Practice 1. Disciplinary Practice and Procedures in Employment*. London: HMSO.

Advisory, Conciliation and Arbitration Service (1987a) *Discipline at Work: the ACAS Advisory Handbook*. London: HMSO.

Advisory, Conciliation and Arbitration Service (1987b) *Employing People: the ACAS Handbook for Small Firms*. London: HMSO.

Analoui, F. and Kakabadse, A. (1989) Defiance at work. *Employee Relations* **11** (3), 1–62.

Analoui, F. and Kakabadse, A. (1993) Industrial conflict and its expressions. *Employee Relations*, **15** (1), 46–62.

Bouwen, R. and Salipante, P. F. (1990) Behavioural analysis of grievances: episodes, actions and outcomes. *Employee Relations*, **12** (4), 27–32.

Croney, P. (1988) An investigation into the management of labour in the hotel industry. MA thesis, University of Warwick.

Edwards, P. K. (1989) The three faces of discipline. In K. Sisson (ed.), *Personnel Management in Britain*, pp. 296–325. Oxford: Blackwell.

Edwards, R. (1979) *Contested Terrain: the Transformation of the Workplace in the Twentieth Century*. London: Heinemann.

Evans, S., Goodman, J.F.B. and Hargreaves, L. (1985) *Unfair Dismissal Law and Employment Practice in the 1980s. Department of Employment Research Paper 53*. London: HMSO.

Fenley, A. (1986) Industrial discipline: a suitable case for treatment. *Employee Relations*, **8** (3), 1–30.

Fox, A. (1985) *Man Mismanagement*, 2nd edition. London: Hutchinson.

Gabriel, Y. (1988) *Working Lives in Catering*. London: Routledge & Kegan Paul.

Kelliher, C. (1984) An investigation into non-strike conflict in the hotel and catering industry. MA thesis, University of Warwick.

Kessler, S. and Bayliss, F. (1992) *Contemporary British Industrial Relations*. London: Macmillan.

Marchington, M. (1992) Managing labour relations in a competitive environment. In A. Sturdy, D. Knights and H. Willmott (eds), *Skill and Consent: Contemporary Studies in the Labour Process*, pp. 149–84. London: Routledge.

Millward, N., Stevens, M., Smart, D. and Hawes, W. R. (1992) *Workplace Industrial Relations in Transition*. Aldershot: Dartmouth Publishing Company.

Osigweh, C.A.B. and Hutchison, W.R. (1990) To punish or not to punish? Managing human resources through 'positive discipline'. *Employee Relations*, **12** (3), 27–32.

Pigors, P. and Myers, C.S. (1977) *Personnel Administration*, 8th edition. Maidenhead: McGraw-Hill.

Price, L. (1993) The limitations of the law in influencing employment practices in UK hotels and restaurants. *Employee Relations*, **15** (2), 16–24.

Price, L. (1994) Poor personnel practice in the hotel and catering industry: does it matter? *Human Resource Management Journal*, **4** (4), 44–62.

Ralston, R. (1989) The changing nature of personnel management in the hotel and catering industry. MSc thesis, University of Manchester.

Randell, G. (1989) Employee appraisal. In K. Sisson (ed.), *Personnel Management in Britain*, pp. 149–76. Oxford: Blackwell.

Rollinson, D. (1992) Individual issues in industrial relations: an examination of discipline, and an agenda for research. *Personnel Review*, **21** (1), 46–57.

Royal Commission on Trade Unions and Employers' Associations 1965–1968, Chairman Lord Donovan (1968) *Report*. London: HMSO.

Salipante, P. F. and Bouwen, R. (1990) Behavioural analysis of grievances: conflict sources, complexity and transformation. *Employee Relations*, **12** (3), 17–22.

Smith, I.G. (1993) Reward management: a retrospective assessment. *Employee Relations*, **15** (2), 45–59.

Storey, J. and Sisson, K. (1993) *Managing Human Resources and Industrial Relations*. Buckingham: Open University Press.

Torrington, D. (1991) *Management Face-to-face*. London: Prentice Hall.

Torrington, D. and Hall, L. (1991) *Personnel Management: A New Approach*, 2nd edition. London: Prentice Hall.

Wynne, J. (1993) Power relationships and empowerment in hotels. *Employee Relations*, **15** (2), 42–50.

EIGHT

Managing Involvement and Commitment

The purposes of this chapter are to:

- outline the variety of methods that are available for managing employee involvement and maximizing commitment;
- consider the role of communications;
- assess the relevance and effectiveness of these mechanisms in hotels and catering.

8.1 INTRODUCTION

Few managers would state publicly that they believed there was little or no point in trying to achieve some degree of employee involvement, although how some of them actually behave towards employees in practice might cast doubt on the sincerity of such beliefs (see Chapter 2). If businesses function more effectively when employees are more involved in what they are doing, and more committed to what the organization stands for, then managers need to be aware of the proliferation of ways in which this can be achieved. The extent to which managers choose to abdicate, or not to abdicate, managerial prerogative is a matter of managerial style and will be the ultimate determinant of the methods of involvement that are chosen. These can be mirrored in Gospel and Palmer's (1993) spectrum of decision-making, discussed in Chapter 6. Thus managers who believe they are there to 'manage' will adopt minimal, more token forms of employee involvement, whereas more 'open' managers are likely to opt for more power sharing and democratic means.

The importance of good communications is a theme underlying all these mechanisms and procedures. Managers and employees need understand the company's philosophy – what the processes are about, why the procedures are there and what they are supposed to achieve. Managers also need to be able to make them work, which is all about communicating effectively on an interpersonal basis (see also Chapter 6). Communicating on whatever level is, to use Gospel and Palmer's (1993) term, 'all-pervasive' since it occurs vertically, up and down, and horizontally, from one end to the other, in and across the organization through people and by other means. All of these issues are now discussed in more detail.

8.2 EMPLOYEE INVOLVEMENT AND PARTICIPATION

This part of the chapter will examine issues such as employee involvement, securing commitment, participation, empowerment, joint consultation and industrial democracy. These 'participative' issues have traditionally been seen as separate from collective bargaining, although in practice, and in much contemporary analysis, this traditional line of demarcation has become increasingly blurred. For instance, Marchington (1992a) sees collective bargaining and joint consultation as 'representative participation' (p. 161). Collective bargaining, according to Farnham and Pimlott (1990), is 'distributive' and 'competitive' and is essentially a process to resolve a 'situation of basic conflict over the ways in which something might be divided or decided' (Farnham and Pimlott, 1990, p. 427), such as pay or profits, often referred to as conflicts of interest. Collective bargaining comes about from worker pressure through the institution of the trade union.

Most, but not all, of the other 'participative' issues are more 'integrative' and 'cooperative', being more closely focused on resolving mutual problems through joint means in pursuit of common goals.

It is possible to place these participative issues on a spectrum which spans the barely democratic (unitarist) to the highly democratic (pluralist), although such classification inevitably oversimplifies the complex nature of the form that each issue may take, and relationship between them (Marchington *et al.*, 1992, pp. 6–11, offer a useful explanation of the differences in terminology).

As a general rule, employee involvement, participation and securing commitment tend to be largely undemocratic or, to use Farnham and Pimlott's (1990, pp. 82–5) terms 'direct' and 'integrative', because they take place directly between managers and individuals and supervisors and teams. Such initiatives tend to be designed and introduced by management, to centre on improving communications, to generate commitment and to enhance the contributions of employees (Marchington *et al.*, 1992, p. ix); they are also often referred to as employee empowerment (Lashley, 1994). For the purposes of this chapter, employee involvement, participation, commitment and empowerment are discussed together under the label *direct/integrative* because they all conform most closely to the direct integrative model.

While some forms of participation and joint consultation can be placed within the less democratic direct/integrative model, most others are more democratic – 'indirect' and 'distributive' in Farnham and Pimlott's (1990, p. 85) terms – because they are conducted through representatives who have been selected by one means or another. Industrial democracy is always democratic. These are discussed under the label *indirect/distributive*.

Given that much of the analysis that has informed earlier chapters in this book has pointed to a style of managing in hotels and catering that is more unitarist than pluralist in tendency, it will be interesting to observe whether this is reflected in the way participative issues are managed. Unfortunately, this is hampered by the lack of empirical work in this area. Hales's (1987) study provides a useful insight into Quality of Working Life measures in the hospitality industry in the mid-1980s, including job redesign and participation practice. The WIRS3 enables a comparison to be made between practices in hotels and catering and elsewhere. Lashley (1994) proposes an analytical framework for fieldwork into empowerment initiatives in hospitality retail operations (see also Lashley, 1995). Merrick (1994) reports on an employee communications system recently put into practice at a motorway service station.

Summary propositions

- Employee involvement and participation cover a broad spectrum of different types of arrangements that are best understood if they are classified in some way.
- Two broad classifications – the direct/integrative approach (less democratic) and the indirect/distributive approach (more democratic) – aid discussion.

Discussion point

- Is there a difference between involvement and participation?

8.3 THE DIRECT/INTEGRATIVE APPROACH

Terminological issues

Much of the current popularity of direct/integrative measures centring on employee involvement, participation and commitment would seem to stem from their 'central position in human resource management policies' (Coopey and Hartley, 1991, p. 18), particularly those proposed by Guest (1987, 1989) and more recently by Iles *et al.* (1990). They are also closely connected to the Quality of Working Life approach to work organization (Hales, 1987) and employee empowerment, which Lashley (1994), from a review of the literature, suggests 'is *the* employment strategy for the next century' (p. 3).

Although Farnham and Pimlott (1990) seek to differentiate employee involvement from employee commitment (some writers use the terms interchangeably), they are inextricably linked (see also Lashley, 1994, 1995; Lashley and McGoldrick, 1994).

> Employee involvement, by aiming to get the support and commitment of all employees in an organization to managerial objectives and goals, thereby reinforcing a sense of common purpose between management and employees, is essentially unitary in its purpose and its methods. It is defensive of the right to manage and the organizational *status quo*. (Farnham and Pimlott, 1990, p. 82)

Employee involvement is about processes set up by management to incorporate employees in decisions, particularly those affecting the way work is done. Employee commitment is about enhancing quality and performance; strong commitment will be manifested in a conscientious, self-directed application to work, regular attendance, minimal disciplinary supervision and a high level of effort, which reflects Torrington and Hall's (1991, pp. 538–9) approach to discipline – 'producing a controlled human performance' – noted in Chapter 7. This link is also expressed by others (Marchington *et al.*, 1992) and, for example, employee involvement is seen as a 'soft' form of participation designed to foster employee commitment (Ramsay, 1991, p. 1). Employee commitment is also integrally related to employee motivation (Farnham and Pimlott, 1990, pp. 77–87).

Lashley (1994) asserts that empowerment operates on a different plane and, therefore, is considerably more problematic:

> If empowerment can be differentiated from other employment initiatives, it engages employees at an emotional level. It is individual and personal. It is about discretion, autonomy, power and control. Whatever the intentions of managers, the effectiveness of empowerment will be determined by the perceptions, experiences, and feelings of the 'empowered'. Fundamentally, these feelings will be rooted in a sense of personal worth and ability to effect outcomes; of having the 'power' to make the difference. (Lashley, 1994, p. 2)

Marchington (1992a, pp. 157–8) implies that empowerment has particular benefits in a service environment where customer service staff place high value of satisfying customers; they are 'getting on' with their work. Others are more circumspect that customer service staff are, or can ever be, completely customer-oriented (for example, Analoui and Kakabadse, 1989; Wynne, 1993).

The substance of involvement

Farnham and Pimlott (1990), Ramsay (1991) and Marchington *et al.* (1992) point to employee involvement being characterized by certain employment practices. The main features of the WIRS3 classification are shown in Table 8.5. Financial involvement can be attempted through profit sharing and share ownership schemes. Ramsay (1991) notes that it is generally accepted that such schemes are too weakly linked to employee performance and are too remote from an employee's control to have a meaningful effect on behaviour (see also Morris *et al.*, 1993). Job involvement, on an individual task or team/workgroup basis, can be sought through suggestion schemes, the use of attitude surveys, job enlargement, job enrichment, job rotation, quality circles and autonomous work groups (for a useful review, see Ramsay, 1991, pp. 5–8; Marchington *et al.*, 1992, pp. 5–21).

Growing interest in involvement

The WIRS series has shown a growth in employee involvement, which Ramsay (1991) attributes to the need of management to gain the acquiesence or support of employees to sustained innovation in highly competitive times, which is also implicit in the HRM approach. Marchington *et al.*'s (1992) research found that '80% of existing schemes had been introduced in the 1980s, reflecting a new wave of interest in the subject' (p. ix). The main reasons for introducing schemes were related to information and education, commitment, securing enhanced employee contributions, recruitment and retention of labour, conflict handling and stability (pp. ix–x).

The findings from the most recent WIRS3 survey, shown in Tables 8.1 and 8.2, are, to an extent, ambiguous. Table 8.1 shows that a sizeable minority of workplaces across all sectors had made changes in the recent past in order to increase employee involvement. Even so, Table 8.2 shows that the nature of these changes was many and varied and that some of the practices associated with the 'new industrial relations' are highly marginal. The very limited evidence of autonomous workgroups, any 'job' change (job rotation, job enlargement) and quality circles casts considerable doubt on the reality of the 'new industrial relations', a point further emphasized by Millward (1994). Changes that have been excluded from Table 8.2 because they were very rarely mentioned by managers in the WIRS interviews include share ownership schemes, management training relating to involvement or participation, quality of work life or job satisfaction initiatives and improvements to fringe benefits. What does seem clear is that managements everywhere have not adopted a universalistic approach to increasing employee involvement, conceivably because there is no single panacea to an issue that is essentially situation specific (this point is implied by Hales, 1987; Coopey and Hartley, 1991). Alternatively, employee involvement may be more talked about than practised.

Table 8.1 *Management change with aim of increasing employee involvement in past three years.*

			PSS (by no. of employees)				HCI (by no. of employees)			
	AIS	SI	All	25–49	50–99	100+	All	25–49	50–99	100+
Unweighted base	2061	1299	702	174	153	375	61	21	18	22
Weighted base	2000	1462	886	518	218	151	99	59	26	14
Yes (%)	45	49	45	40	49	56	41	42	32	52

Table does not include percentage figures for missing cases.
Source: WIRS3 (1990).

Although Morris *et al.* (1993) acknowledged the perceived link between committed employees and competitive advantage, they noted that 'commitment is a well-researched concept, but the human resource literature has tended to assert uncritically that it is a desired goal without demonstrating empirically that it has a positive consequence for organisational performance or that commitment can be managed through human resource policies' (p. 21). Their longitudinal analysis of graduates (not hotel and catering) did suggest, however, that 'human resource policies appeared to influence levels of organizational commitment' (p. 21).

In hotels and catering, much of the basis for securing competitive advantage comes from the need to improve service quality (see, for example, Kane, 1986), often in conjunction with total quality management or customer care programmes. This approach is rendered problematic because of the intangible elements in the consumer purchase (Lashley, 1994). In this regard some hospitality operations, particularly those at the 'luxury' end of the market in the hotel sector (Wynne, 1993) and in retail operations such as McDonald's, are seeking to adopt strategies for managing human resources that ensure greater employee involvement and empowerment (Lashley, 1994, 1995).

Table 8.2 *Changes made to increase employee involvement in past three years.*

			PSS (by no. of employees)				HCI (by no. of employees)			
	AIS	SI	All	25–49	50–99	100+	All	25–49	50–99	100+
Unweighted base	2061	1299	702	174	153	375	61	21	18	22
Weighted base	2000	1462	886	518	218	151	99	59	26	14
More two-way communication (%)	13	15	12	10	13	15	8	6	12	8
Meetings – general consultative (%)	9	10	9	8	12	11	6	6	5	7
Delegation (%)	6	7	7	9	6	4	3	6	–	–
More information to employees (%)	5	6	6	7	2	6	7	11	–	1
Training/briefing groups (%)	5	5	5	2	7	11	4	–	9	13
New consultative committee (%)	4	3	3	3	4	4	2	–	5	5
New incentive scheme (%)	3	4	5	7	2	3	3	6	–	–
New suggestion scheme (%)	2	2	2	2	2	2	[a]	–	–	3
Autonomous workgroups/any job 'change' (%)	2	2	[a]	–	1	–	–	–	–	–
Quality circles (%)	2	1	2	1	2	7	1	–	[a]	8
Reorganized management/ restructuring management responsibilities (%)	2	2	2	1	4	2	–	–	–	–

[a]Fewer than 0.5 per cent.
Table does not include percentage figures for missing cases and changes below 2 per cent.
Any slight discrepancies between figures are due to the rounding of decimal points.
Source: WIRS3 (1990).

Lashley (1994) argues that empowerment can come about through employee involvement, commitment, participation and delayering. He defines involvement as 'those initiatives where the managerial intention is to gain from employees' experiences, ideas and suggestions' (p. 7) and cites the examples of quality circles at Accor, suggestion schemes at McDonald's and the use of team briefing at Hilton Hotels. When employees are empowered through greater commitment to the organization's goals, through, in the main, financial measures such as share ownership, 'employees take more responsibility for their own performance and its improvement' (Lashley, 1994, p. 7). Employee empowerment through participation centres on 'Initiatives which aim to involve employees in some aspect of decision-making and organisational activities which might more traditionally be the domain of management' (Lashley, 1994, p. 6). Measures include workers deciding work rotas and schedules at Harvester Restaurants and the use of enhanced training programmes such as 'Whatever It Takes Training' at Scott's Hotels. Delayering, much publicized by Peters and Waterman (1983) as the hallmark of excellence, is about reducing the number of management tiers in the organization to bring managers closer to the customer (see also Lashley, 1995).

Ephemera or substance?

As already noted, some commentators remain sceptical about management's intentions in regard to employee involvement and its value in organizational terms. Farnham and Pimlott (1990) speak for many by noting that although terms including job enrichment and participative management have increasingly become part of the management vocabulary, 'their long-term beneficial effects in reducing industrial conflict, improving employee morale, and raising organizational productivity have yet to be proved' (p. 87).

In terms of hotels and catering, Lashley (1994) and Wynne (1993) remain sceptical about the value and efficacy of empowerment and customer care programmes. Hales's (1987) study casts doubt on the use of employee involvement measures as 'universally-understood practices instigated coherently and deliberately on an organization-wide basis for the single purpose of improving the quality of working life of lower-level operatives' (p. 270). He found that job redesign measures were 'as much inspired by productivity gains of work rationalisation and labour-saving as by considerations of humanising work ... often the result of an *ad hoc* evolutionary development at unit level rather than purposive, instigated change throughout an organisation' (p. 271).

Much of the analysis that informs this book, and other texts (for example, Gabriel, 1988; Wood, 1992), points to a substantial level of powerlessness or disempowerment, with negative outcomes of high labour turnover, stress, low productivity and illicit practices. If, as Coopey and Hartley (1991) note, orientation to work is an important antecedent to commitment, labour turnover needs to be managed more effectively and hotels and catering needs to review its approach to recruitment and training of the kind of high quality labour that is likely to be associated with committed employees. Reliance on the external labour market does not square readily with the notion of employee commitment and other employee involvement initiatives.

Paradoxically, fiddling or pilfering may actually make employees feel empowered, although not in the way that the protagonists of employee empowerment mean. Empowerment is often presented far too simplistically as being a 'fix-all in the workplace' (Scott and Jaffe, 1991, p. 9). As Lashley (1994) notes, employees will only feel empowered if they are empowered to make significant decisions. Not all employees have the same levels of need for power and introducing and sustaining empowerment has major training implications for managers. 'The feeling of personal power or of pleasure in delighting the customer is fundamental to the effectiveness of empowerment as an employment strategy. For it is in this dimension that empowerment has the potential for transforming service delivery encounters' (Lashley, 1994, p. 13). But getting there successfully is, perhaps, like searching for the Holy Grail. (Ramsay's

(1991) guidance for managers on employee involvement/participation is included at the end of the section on the indirect/distributive approach.)

Summary propositions

- Employee involvement measures of a more unitary, undemocratic kind have become more fashionable across all employment sectors, including hotels and catering.
- There is still scant evidence to link the use of such measures to positive organizational outcomes including increased organizational effectiveness, improved attitudes and higher morale.
- The use of particular 'participative' measures would seem to be situation specific rather than universalistic.
- Such measures imply a high quality approach to labour based on sophisticated recruitment, training and retention measures that are associated with strong internal labour markets. This may be incompatible with the approach of many hotel and catering companies, which seek to perpetuate weak internal labour markets through high reliance on the external labour market.

Discussion points

- Evaluate the means available to enable managers to secure increased employee involvement through non-democratic means.
- Who gains the most from empowerment: managers, workers or customers? What are these gains and are they accompanied by equivalent losses for the other parties?

8.4 THE INDIRECT/DISTRIBUTIVE APPROACH

Joint consultation

Although the indirect/distributive model may be said to hold more relevance for unionized workplaces than non-unionized workplaces, it is, nevertheless, a significant feature in hotels and catering. Having said that, there is, perhaps, an element of doubt about the extent to which schemes that fit within this model are really democratic, put neatly by Hales (1987, p. 271) thus: 'there seems to be greater readiness on the part of management to admit that Consultative Committees are intentionally manipulative, diversionary and designed to increase, not relinquish, managerial control'.

Marchington's (1988) models of joint consultation, outlined below, show how this means of involving employees can span both the direct/integrative and indirect/distributive models.

1. Non-union model. Here management takes a decision and informs employees through representatives. There is no employee influence and the mechanism is purely a vehicle for communications, up and down. This is similar to Farnham and Pimlott's *pseudo-consultation*, where the 'express intention of such an

approach is to prevent the emergence of employee-based power centres in non-union firms' (Farnham and Pimlott, 1990, p. 431).

2 Competitive model. Joint consultative committees are upgraded to render collective bargaining less meaningful. The body becomes more of a mechanism to establish and justify management's actions. Such committees are often accompanied by more direct forms of participation, such as quality circles and briefing groups. This model is similar to Farnham and Pimlott's (1990) *classical consultation*, whereby employees have some influence on matters of common interest, with management taking responsibility for taking and implementing the final decisions.

3 Adjunct model. The joint consultative committees have 'teeth' and discuss 'hard', high-level information, often as a precursor to negotiations. There is mutual influence between managers and the workforce, represented by shop stewards. This is similar to Farnham and Pimlott's (1990) *integrative consultation*, which is an advanced form of employee participation.

4 Marginal model. The committees are symbolic and hold no real substance. They deal with trivial matters (tea and toilets) and represent little more than weak fire-fighting mechanisms.

Croney's study (1988), which mirrors the situation found by Hales (1987), infers that the types of consultative arrangements found in four hotel groups conform most closely to the non-union (Marchington, 1988) and pseudo-participation (Farnham and Pimlott, 1990) models because they were there to make sure that staff knew management's opinion and authority within a unitary style of management. Croney found evidence of both direct/integrative ('click teams' akin to quality circles) and indirect/distributive (committees of management and employee representatives) participation.

However, Croney (1988) identified problems with these consultative arrangements. The espoused policy of all organizations emphasized the importance of consultation, involvement and communication as part of management style and philosophy, although one organization did not have any structure or procedures for consultation. Yet unit management displayed little enthusiasm for managing participation of any kind. Managers held committees involving employee representatives because they 'had to', but not as often as head office would have liked. 'Click teams' were a 'pain to carry out'. This danger of managers not conforming to espoused policy because they choose to work differently through operational policies was identified in Chapter 2 and, in relation to involvement (Ramsay, 1991), is also reviewed at the end of this section.

Reading between the lines of remarks made by Forte (1982) shows an interesting tension in the consultative process. First the onus is on management to participate in day-to-day affairs (rather than staff participating in management) and management remains firmly in charge:

> The nature of our work calls for active participation by management in day to day operations and they need to have a very close relationship and understanding with their staff. Equally, staff need to know that line management has the authority and responsibility for taking decisions which vitally affect them in their work. (Forte, 1982, p. 32)

Forte also acknowledges that staff do have a contribution to make. It is interesting to see how the other half of the committees' time is spent:

> ... the minutes of the staff consultative committees show that on the insistence and initiative of the staff members more than half the time is spent in discussing ways in which business can be improved, customers can be given better service and even where expenses can be saved. (Forte, 1982, p. 33)

Marchington *et al.* (1992, pp. 63–4) report brief details of the consultative arrangements in one hotel which is part of a larger organization.

Evidence from the WIRS3

The extent and constitutional arrangements of joint committees of managers and workers found in the WIRS3 are shown in Table 8.3. Joint committees are a significant minority practice across all sectors. The extent of such committees in hotels and catering does not differ much from the pattern found overall, except in private sector services where the incidence of joint committees is less marked. Larger organizations in hotels and catering do have a greater tendency to have joint committees than most other comparator groups and, given the extent of trade union organization across these groups (see Table 3.1), it may well be the case that such committees conform to the non-union and pseudo-participation models described above. As a general rule, such committees meet less often in the smaller establishments than in the larger establishments.

What is clear is that there is no obvious link between the extent of trade union recognition and density/collective bargaining and the extent of joint committees. Sectors of the economy with relatively high levels of trade union activity – all industries and services and service industries (see Table 3.1) – show a similar pattern of joint committee arrangements to the virtually non-unionized hotel and catering industry. In other words, there is no overwhelming evidence to show that increased incidence of joint committees arises in establishments which neither recognize nor deal with trade unions. However, the moderately unionized private sector services have the least extent of joint committees. Evidence from the WIRS3 shows that a small proportion of joint committees do not actually function. As Ramsay (1991, p. 12) has pointed out, 'At best, schemes may need regular review and revamping if vitality is to be maintained'.

The use of joint committees for negotiating purposes in hotels and catering, inferred in Table 8.4, is brought more sharply into focus in Table 8.5. Table 8.4 shows that such committees' broader role is most marked in smaller hotel and catering establishments, and that these committees are more broad based in terms of worker representation than those operating in other sectors.

The extent of committees' influence in positive terms is slightly less marked in hotels and catering than elsewhere, as shown in Table 8.5. Overall, joint committees tend to be fairly influential, but are more likely to be very influential than not influential. Such influence may, or may not, be related to the relative autonomy of establishment level committees. Table 8.6 shows that joint committees in hotel and

Table 8.3 *Joint committees of managers and workers.*

	AIS	SI	PSS (by no. of employees) All	PSS 25–49	PSS 50–99	PSS 100+	HCI (by no. of employees) All	HCI 25–49	HCI 50–99	HCI 100+
Unweighted base	2061	1299	702	174	153	375	61	21	18	22
Weighted base	2000	1462	886	518	218	151	99	59	26	14
JCC present (%)	29	31	20	14	24	34	28	24	35	38
Functioning committee[a] (%)	26	28	17	12	21	30	24	18	35	30
No functioning committee (%)	74	72	83	88	79	70	76	82	65	70
Functions and negotiates (%)	8	8	3	2	3	6	10	11	5	13
All or some representatives chosen by trade unions (%)	10	12	3	1	3	8	2	–	5	6
No representatives chosen by trade unions (%)	15	16	15	11	18	22	22	18	29	24

[a]'Functioning' committee is one that meets every three months or more often and discussed something important in the past 12 months according to management.
Any slight discrepancies between figures are due to the rounding of decimal points.
Source: WIRS3 (1990).

Table 8.4 *Characteristics of first committee.*

	AIS	SI	PSS (by no. of employees) All	PSS 25–49	PSS 50–99	PSS 100+	HCI (by no. of employees) All	HCI 25–49	HCI 50–99	HCI 100+
Unweighted base	927	592	212	26	40	146	23	5	7	11
Weighted base	571	457	175	73	52	51	28	14	9	5
Negotiation and consultation (%)	31	30	18	19	16	19	48	73	15	38
Worker representatives[a] (%)										
Manual only	13	11	3	5	3	2	12	24	–	–
Non-manual	44	53	41	37	53	36	2	–	–	12
Both	42	36	54	58	40	62	86	76	100	88

[a]Excludes missing cases.
Any slight discrepancies between figures are due to the rounding of decimal points.
Source: WIRS3 (1990).

catering workplaces are more autonomous to the extent that fewer workplaces operate in the context of joint committees operating at a higher level than the establishment.

Table 8.5 *Extent of committees' influence.*

			PSS (by no. of employees)				HCI (by no. of employees)			
	AIS	SI	All	25–49	50–99	100+	All	25–49	50–99	100+
Unweighted base	934	594	212	26	40	146	23	5	7	11
Weighted base	572	464	175	73	52	51	28	14	9	5
Very influential (%)	32	31	28	31	35	16	26	24	31	22
Fairly influential (%)	47	46	53	57	44	58	50	52	46	52
Not very influential (%)	16	17	10	7	4	21	23	24	24	16
Not at all influential (%)	3	3	1	–	3	1	1	–	–	4

Table does not include percentage figures for missing cases.
Any slight discrepancies between figures are due to the rounding of decimal points.
Source: WIRS3 (1990).

Table 8.6 *Joint committee at higher level than establishment.*

			PSS (by no. of employees)				HCI (by no. of employees)			
	AIS	SI	All	25–49	50–99	100+	All	25–49	50–99	100+
Unweighted base	1666	1067	522	117	123	282	50	15	15	20
Weighted base	1511	1190	632	349	171	112	76	40	22	13
Yes (%)	43	50	40	44	34	37	22	30	6	26

Table does not include percentage figures for missing cases.
Any slight discrepancies between figures are due to the rounding of decimal points.
Source: WIRS3 (1990).

Industrial democracy

The concept of industrial democracy, in essence where workers are given considerable formal determining powers in the employment relationship, is less well established in Britain than in some other European countries. This looks set to change in light of the Directive on European Works Councils, due to be implemented by 1996. The Directive will require information arrangements and consultative machinery at European level for all companies or groups with more than 1,000 employees employing at least 150 workers in more than two European Union countries (excluding the UK). Although it will not apply as such to the UK as a result of the Social Chapter opt-out (see Chapter 9), there is considerable uncertainty about the full implications of the Directive for UK-based multinational companies and for the UK employees of multinational companies (see Gold and Hall, 1994; Industrial Relations Services, 1994).

Employees of the big multinationals will have the right to be consulted on strategic corporate decisions, such as mergers, takeovers, technological developments and shifts of production from one country to another. An estimated 300 British-based multinationals are likely to be affected. Uniform compulsory consultation structures have been abandoned in favour of voluntary arrangements between employers and employees. Managements have the right to withhold information if disclosure could be 'seriously prejudicial' to the undertakings concerned.

Aikin *et al.* (1994) provide a useful review of existing consultation requirements under UK law and argue that participation by employees is already inescapable in spite of the Maastricht opt-out, a view affirmed by Welch (1994). In the major cases of *Commission of the European Communities* v *United Kingdom of Great Britain and Northern Ireland*, C-382/92 [1994] IRLR 392 ECJ and *Commission of the European Communities* v *United Kingdom of Great Britain and Northern Ireland*, C-383/92 [1994] IRLR 412 ECJ, the European Commission's complaint that the UK had failed to comply with the Directives on redundancies and acquired rights by restricting information and consultation to representatives of recognized trade unions was upheld. The UK's position had allowed employers to avoid consultation by not recognizing a trade union, when in fact the Directives required information and consultation with representatives of the workforce.

Lessons for management practice

As a final point, Ramsay offers suggestions for all types of involvement, both the direct/integrative and the indirect/distributive models discussed above. 'Employee involvement with substance is expensive, in time and money' (Ramsay, 1991, p. 16). His checklist of lessons for management practice (pp. 16–7) is summarized as follows.

Management issues

Management at all levels must be deeply committed and strong support must come from the top. Do not involve managers who are not in a position to make real decisions. The system must have support throughout; lack of middle and lower level support can undermine a system (see also Croney, 1988).

Making it work

The system must have clearly considered objectives. Training and support must be provided for all those involved. There must be adequate and impartial monitoring through measurement criteria that are identifiable from specified objectives. Information provided and the consultation conducted must be taken seriously. It is easy to expect too much.

Summary propositions

- Joint committees in hotels and catering, even though they exist, may lack real influence because of the way in which they are constituted and because middle and junior managers are not committed to the principle.

- Developments arising from the social dimension of European integration (the Social Chapter) are likely to require UK multinationals to consult more formally on employment issues in the future.

Discussion points

- How can more democratic means of workforce participation be achieved in hotels and catering?
- What are the benefits and drawbacks of joint consultation for managers and workers?

8.5 COMMUNICATING INFORMATION

Communications structure

All commentators reinforce the importance of effective communications structures and information systems to underpin direct/integrative and indirect/distributive employee involvement schemes, (Bland and Jackson, 1990; Farnham and Pimlott, 1990; Ramsay, 1991; Marchington *et al.*, 1992; Marchington, 1992b), although the case is more broadly based. 'For generations of managers, good communication has been accepted as a touchstone of good employee relations' (Ramsay, 1991, p. 9). 'The presence of legitimized and effective channels of communication can ensure that conflict is voiced, expressed and dealt with rather than being suppressed, and the frustrated, aggrieved individual left in isolation can therefore choose from among the less desirable choices of action which are open to him/her' (Analoui and Kakabadse, 1993, p. 57). However, Townley (1989, p. 329) notes that 'British management has not placed a strong emphasis on employee communication programmes, despite many prescriptive statements that communication is of central importance for the effective running of an organization'.

Effective communication is multidimensional and pervasive. Vertical communication occurs downwards from managers to staff, and upwards from staff to managers. Horizontal communication takes place between managers and managers and between staff and staff. Upward communication may be of lesser importance to the managers of some employee involvement schemes, particularly those which seek to reinforce managerial authority. Communication also takes place informally through the 'grapevine' and not necessarily along the formal planes of organization structure. In some cases this may undermine managerial authority and lead to other negative reactions and consequences, if for example, employees find out about a major change to their circumstances from someone else. Communication is also a process of education (see Townley, 1989).

Information can be communicated in writing, orally, audio-visually and electronically (Bland and Jackson, 1990, pp. 94–118; Marchington, 1992b, pp. 34–56) and organizations may not always strike an appropriate balance among these, particularly between written and oral forms. Written communications have been found to be relatively poor but are still used most often (for a useful overview see Bland and Jackson, 1990, pp. 40–78). The main examples include noticeboards, newspapers, letters to employees and suggestion schemes. The Employment Act 1982 requires companies employing more than 250 employees to include in their annual reports statements of actions taken to promote employee involvement arrangements. Oral communication is more time consuming but is likely to prove to be more effective (see Bland and Jackson, 1990, pp. 79–93, and the section on interpersonal communications in Chapter 6). Methods include team briefing, employee opinion surveys, and development and appraisal systems. Marchington (1992a, pp. 169–70) casts doubt on the claim that team briefing increases workers' commitment, or that it is taken seriously by managers.

Evidence from the WIRS3 about methods of communication between management and workforce is shown in Table 8.7. To some extent, these methods overlap with the methods of employee involvement, participation and empowerment discussed above. Arguably regular meetings signify nothing more than 'normal management', as opposed to more specialized managerial initiatives such as quality circles and suggestion schemes, although this difference is not intended to imply that 'normal management' is necessarily the easy option. In terms of regular meetings, what differentiates hotels and catering from elsewhere is the lesser extent of regular meetings between senior managers and all sections of the workforce. Importantly, communicating, by whatever method, is far from being the norm in any workplace.

As regards other methods of communication, hotels and catering is generally less 'communicative' than elsewhere. Indeed, Table 8.7 shows a broadly similar pattern of communications practice in most areas among all industries and services, service industries and private sector services, except in terms of a functioning consultative committee.

Information provision

Communications are about the exchange of information, both the giving and receiving of it. Many contend that the change process is aided by a free exchange of information. Change will be more successfully effected if those involved understand the rationale for moving forward and, therefore, information must be provided to justify the change and to gain the commitment of those to the new order of things. From the WIRS3, shown in Table 8.8, it is clearly evident that different employment issues are managed differently in terms of information provision to the extent that managers appear to be much more willing to impart information about some issues than others. Matters of finance and investment are considerably less 'open' than those of an employee relations nature. Even so, those employee relations issues which have the most overt direct 'cost' connotations, namely staffing and manpower plans, are at least as 'closed' as, if not more than, financial matters.

Although trade unions have the right to request information from employers for collective bargaining purposes (see ACAS, 1977), this has little relevance in the predominantly non-unionized hotel and catering industry.

Summary propositions

- Effective communications incorporating written, verbal and audio-visual methods are based on organizational structures and systems that are multidimensional and pervasive.

Table 8.7 *Methods of communication between management and workforce.*

			PSS (by no. of employees)				HCI (by no. of employees)			
	AIS	SI	All	25–49	50–99	100+	All	25–49	50–99	100+
Unweighted base	2061	1299	702	174	153	375	61	21	18	22
Weighted base	2000	1462	886	518	218	151	99	59	26	14
Regular meetings among work-groups or teams at least once a month (%)	35	39	35	34	37	37	34	37	25	37
Regular meetings between junior managers/ supervisors and all workers for whom responsible (%)	48	55	50	47	48	59	50	46	50	66
Regular meetings between senior managers and all sections of the workforce (%)	41	43	41	39	43	45	29	24	38	35
Systematic use of management chain for communication (%)	60	61	59	56	59	69	50	49	52	50
Suggestion schemes (%)	28	32	33	31	36	36	21	12	28	46
Regular newsletter (all employees) (%)	41	47	45	38	44	66	33	24	34	71
Functioning consultative committee (%)	26	28	17	12	21	30	24	18	35	30
Others[a]	48	53	48	47	46	53	42	43	36	47
Others[b]	25	28	25	24	23	31	18	11	25	29

[a]Includes quality circles.
[b]Excludes quality circles.
Any slight discrepancies between figures are due to the rounding of decimal points.
Source: WIRS3 (1990).

- Informal grapevine communications may undermine formal managerial communications strategies.
- The range of issues about which information is communicated is relatively limited.
- There is considerable scope for improving communications in hotels and catering.

Table 8.8 *Management provision of information before implementation of changes.*

			PSS (by no. of employees)				HCI (by no. of employees)			
	AIS	SI	All	25–49	50–99	100+	All	25–49	50–99	100+
Unweighted base	2061	1299	702	174	153	375	61	21	18	22
Weighted base	2000	1462	886	518	218	151	99	59	26	14
Terms and conditions of employment (%)										
A lot	62	62	64	66	61	64	66	67	63	71
Little	18	17	20	18	24	18	31	31	37	14
None	18	19	16	16	16	18	3	2	–	14
Safety and occupational health arrangements (%)										
A lot	66	66	63	62	63	66	72	71	68	80
Little	18	18	18	17	18	20	19	20	21	14
None	14	15	18	20	16	13	9	9	11	5
Staffing/manpower plans (%)										
A lot	33	36	26	28	22	25	31	34	40	3
Little	27	27	27	29	26	24	30	34	25	21
None	40	37	47	44	52	51	39	32	35	75
Working methods or work organization (%)										
A lot	68	70	63	62	61	66	69	64	79	74
Little	19	18	21	21	23	20	24	33	16	–
None	12	12	16	17	16	14	7	3	5	25
Internal investment plans (%)										
A lot	19	18	20	21	20	16	29	32	29	20
Little	22	20	24	24	21	26	18	21	11	17
None	58	60	56	55	59	57	52	47	60	62
Financial position of establishment (%)										
A lot	28	31	28	29	24	31	30	32	28	24
Little	31	31	29	25	36	34	26	24	31	22
None	40	37	42	46	40	33	43	44	41	44
Financial position of organization (%)										
A lot	23	26	29	30	23	33	13	16	11	6
Little	23	25	21	15	28	32	20	15	23	35
None	29	31	26	24	33	22	44	38	51	57
Single establishment	21	15	23	29	17	12	23	32	16	–

Table does not include percentage figures for missing cases.
Any slight discrepancies between figures are due to the rounding of decimal points.
Source: WIRS3 (1990).

Discussion points

- What information should managers communicate to workers and what information should remain confidential?
- How can managerial communications down and across the organization be improved?

- How can workforce communications down and across the organization be improved?
- How can customer communications be improved?

REFERENCES

Advisory, Conciliation and Arbitration Service (1977) *Code of Practice 2. Disclosure of Information to Trade Unions for Collective Bargaining Purposes*. London: HMSO.

Aikin, O., Mill, C., Burnage, A., Foulds, J., Lines, T., Pope, C., Thomason, G. and Wooldridge, E. (1994) No escape from consultation. *Personnel Management*, October, 54–7.

Analoui, F. and Kakabadse, A. (1989) Defiance at work. *Employee Relations*, **11** (3), 1–62.

Analoui, F. and Kakabadse, A. (1993) Industrial conflict and its expressions. *Employee Relations*, **15** (1), 46–62.

Bland, M. and Jackson, P. (1990) *Effective Employee Communications*. London: Kogan Page.

Coopey, J. and Hartley, J. (1991) Reconsidering the case for organizational commitment. *Human Resource Management Journal* **1** (3), 18–32.

Croney, P. (1988) An investigation into the management of labour in the hotel industry. MA thesis, University of Warwick.

Farnham, D. and Pimlott, J. (1990) *Understanding Industrial Relations*, 4th edition. London: Cassell.

Forte, R. (1982) How I see the personnel function. *Personnel Management*, August, 32–5.

Gabriel, Y. (1988) *Working Lives in Catering*. London: Routledge & Kegan Paul.

Gold, M. and Hall, M. (1994) Statutory European Works Councils: the final countdown? *Industrial Relations Journal*, **25** (3), 177–86.

Gospel, H.F. and Palmer, G. (1993) *British Industrial Relations*, 2nd edition. London: Routledge.

Guest, D. (1987) Human resource management and industrial relations. *Journal of Management Studies*, **24** (5), 503–21.

Guest, D. (1989) Personnel and HRM: can you tell the difference? *Personnel Management*, January, 48–51.

Hales, C. (1987) Quality of Working Life. Job redesign and participation in a service industry: a rose by any other name? *The Service Industries Journal*, **7** (3), 253–73.

Iles, P., Mabey, C. and Robertson, I. (1990) Human resource management practices and employee commitment possibilities. *British Journal of Management*, **1** (3), 147–57.

Industrial Relations Services (1994) The UK and 'European Works Councils'. *European Industrial Relations Review*, 246, July, 14–21.

Kane, J. (1986) Participative management as a key to hospitality excellence. *International Journal of Hospitality Management*, **5** (3), 149–51.

Lashley, C. (1994) Is there any power in empowerment? Paper presented to the Third Annual CHME Research Conference Napier University, April.

Lashley, C. (1995) Towards an understanding of employee empowerment in hospitality services. *International Journal of Contemporary Hospitality Management*, **7** (1), 27–32.

Lashley, C. and McGoldrick, J. (1994) Barriers to employee empowerment. Paper presented to the Third Annual Conference on Human Resource Management in the Hospitality Industry, 'Quality and Human Resources', London, February.

Marchington, M. (1988) The four faces of employee consultation. *Personnel Management*, May, 44–7.

Marchington, M. (1992a) Managing labour relations in a competitive environment. In A. Sturdy, D. Knights and H. Willmott (eds), *Skill and Consent: Contemporary Studies in the Labour Process*, pp. 149–84. London: Routledge.

Marchington, M. (1992b) *Managing the Team: a Guide to Successful Employee Involvement.* Oxford: Blackwell.

Marchington, M., Goodman, J., Wilkinson A. and Ackers, P. (1992) *New Developments in Employee Involvement. Employment Department Research Series No. 2.* London: HMSO.

Merrick, N. (1994) On the road to good relations. *Personnel Management Plus*, **5** (11), 24–5.

Millward, N. (1994) *The New Industrial Relations?* London: Policy Studies Institute.

Morris, T., Lydka, H. and Fenton O'Crevy, M. (1993) Can commitment be managed? A longitudinal analysis of employee commitment and human resource policies. *Human Resource Management Journal*, **3** (3), 21–42.

Peters, T.J. and Waterman, S. (1983) *In Search of Excellence*. New York: Harper and Row.

Ramsay, H. (1991) Reinventing the wheel? A review of the development and performance of employee involvement. *Human Resource Management Journal*, **1** (4), 1–22.

Scott, C.D. and Jaffe, D.T. (eds) (1991) *Empowerment: Building a Committed Workforce.* London: Kogan Page.

Torrington, D. and Hall, L. (1991) *Personnel Management: a New Approach*, 2nd edition. London: Prentice Hall.

Townley, B. (1989) Employee communication programmes. In K. Sisson (ed.), *Personnel Management in Britain*, pp. 329–55. Oxford: Blackwell.

Welch, R. (1994) European Works Councils and their implications: the potential impact on employer practices and trade unions. *Employee Relations*, **16** (4), 48–61.

Wood, R.C. (1992) *Working in Hotels and Catering*. London: Routledge.

Wynne, J. (1993) Power relationships and empowerment in hotels. *Employee Relations*, **15** (2), 42–50.

PART THREE
LEGAL AND SOCIAL ISSUES

NINE

Managing Workplace Employment Law

The purposes of this chapter are to:

- outline the role of individual and collective law in the employment relationship;
- outline how Britain's European Union (EU) membership has changed, and will continue to change, UK employment law;
- review the most significant aspects of workplace employment legislation in relation to hotels and catering;
- assess the effect of the law on employment practice.

9.1 INTRODUCTION

Under common law, employment rights and duties are based on the individual's contract of employment (see Rideout, 1989, pp. 1– 139; Pitt, 1992, pp. 43–73; Smith *et al.*, 1993, pp. 64–146, 249–92). Although the contract is, *prima facie*, freely entered into, in reality the employer is almost always more powerful than the employee. From this, public policy has been largely supportive, at least until relatively recently, of the need for individuals to associate freely into trade unions that can go about their business effectively and of the need to provide a platform of individual employment rights.

Until the 1960s laws which had a direct and indirect effect on regulating the employment of individual workers and the functioning of trade unions were a clearly identifiable, but relatively small, aspect of British industrial relations. Piecemeal health and safety legislation and minimum wage legislation were examples of early 'social protection' legislation that applied to selected groups of 'vulnerable' workers. Trade unions functioned on the basis of 'immunities' which protected them from being prosecuted under civil law for taking industrial action. This 'simple form of a shield rather than an arsenal of legal rights' (Mackie, 1992, p. 295) enabled them to bargain freely with employers.

However, since the mid-1960s, the law has become an increasingly pervasive instrument in British industrial/employee relations, and is commonly differentiated as individual and collective employment law by many industrial/employee relations commentators (see, for example, Farnham and Pimlott, 1990, pp. 242-312), although lawyers do not generally differentiate the law on this basis (see Rideout, 1989; Pitt, 1992, 1994; Smith *et al.*, 1993).

Although these legal distinctions are, in one sense, somewhat arbitrary, individual employment law is essentially concerned with conferring rights on individual employees, whereas collective employment law is largely about regulating the role and behaviours of trade unions and employers. An example of how difficult it is to be precise about splitting collective and individual law lies in the fact that an employer can single out and dismiss individual strikers during collective industrial action. Individual claims for equal pay (see Chapter 10) can trigger off a collective response if an employer decides to alter a pay structure as the result of an individual case. Given the limited presence of trade unions and the passive role of employers in employee relations matters in hotels and catering, the topic of collective law will be discussed only briefly later in the chapter, although the reader is referred elsewhere for more detailed analysis and discussion.

Although successive Conservative administrations since 1979 have pursued a stated employment policy of deregulation, some measures have placed increased regulation on employers. As a general rule collective laws have shifted the balance of power even more strongly in favour of employers, but they have also involved employers in some extra expense. The maintenance of 'check-off' facilities, that is the system where an employer deducts trade union subscriptions directly from wages, is now more costly to the employer because of more cumbersome administrative arrangements. Extra expense also arises with regard to some individual rights, as employers must now fund the whole cost of redundancy payments and statutory sick pay. Individual rights have been weakened and strengthened. On the one hand, the repeal of some social protection legislation, such as the abolition of wages councils and the extension of the service qualification for protection against unfair dismissal, have weakened the platform of individual employee rights. On the other hand, individual workers have increased rights in relation to trade union membership and Sunday working, and all part-time employees are covered by employment protection legislation regardless of hours worked (EPPTER, 1995).

Individual employment law is now arguably of much greater significance than collective law, at least to industries such as hotels and catering, particularly in light of the changes that are being made to British law as a result of the European Social Charter. Individual law forms too big an area to be the subject matter of a single chapter. Indeed, this third part of the book cannot provide a detailed analysis and evaluation of every legal provision and, therefore, the reader is frequently referred elsewhere. It should also be noted that Northern Ireland is subject to different employment legislation that is not reported in this book. The legal framework, within which managers must manage, is subject to continual adjustment as new precedents are established within existing laws and new laws reach the statute book. Every attempt has been made to report key legal issues and leading cases that are as up-to-date as possible (to February 1995).

Classifying the law

For the purposes of this book, individual law, overlapping to some extent with 'social protection' law, is divided into three broad areas which span, to varying extents, the formation, maintenance and termination of the employment contract. The first of these broad areas in this chapter covers the main individual employment rights in the workplace, such as the right to a written statement of terms and conditions of employment, rights relating to officially permitted 'time off' from work and rights in relation to dismissal. The second area, discussed in Chapter 10, covers key legal aspects that relate broadly to the issue of 'equality', including the right not to be discriminated against in employment on grounds of race or sex, the right to equal pay and the right to maternity leave and maternity pay. The third area, embracing legal aspects related to health, safety and welfare and social protection, is discussed in Chapter 11. Data from the WIRS3 on points related to the law are included in each chapter.

Since the European dimension is a highly significant influence on all the legislation discussed in this chapter, and the other two chapters in this part of the book, it is considered next in outline form.

9.2 THE EUROPEAN INFLUENCE

The social dimension

Although Articles of the Treaty of Rome 1957 contained the main provisions on social policy deemed necessary for the European Economic Community (EEC) – later the European Community (EC), now the European Union (EU) – to achieve its economic objectives, it was only later that attempts were made to enhance the social dimension. Treaties are not self-executing and when the UK joined the EEC, the effect of Community law was defined in the European Communities Act 1972. The whole area of the effect of Community law is extremely complex (see, for example, Hepple, 1991; Aikin, 1992; Steiner, 1994). Detailed discussion of this area falls outside the main purpose of this book, but the main instruments used to effect EC law are outlined below, and the principles which govern whether national law or EC law has primacy are summarized in Appendix 3.

The first serious attempt to move social policy forward came when a Social Action Programme was adopted in 1974 by the member states shortly after the UK joined the EEC. Attention was, however, concentrated in a few limited areas, including employment protection, the equal treatment of men and women and health and safety, so achievements were relatively small. British employment laws were introduced during the 1970s to reflect these developments, largely by 'sympathetic' Labour administrations.

During the early 1980s progress on new measures relating to matters such as part-time work and worker participation was frustrated by disagreement and deadlock among the member states. During this period of stalemate, Britain was forced to amend national laws which were found to have been in breach of particular articles of the Treaty of Rome, such as in relation to the transfer of undertakings and equal pay legislation (see Chapter 10). These were anathema to a Conservative administration that was vehemently opposed to over-regulation in the employment relationship (for more details on the earlier history of European social policy, see Gold, 1993b, pp. 10–27; Hepple, 1993, pp. 12–13).

Matters took on a different complexion after the Single European Act 1986 was signed by *all* the member states. The Act was designed 'to give the programme to complete the internal single market by the end of 1992 a legal base; it included clauses on social policy' (Lucas, 1993, p. 90). In order to move social policy even further forward, the Social Charter was adopted in 1989 'as a solemn declaration' by all member states, with the exception of the UK, which maintained that implementation of principles enshrined in the Charter would increase the costs and burdens on business and create unemployment.

The Social Charter does not have the force of law but the Action Programme, the vehicle for implementing the principles of the Charter, is turning some of the proposals into law. Thus the principles of the Charter may be given effect in any of the following ways. Regulations apply directly to all member states, and prevail over and do not have to be incorporated into national law. Directives are binding on the results to be achieved by each member state within a certain time period but leave the method of implementation up to national governments (see Appendix 3). Decisions are binding on the parties to which they apply, but Recommendations and Opinions have no legal force.

Interestingly, the Action Programme did not propose any initiatives related to the right of freedom of association and collective bargaining, where matters were deemed to be the preserve of both sides of industry or the member states (subsidiarity). The decision-making process and legal base (see Appendix 3) is complex, but as a result of the Single European Act most decisions on social policy can be taken on a majority vote, particularly those that are concerned with health and safety as a result of the introduction of Article 118A into the Treaty of Rome (for more details, see Chapter 11). For further discussion on these issues, see Wedderburn (1989), Lucas (1991c, 1993), Gold (1993a) and Hepple (1993).

The UK's opt-out from the social provisions of the Political Union Treaty (the Social Chapter) at Maastricht in 1991 left the other 11 (at that time) member states free to take a 'fast track' approach to certain social proposals, with no participation by the UK (see for example, Towers, 1992; Gold, 1993b, pp. 27–38; Hepple, 1993). Where proposals are subject to qualified majority voting, largely health and safety-related, the position remains unchanged. Thus the Directive on the protection of young people at work has proceeded under the Action Programme post-Maastricht.

The importance of the opt-out, often referred to as the Maastricht or social protocol, relates to some of the more contentious proposals that are based on unanimity; progress has not been possible where one member state blocks the proposals. Thus the UK has not been involved in discussions on the more controversial Directive on European Works Councils, and will not be bound by its requirements except to the extent that multinational companies with more than 1,000 employees employing at least 150 workers in more than two member states will be so bound (see Chapter 8). Adoption of the Directive occurred in September 1994, with implementation due within two years. In other words, the intention of the UK's opt-out – to avoid being bound by certain legally enforceable requirements – has not fully materialized, and the UK has had no moderating voice in the formulation process. Another controversial Directive on atypical workers, which includes the requirement for National

Insurance contributions to be levied at a lower threshold than at present, has made little progress to date. The UK remains opposed to this Directive, which looks set to become another candidate for the social protocol route.

There are now moves to challenge the basis of Britain's opt-out, which have been set out in a White Paper produced by the European Commission in July 1994 (see Industrial Relations Services, 1994a). Other member states have complained 'that the two-speed social Europe is legally unworkable and politically damaging' (Carvel, 1994). It is possible that Britain's social policy opt-out will not be renewed in 1996. The White Paper also contains proposals to complete implementation of the Action Programme, including measures on atypical workers, although these have been watered down quite significantly.

Significance: for whom?

The reality and potentiality of the effect of European developments on British individual employment law and industrial relations are undoubtedly significant. Gill (1994, p. 427) envisages a 'weaving of a European dimension into national industrial relations systems rather than the imposition of a new EC-wide model'. At present Directives are directly enforceable on public sector employees only, while private sector employees have to rely on governments to implement Directives in domestic legislation (private sector individuals can sue the government as per *Frankovich* – see Appendix 3). In spite of a recommendation by the European Court of Justice's Advocate General to end this distinction, in the full ruling of *Paola Faccini Dori* v *Recreb Sri*, C-91/92 [1994] *The Times*, 4 August, ECJ, this point was not accepted.

Changes through the UK's domestic legislative programme, and as a result of decisions reached in the European Court of Justice (ECJ) – the ultimate court for decisions on European law – will continue to occur. Such developments should, in theory at least, be of particular benefit to employees in industries such as hotels and catering, provided employers do not find ways round the legislation and the law is monitored and enforced effectively. However, there is virtually nothing in the social policy provisions that will require the UK government to legislate domestically to change the nature of collective law in Britain. This does not mean that change will not occur, but it looks as though the trade unions will have to rely rather more on the ECJ if their lot is to be changed.

The decisions from *Commission of the European Communities* v *United Kingdom of Great Britain and Northern Ireland*, C-382/92 [1994] IRLR 392 ECJ and *Commission of the European Communities* v *United Kingdom of Great Britain and Northern Ireland*, C-383/92 [1994] IRLR 412 ECJ are important not only to union members but also to non-union members. The ECJ affirmed that the redundancy and acquired rights Directives require information provision and consultation to take place with workforce representatives. Consultation is not restricted to recognized trade unions so the UK cannot avoid such obligations by derecognizing trade unions. Thus some extension of 'collective bargaining' by a non-trade union route looks likely to follow from these important judgements (see also the collective dimension below).

The potential implications of the European Social Charter in relation to employment practice remain to be seen, particularly the extent to which employers attempt to circumvent the law through employment strategies, work organization and other methods. This issue has already been considered in relation to hotels and catering (see Lucas, 1991c, 1993). The broad conclusion reached at the time of writing was that

> Although much of its content and implications remain an unknown quantity, the Social Charter appears to pose less of a threat to employment practice than was predicted. Some controversial proposals have been diluted and others have been effectively shelved thus presenting the hospitality industry with the opportunity to continue with its 'cost-cutting' employment practices, at least in the short-to medium-term. (Lucas, 1993, p. 89)

This remains largely true to the extent that little has changed to divert employers from perpetuating the creation and maintenance of low paid, part-time jobs of fewer hours, and it is these direct, tangible 'real' employment costs that dominate the employment decision rather than the indirect, intangible 'perceived' costs of employment legislation (Lucas, 1993, pp. 95–7). This is not to deny that changes elsewhere, particularly in relation to equal pay and sex discrimination, are unimportant, but these, and others relating to health and safety, are discussed separately in the next two chapters.

Summary propositions

- The UK's EU membership has led to significant changes in British individual employment law, but not in collective law. This trend is likely to continue, but the extent of future changes is not yet known.
- The real effects of the law on the management of the employment relationship, and on employment practice, remain to be seen.

Discussion point

- To what extent does the European influence pose a threat or offer an opportunity to managing the employment relationship in hotels and catering?

9.3 INDIVIDUAL WORKPLACE LAW

The history and paradoxes of individual law in the employment relationship in British industrial/employee relations have been discussed more fully elsewhere (see, for example, Hendy, 1993, pp. 7–19; Smith *et al.*, 1993, pp. 1–27). Important early statutes included the Contracts of Employment Acts 1963 and 1972, the Redundancy Payments Act 1965, the Industrial Relations Act 1971 (introducing unfair dismissal provisions) and the Employment Protection Act 1975. As noted earlier, such legislation was perceived to be necessary in order to redress an imbalance in the employment relationship (individual workers were disadvantaged compared to more powerful employers), most marked among unorganized industries such as hotels and catering. However, there has been an acknowledgment that small firms should be exempted from some legal provisions, but this has not been consistently applied across the range of legislative measures (see Smith, 1985).

The main body of individual law is now contained in the Employment Protection (Consolidation) Act (EPCA) 1978, although a number of changes effected since 1979, particularly under the Trade Union Reform and Employment Rights Act (TURERA) 1993, point to the need for a new consolidating act to be passed. The consolidation of all individual measures seems all the more necessary because the extent and influence of this instrument looks set to grow, most particularly because of the influence of EU membership. Collective law was consolidated by way of the Trade Union and Labour Relations (Consolidation) Act (TULRCA) 1992, although there are some important new changes contained in TURERA 1993.

A summary of the main provisions of individual workplace law, and examples of some useful leading cases, are shown in Appendix 4. These are broad and basic outlines to guide the reader through the attendant analysis and evaluation that follows, and do not constitute a comprehensive account of the law. Any manager or student who needs more specific advice or information on any aspect, particularly to deal with an actual case, must check the details of the law more closely. More detailed guidance is given in Employment Department leaflets, available from Job Centres, Employment Offices, Unemployment Benefit Offices and regional offices of the Employment Department. ACAS and Citizens Advice Bureaux may be able to provide advice and assistance. Other texts offer more detailed analysis and guidance, including Farnham and Pimlott (1990, pp. 242–76), HCTC (1990), Pitt (1992, 1994) and Smith *et al.* (1993).

The tribunal system

The majority of individual employment law is enforceable through the industrial tribunal system (for details on scope and procedures, see Torrington and Hall, 1991, pp. 556–68) and cases can be appealed on a point of law to the Employment Appeal Tribunal (EAT) (for more details, see Smith *et al.*, 1993, pp. 293–323). Thereafter cases can proceed to the Court of Appeal (CA) and then to the House of Lords (HL). The European Court of Justice (ECJ) is the ultimate decision-making court in relation to European law; its decisions have the effect of overriding all UK law. The importance of this court will become clearer in relation to decisions about equal pay and sex discrimination, discussed in Chapter 10.

Industrial tribunals and the EAT are chaired by a legally qualified person and a High Court judge respectively, who are assisted by two lay members (four in some EAT cases). The lay members comprise one nominee from a panel drawn from employers' organizations and the other from a panel drawn from employees' organizations. They are required to act independently as full tribunal members, although their industrial experience is meant to help the tribunals to reach sensible and practicable decisions.

Compensation or redress to the aggrieved party (most usually a former employee) following a tribunal case is normally in the form of monetary compensation (a tribunal may also order re-engagement or reinstatement). Here, different statutory provisions impose different requirements, and these are mentioned, as necessary, at the appropriate point in the text (see also Appendices 4 and 5). The one common element shared by a number of statutory provisions is that compensation is frequently related to a week's pay, to which there is a maximum limit. This currently stands at

£205 per week, a level unchanged since April 1992. The shape of things to come points to more pressure on employers to take workers back rather than awarding compensation (Gibb, 1994), an option that has been possible, but rarely utilized, for many years.

Although some modifications have been made to the jurisdictions of industrial tribunals over time, most recently TURERA 1993 (Sections 36 to 42) has made some important changes to the constitution and jurisdiction of industrial tribunals and the EAT. Pre-hearing reviews (PHR) have replaced the rarely used pre-hearing assessment (introduced in 1980; see Smith *et al.*, 1993, pp. 317–18), the general principle being to dispose of claims that stand virtually no chance of success at tribunal, such as those which fall outside the jurisdiction of the tribunal system. As no evidence is to be considered in a PHR, its potential efficacy must still remain questionable. However, the power of a tribunal to order advance payment of up to £150 may be sufficient to deter some 'vulnerable' applicants from proceeding with a justifiable case (consider what that represents in terms of a week's pay to many hotel and catering workers – see Chapters 3 and 5).

It is important to note that ACAS maintains a rather more important role in pre-tribunal proceedings than the PHR or its predecessor. In 1990–1 there were 381 pre-hearing assessments, which marked a rapid decline from a peak of 3,555 in 1983 (Smith *et al.*, 1993, p. 318). In 1993 ACAS received nearly 72,000 notified tribunal cases for individual conciliation, the majority of which (45,000) related to unfair dismissal (ACAS, 1994, p. 76). In terms of completed cases (66,000), around two-thirds were either settled or withdrawn, so that only around one-third of cases proceeded to industrial tribunal.

Certain cases, such as those that are non-contested or relate to the Wages Act 1986, can now be heard by the chair alone, and if these are appealed the EAT judge can also sit alone. The power to extend the jurisdiction of industrial tribunals to certain contract claims (Section 38) has been invoked and effected by the Industrial Tribunals Extension of Jurisdiction (England and Wales) Regulations 1994 (see Aikin, 1994); this largely follows from *Delaney* v *Staples* [1992] IRLR 191 HL, [1991] 112 CA and [1990] IRLR 86 EAT (see wages and pay, and notice below). Further changes to industrial tribunals are pending in the light of the potential for huge compensation awards to be made under the Sex Discrimination Act 1975 (as amended) as a result of *Marshall* v *Southampton and South-West Hampshire Area Health Authority (Teaching) (No. 2)* 271/91 [1993] IRLR 455 ECJ (see Chapter 10), and the burgeoning workload of the industrial tribunal system (Gibb, 1994). A number of these 'procedural' issues are touched upon later in the chapter.

Summary proposition

- Individual workplace law is becoming an increasingly complex area that impinges on the management of the employment relationship.

Discussion points

- What are the potential effects of more or fewer individual employment rights on employment opportunities and employment practice?
- Which aspects of the employment relationship should be regulated by individual employment rights?

9.4 BEGINNING THE EMPLOYMENT CONTRACT

Written statement of terms and conditions of employment

The written statement of terms and conditions of employment (referred to as the written statement) is often referred to incorrectly as the contract of employment. A contract of employment exists once an employee has accepted an offer of employment, and may not even be written down in the early stages of employment; here it is governed by the common law. The written statement has the effect of extending the employment contract through the specification of certain statutory requirements in writing (see Appendix 4, Section A4.1). There is no statutory requirement for an employee to sign, or sign for, the written statement; signing does not constitute an agreement or acceptance of terms because the law overrides such circumstances. A more detailed legal analysis of the written statement is given in Smith *et al.*, (1993, pp. 71–80).

Important changes to the documentation that employers must now provide on the written statement, while cumbersome, have the effect of making the contract more explicit (TURERA 1993, Section 26 and Schedule 4). This implements the Directive on an employer's obligation to inform employees of the conditions applicable to the contract or employment relationship deriving from the Social Charter.

Employers must now give all employees (EPPTER 1995) all the principal statements (these may be given in 'instalments') relating to these terms and can no longer refer employees to accessible documents. There are some exceptions to this. Employees may be referred to accessible documents for certain aspects relating to grievance procedures and disciplinary rules and procedures, particulars of sick pay and other terms relating to sickness or injury, and of pensions. For notice periods, reference may be made to the law (EPCA 1978, Section 49) or the provisions of a suitably accessible collective agreement (see Appendix 4, Section A4.1). Where there are no particulars of matters 4 to 7 and 9 to 15 listed in Appendix 4, Section A4.1, the fact must be stated.

The consequences of non-compliance remain unchanged by TURERA 1993; an industrial tribunal can only determine the correct particulars. Particulars cannot be invented where none are agreed, unless implied by law, such as the right to reasonable notice.

Other contractual points

There may be other contractual arrangements, relating to, for example, bereavement leave, deductions from pay or the right of search, that do not need to be specified in the written statement. However, if terms are made as explicit as possible this minimizes the possibility of lengthy disputes about what contractual obligations actually apply, particularly at an industrial tribunal. Thus contractual terms such as those relating to the provision of accommodation and its vacation, the provision of meals and uniforms, joint contracts (managing spouses or partners in public houses or hotels), deductions from pay, the right to suspend employees (with and without pay), safety responsibilities, maternity rights (see Chapter 10), continuity of employment and overtime working should be made as explicit as possible in writing. However, explicit terms may still be subject to disputed interpretation as to their effect.

Once agreed, unless the contract provides for it, the terms of the contract cannot be changed without the consent of both parties. Consultation and agreement are sensible precursors to any changes, because unilateral employer change could lead to a claim for constructive dismissal (see dismissal below).

Apart from the requirements set out in the written statement and elsewhere, in the employment relationship employers and employees have common law duties to one another. An employer must take reasonable care for the employee's safety, pay the agreed wages, not require the employee to do unlawful acts and provide work for workers paid by results or commission if work is available. Employees must obey all lawful and reasonable orders, not commit misconduct, give faithful and honest service and use reasonable skill and care in their work. In some areas, common law has been 'enhanced' by statute law, such as health and safety legislation and the law on unfair dismissal. Disputes about the employment contract may be tested through the civil and criminal law and, where specified, through the industrial tribunal system.

Summary proposition

- A clear and explicit statement of employment terms, preferably in writing, is likely to give clarity to the employment contract, although the meaning of those terms may still be disputed.

Discussion point

- What are the main features of an employment contract that should be specified in writing for management jobs and operative jobs? Consider these in relation to an hotel and catering organization of your choice.

9.5 MAINTAINING THE EMPLOYMENT CONTRACT

As a significant body of workplace rights that are part of 'contract maintenance' fall within the spheres of equality of opportunity and health and safety, they are discussed in the next two chapters. Regulating the employment relationship through the grievance and disciplinary process has already been discussed in Chapter 7. This section discusses rights relating to pay, time off work and the transfer of undertakings, although this last area may also have implications for contract termination.

Wages and pay

In terms of hotels and catering, the main body of rights relating to pay, with the exception of equal pay and maternity pay (see Chapter 10) and sick pay (Chapter 11), is contained in Appendix 4, Section A4.2. A more detailed legal analysis of wages and pay is given in Smith *et al.*, (1993, pp. 147–80).

With the exception of deductions from pay, there is virtually no known information about the significance of these rights in hotels and catering. It is possible that itemized pay statements may not always be issued, particularly if employers are circumventing Inland Revenue requirements with regard to income tax. The recent change to the hours qualification with regard to itemized pay statements (EPPTER 1995) means that *all* employees must be given a statement. Such rights are not generally the subject of industrial tribunal proceedings (Employment Department, 1993, p. 528) or individual conciliation with ACAS (industrial tribunal and non-industrial tribunal cases) (ACAS, 1994, p. 76).

The number of cases brought under the Wages Act 1986 (Part I) for unlawful deductions, which include the non-payment of wages and holiday entitlements, has been increasing generally (Employment Department, 1993, p. 528). Claims under the Wages Act now constitute the second largest number of registered tribunal cases (7,510 in 1992–3) and the second largest area for individual conciliation by ACAS; unfair dismissal claims remain the predominant issue (see below). In a leading case in this area, *Delaney* v *Staples* [1992] IRLR 191 HL, it was confirmed that wages in lieu of notice were not pay under the Wages Act but constituted compensation (damages) for breach of contract. As a result the employee had to bring a claim for non-payment of wages in lieu before the ordinary courts, although non-payment of wages went before a tribunal. This anomaly contributed to the decision to widen industrial tribunals' jurisdiction in relation to certain contact claims, noted above (see also notice). The effect of this extension is that every employee, regardless of length of service, will be able to bring a claim relating to termination of employment. Maximum compensation is £25,000.

Claims for unlawful deductions are most likely to occur in smaller workplaces (median size 19 employees) in the private sector in distribution, hotels and catering or repairs where unions are not recognized (Employment Department, 1994, p. 22). Such claims are almost twice as likely to be lodged by males as by females, the majority of whom are not union members. There is a greater chance of a claim for unlawful deductions being upheld in favour of the applicant at tribunal (68 per cent of cases) than is the case in relation to claims for unfair dismissal (41 per cent of cases) (Employment Department, 1994, p. 24). The median award made by tribunals (£265) is higher than the median award made in pre-tribunal proceedings (£166) (Employment Department, 1994, p. 25). For a useful review of developments under Part I of the Wages Act 1986 see Industrial Relations Services (1994b).

As noted by Smith *et al.*, (1993, p. 177), the provision relating to deductions to make good cash shortages or stock discrepancies in retail employment is open to some abuse. Some hotel and catering employees fall within retail employment, and this also has implications in relation to the Sunday Trading Act 1994, mentioned in Chapter 11. There is no requirement that the grounds for deduction should be fair or reasonable, and the 10 per cent limit on deductions does not apply to a worker's final payment of wages on termination (see Appendix 4, Section A4.2).

Minimum wages

Important changes in relation to hotel and catering workers' statutory rights to pay have taken place recently and cannot go unmentioned. In 1986 the wages council system was radically reformed. Workers under the age of 21 were removed from the scope of the councils, and wages councils were restricted to setting a single minimum rate and overtime rate, and a maximum accommodation charge – previously scales of pay rates and other terms had been set. In 1993 the councils were abolished in their entirety through the repeal of Part II of the Wages Act 1986, and around one million hotel and catering workers lost the right to safety net minimum wage protection.

The effects of these changes are difficult to assess overall, although the decision to abolish certain wages councils in the 1970s, including the council covering workers employed in industrial and staff canteens, was found to have led to adverse effects on the workers and the economic performance of organizations (Craig *et al.*, 1982). The vast majority of hotel and catering workers are now no longer protected by *any* form of collective bargaining, inadequate though wages councils might have been (Lucas, 1991a). While some deterioration of pay and conditions undoubtedly occurred after 1986 (Lucas, 1991b, 1992; Radiven and Lucas, 1994), other evidence suggests that, for young workers, pay levels may not have fallen to the low levels that could have been expected (Lucas and Bailey, 1993). It is too soon to make any realistic assessment about the effects of total abolition, although it is probable that some workers will become worse off, and the case that abolition will create more employment is far from established (see, for example, Dickens *et al.*, 1993). Across the economy as a whole, the earnings gap between the highest paid and the lowest paid has been widening over time (Gosling *et al.*, 1994). It is conceivable that the abolition of wages councils could exacerbate this trend and, in particular, widen the gap between women's and men's pay (see Chapter 10, equal pay).

The Social Charter provisions relating to an 'equitable' wage are in the form of an Opinion which has no legal force, so any change in government policy would require a Labour government to be re-elected. The topic of a Statutory National Minimum Wage (SNMW) (see, for example, Bayliss, 1991; Wilkinson, 1992) remains high on the agenda of the Labour Party, which, were it to be elected, would reintroduce a statutory minimum wage that would cover more hotel and catering workers than had been the case under the wages council system (excluding young workers). However, such a measure would be more corporatist than the semi-democratic wages councils because those determining the wage would not be drawn from among the employers and workers. There would also not appear to be any plans for other statutory minimum terms, such as holidays, shift premia and overtime rates, so the SNMW system would be less comprehensive (see Lucas, 1989, 1994). However, the working time Directive does provide for a statutory minimum period of paid holiday entitlement and lays down other minimum requirements relating to matters such as rest periods and meal breaks (see Appendix 6, Section A6.4). These requirements, subject to any exemptions, will apply to hotel and catering workers, if and when the Directive becomes enforceable (there is further discussion of this Directive in Chapter 11).

Time off

Employment protection legislation provides for 'reasonable' time off work, with pay and without pay (see Appendix 4, Section A4.3). More details about trade union issues are given in the ACAS Code of Practice 3, *Time off for Trade Union Duties and Activities* (ACAS, 1991), and in Smith *et al.*, (1993, pp. 466–70). Time off in relation to equality and health and safety is dealt with in Chapters 10 and 11 respectively.

Transfer of an undertaking

Under the EPCA 1978 an employee who changes employer may maintain continuity of employment from the date he or she commenced employment with the old employer in certain circumstances (Smith *et al.*, 1993, pp. 138–40). The transfer of the tenancy of an hotel would now fall within such regulations, as in *Young* v *Daniel Thwaites and Co* [1977] ICR 877. Additional rules under the Transfer of Undertakings (Protection of Employment) Regulations 1981 (TUPE) contain provisions relating to the dismissal of employees as a consequence of the transfer and continuity of contracts of employment and trade unions (see Appendix 4, Section A4.4). The issue of dismissal during a transfer of undertaking is featured in *McGrath* v *Rank Leisure* [1985] IRLR 323 EAT.

The concept of automatic transfer has created particular difficulties, such that the status of the application and effect and scope of the regulations are very uncertain, although amendments under TURERA 1993 (Section 33) were designed to correct some of the defects. The TUPE provisions do apply to the contracting out of a peripheral service such as a canteen – see *Rask and another* v *ISS Kantineservice A/S* 209/91 [1993] IRLR 133 ECJ – a point that would seem to have important implications for catering operations in the catering services sector, rather than those in the commercial sector that is the focal point for this book.

As noted above, the important judgements given in *Commission of the European Communities* v *United Kingdom of Great Britain and Northern Ireland*, C-382/92 [1994] IRLR 392 ECJ and *Commission of the European Communities* v *United Kingdom of Great Britain and Northern Ireland*, C- 383/92 [1994] IRLR 412 ECJ ruled that the UK had failed to implement the acquired rights Directive 1977 and the collective redundancies Directive 1975. TUPE was held to cover contracting out under the government's compulsory competitive tendering programme (Thatcher, 1994), and employees' representatives should have been consulted before the transfer. This means that organizations that have derecognized trade unions will have to consider the introduction of some form of 'worker representation'. For a fuller account of the increasingly complex rules and regulations that govern continuity of employment, see Smith *et al.*, (1993, pp. 129–46).

Summary propositions

- The platform of legally enforceable workplace employment rights has changed since the mid-1980s, to both the benefit and the detriment of employees.
- While the number of instances of workers claiming unlawful deductions from pay has increased, substantially more workers have lost the right to a statutory minimum wage.

Discussion points

- What are the pros and cons of making deductions from pay to make good cash shortages or stock discrepancies, from a managerial and a worker's point of view?
- Argue the case for and against a Statutory National Minimum Wage.

9.6 TERMINATING THE EMPLOYMENT CONTRACT

A number of different issues are involved in terminating the employment contract, of which, undoubtedly, unfair dismissal has been, and continues to be, the most significant item. Therefore, the bulk of the discussion falls on this matter, although the closely related issue of redundancy is highly topical in times of recession and rapid change.

Notice

Until the mid-1960s common law rules on notice made no distinction for the long-serving employee. Thus the fact that an employee on a weekly contract with 20 years' service was only entitled to one week's notice was deemed to be deficient in times of large-scale redundancies (Smith *et al.*, 1993, p. 263). Statutory minimum periods of notice contained in Section 49 of the EPCA 1978 have remained unchanged since 1972 (see Appendix 4, Section A4.5), although many contracts may provide for longer periods of notice to be given by either party, particularly at managerial level (see Smith *et al.*, 1993, pp. 262–7). *All* employees now have the right to notice.

Although the employee may work out the notice period, many employers prefer to pay wages in lieu. Following the case of *Delaney* v *Staples* [1992] IRLR 191 HL, brought under the Wages Act 1986, Smith *et al.*, (1993, p. 265) suggest that 'the employer now has much to gain from putting into the contract of employment a term expressly permitting dismissal with wages in lieu, since it provides the optimum position of a lawful dismissal, no challenge to the payment in tribunal proceedings, an early effective date of termination and the preservation of any restraint of trade clause'.

Wrongful dismissal may arise most typically where an employee is dismissed with no or inadequate notice (for more details on wrongful dismissal, see Smith *et al.*, 1993, pp. 270–92). It has already been noted that under TURERA (Section 38), which amends the EPCA 1978 (Section 131), the power to extend the jurisdiction of industrial tribunals to certain contract claims, including wrongful dismissal claims which have traditionally been matters of civil proceedings, has now been invoked.

Dismissal

Termination of employment is a complex subject that is regulated by a range of different statutes and regulations (for discharge at common law, see Smith *et al.*, 1993, pp. 249–92; for the law and procedure of unfair dismissal, see Smith *et al.*, 1993, pp. 293–401). The subject of dismissal cannot be fully understood without background knowledge of disciplinary issues and the ACAS guidance on discipline (ACAS, 1977, 1987), discussed in more detail in Chapter 7. The main legal parameters of dismissal are given in Appendix 4, Section A4.6, including examples of some leading cases, although equality and health and safety dimensions are also addressed in Chapters 10 and 11 respectively.

In practice, unfair dismissal continues to constitute the most important individual employment right overall in terms of industrial tribunal cases (Employment Department, 1993, 1994) and individual conciliation by ACAS (ACAS, 1994, p. 76). In 1992–3, unfair dismissal cases accounted for 63 per cent of all registered tribunal cases (Employment Department, 1993, p. 528). The significance of the unfair dismissal provisions is especially marked in employment sectors that have been traditionally viewed as 'vulnerable', including non-unionized small firms. Thus private sector employers in distribution, hotels and catering and repairs, over a third of which employ fewer than 25 employees, are most likely to be involved in tribunal cases concerning unfair dismissal (Employment Department, 1994, p. 22). Men are more than twice as likely as women to be involved in tribunal cases (Employment Department, 1994, p. 23).

However, at tribunal the majority of unfair dismissal cases (59 per cent) do not find in favour of the applicant (Employment Department, 1994, p. 24). A 1992 survey of tribunal applications found that the median award made by tribunals (£1,923) was more than double the median award made in pre-tribunal proceedings (£959), which include conciliated settlements (Employment Department, 1994, p. 25). The level of the median award of all tribunal cases had risen to £2,616 in 1992–3 from £1,773 in 1990–1 (Employment Department, 1993, p. 529).

More closely refined figures relating to dismissals in hotels and catering, drawn from the WIRS3 (workplaces employing 25 or more employees), show conflicting results. Table 9.1 shows that the rate of dismissals in hotels and catering is higher than elsewhere, from which it could be posited that there was more likely to be a higher incidence of industrial tribunal claims related to dismissal; this does not seem to follow.

Table 9.1 *Number dismissed (excluding redundancy).*

			PSS (by no. of employees)				HCI (by no. of employees)			
	AIS	SI	All	25–49	50–99	100+	All	25–49	50–99	100+
Unweighted base	2061	1299	702	174	153	375	61	21	18	22
Weighted base	2000	1462	886	518	218	151	99	59	26	14
Number dismissed (%)										
0	57	62	51	58	45	35	35	43	21	30
1–4	33	31	39	37	42	43	45	48	48	29
5–9	5	4	6	4	5	12	7	6	5	18
10–19	2	2	2	1	4	5	7	3	15	12
20–49	1	[a]	1	–	1	2	2	–	5	1
50+	[a]	[a]	[a]	–	–	[a]	–	–	–	–
Dismissals per 1000 employees	15	13	20	28	28	12	45	42	75	25
Mean number of dismissals	1.4	1.2	1.6	1.1	1.9	2.8	2.8	1.6	5.0	3.8

Table does not include percentage figures for missing cases.
[a]Fewer than 0.5 per cent.
Any slight discrepancies between figures are due to the rounding of decimal points.
Source: WIRS3 (1990).

Claims connected to dismissal may relate to matters other than the more routine aspects of unfair dismissal relating to capability and conduct (see Appendix 4, Section A4.6), but virtually all are potentially subject to industrial tribunal proceedings, subject to appropriate qualifications. However, not all dismissals are potentially unfair. The data shown in Table 9.2 show a less active pattern of all tribunal actions occurring in hotels and catering compared to elsewhere in the year prior to the WIRS3, in spite of the high rate of dismissals in the industry. Hotels and catering may have more dismissals that are terminations at the end of a short-term contract or casual engagement.

Table 9.2 *Number of industrial tribunal actions notified in past year.*

	AIS	SI	PSS (by no. of employees) All	25–49	50–99	100+	HCI (by no. of employees) All	25–49	50–99	100+
Unweighted base	2061	1299	702	174	153	375	61	21	18	22
Weighted base	2000	1462	886	518	218	151	99	59	26	14
Actions (%)										
Any	9	7	8	7	8	12	6	3	15	–
1	7	6	7	6	7	10	6	3	15	–
2–4	1	1	1	1	1	2	–	–	–	–
5–49	[a]	[a]	[a]	–	–	[a]	–	–	–	–
50–409	[a]	[a]	[a]	–	–	[a]	–	–	–	–
Tribunal actions per 1000 employees	1.3	1.2	1.9	2.3	1.3	1.9	0.9	0.8	2.2	–

[a]Fewer than 0.5 per cent.
Any slight discrepancies between figures are due to the rounding of decimal points.
Source: WIRS3 (1990).

Paradoxically, in light of the higher rate of dismissals in hotels and catering, the data contained in Table 9.3 suggest a rather less active pattern of behaviour in actually pursuing dismissal on grounds of unfair dismissal than is the case elsewhere. While to some this might be interpreted to mean that employees are less disgruntled in hotels and catering than elsewhere, it is equally plausible that workers in hotels and catering accept dismissal as a constituent part of the employment experience that is not worth challenging because work is generally readily available elsewhere. They are also very unlikely to be union members and to have access to union support at an industrial tribunal.

Table 9.3 *Number of actions on grounds of unfair dismissal.*

	AIS	SI	PSS (by no. of employees) All	25–49	50–99	100+	HCI (by no. of employees) All	25–49	50–99	100+
Unweighted base	2061	1299	702	174	153	375	61	21	18	22
Weighted base	2000	1462	886	518	218	151	99	59	26	14
None (%)	93	95	93	94	94	90	94	97	85	100
Number of actions	245	161	136	36	16	83	6	2	4	–
Unfair dismissal actions per 1000 employees	1.2	1.2	2.0	1.9	1.1	2.4	0.9	0.8	2.2	–

Any slight discrepancies between figures are due to the rounding of decimal points.
Source: WIRS3 (1990).

From an employer's perspective, a high rate of dismissals is also to be expected. Many hotel and catering firms rely on the external labour market, recruiting through unsophisticated means, offering low wages and relying on dismissal to weed out unsuitable workers; this can be seen as a deliberate approach to quality control, and the opposite to paying high wages to attract and retain high quality staff. 'Dismissal or the threat of it remains an important form of control' (Edwards, 1989, p. 306). Others have noted that the chances of a worker being involved in an unfair dismissal claim increase as establishment size decreases. As the great majority of dismissals do not lead to claims for unfair dismissal, employers' fears about the impossibility of firing workers have been overstated (Dickens *et al.*, 1985).

Edwards (1989) also argues that in the broad issue of labour control, the distinction between a dismissal for disciplinary reasons and normal labour turnover breaks down. Even though the conclusion reached in the following citation was drawn from a study in clothing manufacturing, such management style and employee behaviour is clearly identifiable in hotels and catering, as noted throughout earlier chapters.

> ... people who quit 'voluntarily' during the first few months of service with a firm may do so as much for reasons connected with the work that they are expected to do and the form of discipline to which they are subject, as for reasons associated with their personal characteristics ... Such quitting helped enforce managerial authority because those who might question it did not stay long ... The conduct of discipline cannot be separated from overall patterns of workplace relations. (Edwards, 1989, p. 307)

Redundancy

The notion of the state compensating a worker for being dismissed for redundancy by an organization was first established in the mid-1960s during conditions of relatively low, but rising, unemployment. The Redundancy Payments Act 1965 required employers to compensate workers for dismissals in quite specific circumstances according to statutorily defined rules (see Appendix 4, Section A4.7). That compensation was the only 'solution' marked a certain inevitability that most job losses were unavoidable. The state contributed to such compensation by way of a rebate scheme to employers (for more details, see Smith *et al.*, 1993, pp. 402–26). With the benefit of hindsight, redundancy pay can be said to represent an early state-sponsored model of making employment more flexible.

Since then, the possibility that job losses could be avoided, or at least minimized, has been recognized, and consultation requirements were added to the redundancy agenda during the mid-1970s (see Simpson, 1991). Even so, unless trade unions had an agreement, these measures did not lead to workforces having any significant influence over managerial decisions about redundancy (Simpson, 1991, p. 10). Consultation requirements have subsequently been extended by TURERA 1993 (see Appendix 4, Section A4.7), which implements the Directive on collective redundancies deriving from the Social Charter. However, although they would appear to be limited to unionized workplaces, the cases of *Commission of the European Communities* v *United Kingdom of Great Britain and Northern Ireland*, C-382/92 [1994] IRLR 392 ECJ and *Commission of the European Communities* v *United Kingdom of Great Britain and Northern Ireland*, C-383/92 [1994] IRLR 412 ECJ, mentioned above, have generated ambiguity in this regard. For more details on special redundancy procedures, see Smith *et al.*, (1993, pp. 55–9).

This development for more worker involvement has occurred when the issue of reducing the extent of 'state welfare' in circumstances of job loss has been a key plank of Conservative government policy for well over a decade. Although the principle of being compensated for job loss has not been abandoned entirely, the monetary value of redundancy compensation (Smith *et al.*, 1993, p. 418) and unemployment benefit (Smith *et al.*, 1993, pp. 426–46) has been eroded, and employers are now entirely liable for funding redundancy payments. One possible outcome is that employers will now 'think twice' before dispensing with some jobs; conversely such a move could encourage more employers to rely more heavily on more flexible forms of atypical labour that are more readily disposable at virtually no cost.

Redundancy issues are more likely to be in the news in times of recession. In 1993 one-fifth of collectively conciliated disputes involving ACAS were about redundancy. Evidence from the WIRS3 shown in Table 9.4 indicates that a significant proportion of workplaces suffered workforce reductions in 1990, although this was less marked in hotels and catering compared to elsewhere.

Table 9.4 *Workforce reductions in past 12 months.*

			PSS (by no. of employees)				HCI (by no. of employees)			
	AIS	SI	All	25–49	50–99	100+	All	25–49	50–99	100+
Unweighted base	2061	1299	702	174	153	375	61	21	18	22
Weighted base	2000	1462	886	518	218	151	99	59	26	14
Yes[a] (%)	32	30	27	25	27	37	21	21	20	25
Most important reasons (%)										
Lack of demand for products and services	37	29	32	34	26	34	60	52	73	68
Reorganized working methods/ relocation/ integration	37	39	44	39	48	54	13	20	–	8
Improved competitiveness/ efficiency/cost reduction	29	29	34	23	46	45	11	16	–	12
Staff shortages	7	7	9	11	11	4	23	28	27	–

[a]Numbers of cases form the base for most important reasons. These are given in Table 9.5.
Column figures do not sum to 100 because not all the reasons given have been included, and any slight discrepancies in row totals are due to the rounding of decimal points.
Source: WIRS3 (1990).

However, the most important reasons for workforce reductions differ markedly between hotels and catering and elsewhere. The importance of lack of demand for products or services is much more marked in hotels and catering than in other employment sectors. Relatively few hotel and catering workplaces suffered job losses because of work reorganization and improved competitiveness, efficiency and cost reduction measures. Surprisingly, perhaps, staff shortages were given as a reason for workforce reductions in around one-quarter of hotel and catering workplaces, although this was limited to workplaces employing fewer than 100 employees. These differences are not easy to explain at face value. However, based on what has already been established about the industry in earlier chapters, it is possible to infer that hotels and catering are already operating flexibly (a lesser extent of workforce reductions), are more susceptible to product market pressures, are not particularly innovative in terms of 'efficiency' measures and are more prone to the vagaries of the labour market. This begs a number of questions, not least: 'Could not this situation be managed more effectively?'

Compulsory redundancy is not the first, or main, option in terms of reducing workforce numbers, and occurs to a lesser extent in hotels and catering than elsewhere, confirmed by data from the WIRS3 shown in Table 9.5. Natural wastage and redeployment in the establishment are the preferred options across all employment sectors. Termination of temporary staff, though not a significant factor, is more marked in hotels and catering than elsewhere. Early retirement is rare in hotels and catering, probably related to a younger than average workforce, and a lower level of membership of pension schemes than elsewhere (see Table 11.3).

Table 9.5 *Methods used to reduce workforce.*

			PSS (by no. of employees)				HCI (by no. of employees)			
	AIS	SI	All	25–49	50–99	100+	All	25–49	50–99	100+
Unweighted base	917	531	255	49	44	162	17	6	3	8
Weighted base	639	436	242	129	59	55	21	12	5	3
Natural wastage (%)	67	70	70	72	59	78	61	60	47	89
Redeployed in establishment (%)	45	42	43	30	42	75	36	32	27	65
Early retirement (%)	26	27	17	9	22	32	3	–	–	17
Voluntary redundancy (%)	21	19	16	17	9	20	17	24	–	19
Compulsory redundancy (%)	30	20	33	30	40	35	11	16	–	13
Termination temporary/ casual staff (%)	2	2	1	–	–	3	9	–	–	52
None	9	8	10	12	12	3	16	16	27	–

Column figures do not sum to 100 because not all the methods given have been included, and any slight discrepancies in row totals are due to the rounding of decimal points.
Source: WIRS3 (1990).

In terms of registered tribunal cases, redundancy pay is the third most significant issue (7,084 cases in 1992–3) after unfair dismissal (33,683) and Wages Act (7,510) claims, with other redundancy provisions constituting a much smaller body of applications: 490 cases in 1992–3 (Employment Department, 1993, p. 528). Redundancy claims are most likely to be made against smaller private employers (median size 19 employees) in distribution, hotels and catering or repairs (Employment Department, 1994, p. 22). As in Wages Act cases, at industrial tribunal the outcome is most likely to find in favour of the applicant (67 per cent of cases) (Employment Department,

1994, p. 24). There is little difference in the value of the median award made by a tribunal (£688) and the median payment agreed in pre-tribunal proceedings (£646) (Employment Department, 1994, p. 25), ostensibly because the rules of compensation follow a more closely defined formula than in cases of unfair dismissal or those falling under the Wages Act.

The effects of the law

As Price (1993, p. 16) has noted, employment legislation has the dual purpose of providing individual employee protection against unscrupulous employers and encouraging the development of good employment practice, governed more often than not by appropriate procedural arrangements. Price (1994, p. 52) has also observed that managers can only develop and implement good personnel practice if they know about recommended practice and its applicability to their establishments. Perceived levels of knowledge about the contract, fair recruitment and discipline were highest among personnel specialists and lowest among proprietors and partnerships. Managers will find it more difficult to comply with the law if they do not know about it, although ignorance is no defence.

As noted in Chapter 7 from the WIRS3, there has been an undoubted increase in the extent of formalized employment procedures, particularly for dismissals and discipline, in workplaces employing 25 or more employees. Thus the existence of unfair dismissal legislation and the threat of legal proceedings has provided sufficient impetus for most employers in this category to have procedures and to attempt to manage within the law.

By contrast, a subsequent Employment Department survey (1992) covering smaller workplaces (mean size 35 employees) than those characteristic of the WIRS3 survey (mean workplace size 102 employees) showed that formal procedures with regard to the major areas of tribunal jurisdiction were far from being the norm. The most regulated areas were unfair dismissal and race discrimination, where 66 per cent of employers had procedures; 50 per cent of employers had redundancy procedures and only 43 per cent had procedures relating to sex discrimination and matters arising from the Wages Act.

However, some procedures may not conform exactly to recommended guidelines, a point that is unlikely to be discovered until legal proceedings take place. For example, inadequate procedures may contribute to the unsuccessful defence of unfair dismissal claim. The Employment Department survey (1992) of industrial tribunal cases found that subsequently a quarter of employers had made changes to the way in which they dealt with such matters at the workplace. Employers were most likely to have made changes if they had made a settlement with the applicant or had 'lost' the case (see Employment Department, 1994).

The real influence of the law can be viewed from two perspectives. It can be said to be at its most influential when employers, employees and unions go along with it rather than contesting it in the courts. Alternatively its effectiveness is also contingent upon the extent to which it is tested through the legal system and in the outcomes of those judgements. As noted earlier, in the UK the route to satisfaction is mainly via the industrial tribunal system (for an alternative approach, see Lewis and Clark, 1993), although only around one-third of cases go to a full tribunal hearing. Ultimately the most important decision-making body is the European Court of Justice.

As a result of such legal proceedings, reading some legislation as originally published may give little, or no, insight into the current way in which the law is being interpreted in practice. Inevitably this kineticism makes for increasing complexity in the field of employment law, at least for those who are trying to do things correctly. Thus, for managers, 'managing' the law becomes an increasingly difficult and potentially costly area (perceived and real) and, as a consequence, more lawyers are likely to become involved. This is not entirely consistent with the founding spirit of the industrial tribunal system, which was intended to be rather more quasi-legal than legal, relying largely on inputs from practitioners and not lawyers, albeit within a legal framework.

The 1992 survey of industrial tribunal applications sheds some light on these issues (Employment Department, 1994). Initially the Citizens Advice Bureau provided the most important source of advice to applicants, while solicitors and barristers were the most important sources of advice to employers. Although roughly equal proportions of applicants and employers represented themselves at tribunal hearings (around 30 per cent), legal representation was engaged by over one-third of employers, but by only one-quarter of applicants. Citizens Advice Bureaux and friends or relatives (22 per cent) provided more assistance to applicants than trade unions (18 per cent). Having said all of this, cases involving applicants themselves and friends or relatives had the highest success rate in favour of the applicant (42 per cent) and those involving a private solicitor or barrister were the least successful (35 per cent). From an employer's point of view, private solicitors and barristers provided the most successful form of representation. The median overall employer cost, including time spent and fees related to representation or compensation, amounted to £1,486; the median applicant cost came to a more modest £49.

Summary propositions

- Relatively frequent contract termination is an important way of life for both managers and workers in hotels and catering.
- Unfair dismissals provisions would seem to offer a greater extent of potential protection to workers in hotels and catering than to workers elsewhere, although the evidence on this point appears to be contradictory.
- Redundancy is not the main method used to reduce workforce numbers, but the use of alternative methods differs among different sectors of the economy.
- Unfair dismissal and redundancy remain two of the three most important areas of workplace law.
- Workplace law issues will continue to change, and will remain an important factor underpinning the management of the employment relationship.

Discussion points

- What effects does a high rate of dismissals have on the organization, managers and workforce?
- How can fluctuations in demand be managed most effectively to the mutual benefit of the organization and the workforce?

9.7 THE COLLECTIVE DIMENSION

Collective employment law governs the rights, actions, duties, roles, responsibilities and behaviours of both employers and trade unions. However, the vast majority of the legal provisions centre on trade unions rather than on employers, a point reinforced in relation to the titles of the relevant extant legislation, mainly TULRCA 1992 and TURERA 1993. For a fuller account of the law on trade unions and industrial action, see Smith *et al.*, (1993, pp. 447–585) and Hendy (1993).

Briefly, legislation from the 1870s onwards, culminating in the Trade Disputes Act 1906, was concerned to allow workers to organize in free trade unions that could function effectively. The 'traditional' system of British industrial relations was based on the notion of 'voluntarism' and free collective bargaining – collective agreements were not legally enforceable (see Chapter 4). Trade unions were given immunities in civil law to enable them to act legitimately 'in contemplation or furtherance of a trade dispute'. Although the Industrial Relations Act 1971 sought to effect radical changes to British industrial relations (see for example, Armstrong, 1973), its life was short and successive Trade Union and Labour Relations Acts (1974, 1976) restored matters to the 1906 position. Additional measures, contained mainly in the Employment Protection Act 1975, including provisions relating to the statutory recognition of trade unions and the extension of voluntary collective bargaining in wages council industries, were also relatively short-lived (for a useful account of the 1970s, see Kessler and Bayliss, 1992, pp. 21–39).

Since 1979, successive Conservative governments have pursued an unrelenting strategy designed to weaken excessive trade union power (for a good analysis of the changes, see Hendy, 1993). Such a strategy has been implemented by accretion rather than in one fell swoop, and has brought about changes to the autonomy of the trade unions and their activities, the use of union funds, union membership, and the right to strike and take industrial action (see Appendix 4, Section A4.8). In short, it is more difficult for unions to function lawfully and survive (see Kessler and Bayliss, 1992, pp. 67–91; Hendy, 1993).

While it is inevitable that such an 'anti-trade union' scenario could be expected to exert a negative effect on trade union influence in the economy, this point should not be overstated in relation to hotels and catering. Although there are a few reported cases of trade union de-recognition, trade union density has remained low in the industry regardless of whether public policy has been supportive or unsupportive of trade unions and collective bargaining (for example, CIR, 1971; ACAS, 1980; Lucas, 1991a). At the time of the Donovan Report in the late 1960s trade union density was estimated to be around 4 per cent; in 1990 it was 3 per cent (see Table 4.1).

In terms of industrial action and picketing, evidence from the WIRS3 shows a generally less active situation in hotels and catering compared to elsewhere. In hotels and catering, 88 per cent of workplaces had not experienced any forms of industrial action among manual and non-manual workers, compared to 83 per cent of workplaces in all industries and services and 82 per cent of workplaces in all service industries. Industrial action was least marked in private sector services although there is higher trade union density here than in hotels and catering (see Table 4.1). In hotels and catering and private sector services, 98 per cent of workplaces had not been picketed in the past 12 months, compared to 94 per cent of workplaces in all industries and services, and 92 per cent of workplaces in all service industries.

In spite of the undoubted difficulties that face the trade union movement in terms of reasserting its authority as we move towards the twenty-first century, history suggests that industries such as hotels and catering may prove to be very unfertile ground for them, even if substantial parts of the 1980s and 1990s legislation were to be repealed.

Summary propositions

- Trade union reforms of the past 15 years are undoubtedly of considerable importance to the trade union movement in terms of its capacity to function and maintain membership.
- Even if all these reforms were to be repealed it is extremely doubtful whether this would make any difference to the role of trade unions in the hotel and catering industry.

Discussion point

- Can trade unions ever hope to play a role in hotels and catering given the legal restrictions on how they function?

REFERENCES

Advisory, Conciliation and Arbitration Service (ACAS) (1977) *Code of Practice 1, Disciplinary Practice and Procedures in Employment*. London: HMSO.

Advisory, Conciliation and Arbitration Service (1980) *Licensed Residential Establishment and Licensed Restaurant Wages Council Report No. 18*. London: HMSO.

Advisory, Conciliation and Arbitration Service (1987) *Discipline at Work: the ACAS Advisory Handbook*. London: HMSO.

Advisory, Conciliation and Arbitration Service (1991) *Code of Practice 3, Time off for Trade Union Duties and Activities*. London: HMSO.

Advisory, Conciliation and Arbitration Service (1994) *Annual Report 1993*. London: HMSO.

Aikin, O. (1992) A matter of precedence. *Personnel Management*, April, 56–7.

Aikin, O. (1994) Another forum for disputes. *Personnel Management*, August, 46.

Armstrong, E.G.A. (1973) *Straitjacket or Framework? The Implications for Management of the Industrial Relations Act*. London: Business Books.

Bayliss, F.J. (1991) *Making a Minimum Wage Work, Fabian Pamphlet 545*. London: Fabian Society.

Carvel, J. (1994) Brussels guns for UK opt-out. *The Guardian*, 27 July, 1.

Commission on Industrial Relations (1971) *The Hotel and Catering Industry, Part 1. Hotels and Restaurants, Report No. 23*. London: HMSO.

Craig, C., Rubery, J., Tarling, R. and Wilkinson, F. (1982) *Labour Market Structure, Industrial Organisation and Low Pay*. Cambridge: Cambridge University Press.

Dickens, L., Jones, M., Weekes, B. and Hart, M. (1985) *Dismissed: a Study of Unfair Dismissal and the Industrial Tribunal System*. Oxford: Blackwell.

Dickens, R., Gregg, P. , Machin, S., Manning, A. and Wadsworth, J. (1993) Wages councils: was there a case for abolition? *British Journal of Industrial Relations*, **31** (4), 515–29.

Edwards, P. K. (1989) The three faces of discipline. In K. Sisson (ed.), *Personnel Management in Britain*, pp. 296–325. Oxford: Blackwell.

Employment Department (1993) Industrial and employment appeal tribunal statistics 1991–92 and 1992–93. *Employment Gazette*, November, 527–31.

Employment Department (1994) The 1992 survey of industrial tribunal applications. *Employment Gazette*, January, 21–8.

Farnham, D. and Pimlott, J. (1990) *Understanding Industrial Relations*, 4th edition. London: Cassell.

Gibb, F. (1994) Jobs tribunal reform to cut huge awards. *The Times*, 8 July, 1.

Gill, C. (1994) British industrial relations and the European Community. *The International Journal of Human Resource Management*, **5** (2), 427–55.

Gold, M. (ed.) (1993a) *The Social Dimension – Employment Policy in the European Community*. Basingstoke: Macmillan.

Gold, M. (1993b) Overview of the social dimension. In M. Gold (ed.) *The Social Dimension – Employment Policy in the European Community*, pp. 10–40. Basingstoke: Macmillan.

Gosling, A., Machin, S. and Meghin, C. (1994) *What has Happened to Wages?* London: Institute of Fiscal Studies.

Hendy, J. (1993) *A Law unto Themselves. Conservative Employment Laws: a National and International Assessment*, 3rd edition. London: The Institute of Employment Rights.

Hepple, B. (1991) Institutions and sources of labour law: European and international standards. In *Encyclopedia of Employment Law Volume 1*, pp. 1052–61. London: Sweet and Maxwell.

Hepple, B. (1993) *European Social Dialogue – Alibi or Opportunity?* London: The Institute of Employment Rights.

Hotel and Catering Training Company (1990) *Employee Relations in the Hotel and Catering Industry*, 7th edition. London: HCTC.

Industrial Relations Services (1994a) Social Policy White Paper part one. *European Industrial Relations Review*, 248, September, 13–18.

Industrial Relations Services (1994b) Wages Act 1986: the story so far. *Industrial Relations Law Bulletin*, 504, September, 2–11.

Kessler, S. and Bayliss, F. (1992) *Contemporary British Industrial Relations*. London: Macmillan.

Lewis, R. and Clark, J. (1993) *Employment Rights, Industrial Tribunals and Arbitration: the Case for Alternative Dispute Resolution*. London: The Institute of Employment Rights.

Lucas, R.E. (1989) Minimum wages – straitjacket or framework for the hospitality industry into the 1990s? *International Journal of Hospitality Management*, **8** (3), 197–214.

Lucas, R.E. (1991a) Promoting collective bargaining: wages councils and the hotel industry. *Employee Relations*, **13** (5), 3–11.

Lucas, R.E. (1991b) Remuneration practice in a wages council sector: some empirical observations in hotels. *Industrial Relations Journal*, **22** (4), 273–85.

Lucas, R.E. (1991c) Some thoughts on the European Social Charter. *International Journal of Hospitality Management*, **10** (2), 174–7.

Lucas, R.E. (1992) Minimum wages and the labour market – recent and contemporary issues in the British hotel industry. *Employee Relations*, **14** (1), 33–47.

Lucas, R.E. (1993) The Social Charter – opportunity or threat to employment practice in the UK hospitality industry? *International Journal of Hospitality Management*, **12** (1), 89–100.

Lucas, R.E. (1994) Industrial relations theory, discourse and practice – are hotels and catering merely a case of oversight? Paper presented to the British Universities Industrial Relations Association Conference, Worcester College, Oxford, July.

Lucas, R.E. and Bailey, G. (1993) Youth pay in catering and retailing. *Personnel Review*, **22** (7), 15–29.

Mackie, K. J. (1992) The legal background: an overview. In B. Towers (ed.), *A Handbook of Industrial Relations*, 3rd edition. London: Kogan Page.

Pitt, G. (1992) *Employment Law*. London: Sweet and Maxwell.

Pitt, G. (1994) *Cases and Materials in Employment Law*. London: Pitman.

Price, L. (1993) The limitations of the law in influencing employment practices in UK hotels and restaurants. *Employee Relations*, **15** (2), 16–24.

Price, L. (1994) Poor personnel practice in the hotel and catering industry: does it matter? *Human Resource Management Journal*, **4** (4), 44–62.

Radiven, N.A. and Lucas, R.E. (1994) Wages councils – did they matter and will they be missed? Paper presented to the Third CHME Research Conference, Napier University, April.

Rideout, R. (1989) *Rideout's Principles of Labour Law*, 5th edition. London: Sweet and Maxwell.

Simpson, D.H. (1991) Redundancy rights – rights for whom? *Employee Relations*, **13** (1), 4–11.

Smith, I.T. (1985) Employment laws and the small firm. *Industrial Law Journal*, **14** (1), 18–32.

Smith, I.T., Wood, Sir J.C. and Thomas, G. (1993) *Industrial Law*, 5th edition. London: Butterworths.

Steiner, J. (1994) *Textbook on EC Law*. London: Blackstone Press.

Thatcher, M. (1994) Has the Government finally been defeated over TUPE? *Personnel Management*, July, 13.

Torrington, D. and Hall, L. (1991) *Personnel Management: a New Approach*, 2nd edition. London: Prentice Hall.

Towers, B. (1992) Two speed ahead: social Europe and the UK after Maastricht. *Industrial Relations Journal*, **23** (2), 83–9.

Wedderburn, Lord (1989) *The Social Charter, European Company and Employment Rights*. London: The Institute of Employment Rights.

Wilkinson, F. (1992) *Why Britain Needs a Minimum Wage*. London: Institute for Public Policy Research.

TEN

Managing Equality and Workplace Diversity in the Employment Relationship

The purposes of this chapter are to:

- outline the issue of gender in the employment relationship;
- consider issues in equalizing the employment relationship by legislative means in regard to unlawful discrimination and equal pay;
- review recent developments in statutory pregnancy and maternity rights;
- discuss how managers can manage equality and workplace diversity more effectively;
- assess the state of equality management in hotels and catering.

10.1 INTRODUCTION

The greatest inequality divide in employment exists between men and women (Beck and Steel, 1989), and this is true to the extent that the size of the disadvantaged group – women – is considerably greater than that of any of the others, which includes ethnic minorities, disabled people and older people. Women's predominance in numerical terms is particularly marked in hotels and catering (see Tables 2.1 and 2.2) but, as will be argued later, this dimension has been largely overlooked in terms of issues in the employment relationship and in terms of what contemporary industrial/ employee relations means as the twenty-first century draws near (Lucas, 1994).

Studies of men and women have tended to examine the areas of difference and disparity between the sexes in areas such as pay and forms of employment (for example, Robinson and Wallace, 1983, 1984) and occupations (for example, Lashley, 1984). The public prescription for women's relative disadvantage has been to legislate, such that discrimination between the sexes is unlawful, and women are entitled to equal pay with men in particular circumstances. What is still lacking, with the consequence that there may be a gap in the understanding of contemporary employee relations in 1990s Britain, is the role that forces such as women working part-time play in industrial relations terms (Lucas, 1994, pp. 11–12). In other words, more studies of how gendered relations in the workplace affect issues such as power, conflict, cooperation, trust, authority, solidarity and defiance are needed. Legal and institutional reforms have had limited success: men still retain power and make the rules to their advantage.

Before we discuss the equality legislation and how managers can manage equality and diversity in the workplace, it is important to try to understand some of the fundamental issues that underlie the relative disadvantage of particular sectors in the labour market. The examples given relate largely to women because they are ostensibly the most significant disadvantaged section of the labour force in hotels and catering in terms of the employment relationship. Nevertheless, the principles of disadvantage extend to other significant, but less important, groups in terms of their relative size, such as ethnic minorities and the disabled. This is not meant to imply that such groups are any the less worthy of more equal treatment. Indeed, ethnic minority women may be the most disadvantaged sector of the female labour force, but less is known about the situations of these other groups. Their circumstances, though mentioned, are more difficult to report in any detail.

Women and ethnic minorities are the main disadvantaged groups discussed in this chapter (workforce characteristics from the WIRS3 are contained in Appendix 1), although disabled people, older workers, ex-offenders and issues relating to sexual orientation are also mentioned. Issues connected to these groups that have a potential health, safety and welfare implication, such as HIV/AIDS, are discussed in Chapter 11.

Employers also have legal obligations to customers, but many of these are not related to the employment relationship (see Peters, 1992, pp. 75–108). One notable exception is the issue of discrimination on grounds of race or sex, which is mentioned in this chapter.

10.2 GENDERED RELATIONS IN THE WORKPLACE

Although often used synonymously, sex and gender have distinct meanings that need to be appreciated if some of the complexities and subtleties of women's disadvantage in employment are to be fully understood. Sex is a biological fact that differentiates men and women according to their chromosomes, whereas 'Gender ... refers to a classification that societies construct to exaggerate the differences between females and males and to maintain sex inequality' (Reskin and Padavic, 1994, p. 3).

Sex and gender differentiation ensure that females differ from males in easy to spot ways, such as by the clothes they wear or by their hairstyles. Dressing baby girls in pink and baby boys in blue is a relatively recent social construct in the United States (1950s), and an exact reverse of the practice of gender differentiation adopted in the earlier part of the century, whereby girls wore blue and boys wore pink. Prior to that, all babies wore white dresses. To differentiate between the sexes is to justify unequal treatment between them.

The main point is that gendering takes place to maintain male advantage, such that much gendered work assigns tasks on the basis of workers' sex, the higher value placed on men's work and the employers' and workers' construction of gender on the job. One example of the sexual division of labour is that men cook and women serve.

As Whyte (1948) noted, the use of written orders in the restaurant removed the need for women (waiting) to tell men (cooks) what to do, thereby undermining their power base.

'The highest degree of sex segregation in hotel and catering occupations are found in four of the relatively routine and "unskilled" occupational groups of counter/kitchen hands, domestic staff, porters/stewards and caretakers/cleaners' (Bagguley 1987, p. 29). Although over 90 per cent of jobs in each category were held by one sex, interestingly males dominated in the porters/stewards category. However, as this only accounted for 1 per cent of all occupations, it is not significant when compared to the predominance of women in the other three occupational groupings, which comprised 27 per cent of all occupations. Bagguley's (1987, p. 3) point is that the restructuring of catering employment towards more flexible working can only be understood or explained in the context of gender relations in the workforce.

It is argued that such workforce construction policies are made by men to reduce competition for their more highly graded jobs, and are also connected to a seemingly invisible law forbidding women to supervise men (see, for example, Cockburn, 1983; Buswell and Jenkins, 1994). Reskin and Padavic (1994, pp. 97–9) suggest that organizations that wish to pursue the promotion of equality more seriously could reconstruct jobs to broaden access for women to be promoted, which would have the effect of shifting the balance of power by increasing women's power.

The devaluation of women's work is deeply embedded in the major cultures and religions of the world, and has existed for so long that its origins cannot be explained (Reskin and Padavic, 1994, p. 9). When employers create jobs, they often have a particular sex in mind. If the job is male, there is an assumption that the job holder will accept shift work and overtime. If the job is female, it is likely to be part-time, with associated remuneration that discourages long-term employment (for further details of employers' actions, see Reskin and Padavic, 1994, pp. 127–34; see also Buswell and Jenkins, 1994; Kitching, 1994).

Perhaps one of the most highly visible examples of a gendered job was that of a cocktail waitress in the now defunct Playboy Clubs. To accent sex and sexuality, a waitress was required to dress in a skimpy bunny costume and to adopt a particular pattern of behaviour towards customers. In jobs that are gendered in order to control workers – emotion work – employees must manage their own feelings in order to create certain feelings in customers (Hochschild, 1983). Making people feel good is a task that has stereotypically been assigned to women.

Workers also construct work on a gendered basis in order to control one another, to exclude workers of the 'wrong' sex and to get back at employers, although the actions used by males and females may differ (for more details, see Reskin and Padavic, 1994, pp. 134–41). Men may adopt 'macho' behaviour, such as flouting safety regulations to signify the importance of muscle or bravado and to imply that women have no place. Women may use an outbreak of 'hysteria' or tears to resist an employer.

Sexual harassment (that is demanding a sexual favour from a worker as a condition for promotion or a pay increase) and the perpetuation of a hostile working environment where the individual finds it difficult to work are now becoming increasingly accepted as constituting unlawful sex discrimination. In Britain, the Equal Opportunities Commission (EOC) has reported a 58 per cent increase in complaints of sexual harassment (EOC, 1994). The problem is greater in sectors where women are under-represented in the workforce, such as in the police, although of course women can sexually harass men. Thus one method of exerting control over workers is becoming increasingly unacceptable. The exercise of predominantly male power over females in this way is an illegitimate use of power, and this represents a small step forward in reducing men's power over women in the workplace.

An aspect of particular interest is how workers use gender to create solidarity among themselves, although this solidarity may be more defensive than attacking and not necessarily be used to get back at the employer. Reskin and Padavic (1994, p. 11) cite the example of how an almost exclusively male crew tried to use its only female member to confirm the sexuality of an unmarried man whose sexual orientation was unclear by locking them in a room. She was not aware of the plot; nor was there any guarantee of her cooperation. Gender was invoked both to try to get the unmarried male to display stereotypically aggressive masculine behaviour and to cast the female in the role of sex object.

One interesting difference between male and female group behaviour is how women tend to integrate their private lives at work through parties and celebrating birthdays and weddings. By developing this sense of community, are women using it to facilitate acceptance of poor working conditions or to resist them? If such behaviour is a kind of escape route, then it is likely to detract from fighting for justice at work. The notions of accommodation and resistance merit further development in terms of women's propensity to join, or not to join, trade unions. In forging bonds among themselves, men tend to rely on collective solidarity and strength, characteristics that are more closely associated with traditional trade unionism. For a more detailed discussion about 'naming men as men' as part of a critical analysis in gendered power relations in organizations, see Collinson and Hearn (1994).

Summary proposition

- Gendered relations in the workplace is an overlooked topic that requires more systematic analysis and evaluation in terms of contemporary industrial/employee relations in Britain.

Discussion points

- Identify and explain examples of gendered employment in hotels and catering.
- What are the benefits and drawbacks of gendered employment for the organization, managers and workforce?

10.3 DISCRIMINATION

In employment, where certain groups are perceived to be disadvantaged in some way, society has deemed that they should be afforded special legislative protection in order that their relative disadvantage may be reduced and they do not suffer personal detriment. Refusal to employ a particular individual, for whatever reason, is not unlawful under common law. As yet, statutory anti-discrimination protection is restricted to circumstances related to sex and marital status (including pregnancy) or race, colour, nationality, ethnic or national origins. However, many employers operate voluntary non-discrimination policies against other identifiable groups, such as disabled people or on the grounds of age (see Appendix 5, Section A5.1).

The Equal Pay Act 1970 (phased in over a five-year period) and the Sex Discrimination Act 1975 came into effect together at the end of 1975. The Equal Pay Act affects contractual matters while the Sex Discrimination Act affects non-contractual matters, and, although their provisions are largely mutually exclusive, in practice they may be viewed as 'a harmonious whole'. There is no provision for indirect claims under the Equal Pay Act (see sex discrimination below). A year later the Race Relations Act 1976 reached the statute books.

European Community (EC) law has had a profound influence on the development of equal treatment, deriving from Article 119 of the Treaty of Rome, from which the 76/207/EEC equal treatment Directive and the 75/117/EEC equal pay Directive came into being, the implications of which will be developed below. There is still considerable uncertainty about the precise implications of many decisions in relation to equal treatment, also mentioned below. In the first part of this section, discrimination is discussed first, then equal pay (for more background details to equality issues, see Rideout, 1989, pp. 227–52; Bourn, 1992, pp. 5–16; Cox, 1993, pp. 41–63; Smith *et al.*, 1993, pp. 181–9; Earnshaw, 1994).

Sex discrimination

Although women are deemed to be disproportionately disadvantaged in the labour market compared to men, the Sex Discrimination Act 1975 (amended 1986) applies equally to men. Its scope extends to individuals seeking work, as well as those in employment, including some self-employed workers who are not engaged under a contract of employment. Customers come within the scope of the law because it is unlawful for an employer to discriminate on sex grounds in the provision of goods, facilities and services as part of a business, such as service in a restaurant or the letting of hotel rooms. A summary of the Act's provisions is contained in Appendix 5, Section A5.1. Discrimination in relation to death or retirement was excluded but has been subject to considerable litigation, mentioned below. Although there is no express prohibition of sexual harassment, this may amount to unlawful sex discrimination where an employer subjects the woman to any 'detriment'. The EOC has overall responsibility for sex discrimination matters, including equal pay (see Appendix 5, Section A5.1). More details about the EOC's activities can be gleaned from its Annual Report (EOC, 1994).

Direct and indirect

Direct discrimination is treating a woman less favourably than a man or vice versa, or a married person less favourably than a single person. This could arise if an employer refused to engage, promote, transfer or train a woman (or man) because she (he) was married. Refusal to serve a woman at a bar, as in *Gill and Coote* v *El Vino Company Ltd* [1983] IRLR 206 CA, and barring a man from a public house for wearing earrings, as in *McConomy* v *Croft Inns Ltd* [1992] IRLR 561 NIHC, are cases of direct sex discrimination.

Indirect discrimination is more subtle and difficult to identify. It involves applying a requirement or condition to both men and women which, in practice, has the effect that a significantly smaller proportion of men, women or married persons are able to comply it with as compared with persons of the opposite sex or single persons, and which is to his or her detriment and cannot be justified. Examples of unlawful indirect discrimination could be the requirement for a job candidate not to have young children and imposing an age requirement of 20 to 30 for a post. Both requirements are likely have a more detrimental effect on women than men. As a rule positive discrimination is unlawful (for further details on discrimination, see Bourn, 1992, pp. 32–7; Smith *et al.*, 1993, pp. 189–203; Earnshaw, 1994, pp. 114–17).

Remedies

Proving discrimination in practice is extremely difficult; the burden of proof is on the applicant. Although there has been no statutory reversal of the burden of proof (possible under the Social Charter, but looking increasingly unlikely), the Employment Appeal Tribunal (EAT) has 'evolved a potentially useful approach which is that if the applicant shows less favourable treatment in circumstances consistent with discrimination, the tribunal may properly infer unlawful discrimination unless the respondent employer can satisfy it that there is an innocent explanation ' (Smith *et al.*, 1993, p. 203).

Complaints to an industrial tribunal must be made within three months of the date of the act complained of. There are three remedies available to the tribunal: a declaration that rights have been infringed; a recommendation that the employer take action within a specified time to remove the discrimination; and an order for compensation. The upper limit for damages for injury to feelings (originally the same as the compensatory award for unfair dismissal, £11,000) has been subsequently removed following *Marshall* v *Southampton and South-West Hampshire Health Authority (Teaching) (No. 2)* 271/91 [1993] IRLR 455 ECJ by the Sex Discrimination and Equal Pay (Remedial) Regulations 1993 (see Appendix 4, Section A4.6).

The number of sex discrimination complaints has increased recently from 1,078 cases (1990–1) to 1,386 cases (1992–3). The bulk of cases were settled by ACAS conciliation or withdrawn. Of 379 cases that went to tribunal in 1992–3, only 127 cases were successful (9 per cent of all complaints made). The median award increased from £1,142 to £1,416 over the same period (Employment Department, 1993). Most sex discrimination cases (85 per cent) are brought by women, who are most likely to work in small private firms (median size 30 employees) in distribution, hotels and catering or repairs (Employment Department, 1994).

This pattern of litigation and its effects will inevitably be altered as a result of the *Marshall* case, particularly from the recent rash of retrospective claims made by ex-service personnel. In *Ministry of Defence* v *Cannock and others* [1994] IRLR 509 EAT, the EAT laid down guidelines for assessing sex discrimination compensation in the armed services, and said that awards of tens or hundreds of thousands of pounds were disproportionate to the wrong done. This is likely to set a precedent for future non-armed forces claims.

Part-timers: a major breakthrough

In a milestone case, the House of Lords overturned a Court of Appeal decision that only the European Commission could bring proceedings to challenge (by judicial review) the legality under Community law of the exclusion of part-time workers from certain statutory employment rights under the EPCA 1978 (*R* v *Secretary of State for Employment, ex parte EOC and another* [1994] IRLR 176 HL reversing [1993] IRLR 10 CA). It was decided that since the EOC was established with a duty to work towards eliminating discrimination, it had the *locus standi* to seek judicial review of any discriminatory legislation, and that such a review was available whenever UK law did not comply with EC legislation. The requirement to work for two years at 16 hours a week or five years at eight or more hours a week discriminated against a part-time workforce that was predominantly female.

For a while it was uncertain if there would be a cut-off point at eight hours; early tribunal cases tended to ignore the eight-hour minimum. EPPTER, effective from 6 February 1995, brought all part-time employees, regardless of hours worked, into the scope of all employment protection legislation. Qualifying length-of-service thresholds continue to apply in appropriate cases (Employment Department, 1995). Deakin (1994) believes that the enforced removal of the 16-hour threshold should be taken as an opportunity for a more wide-ranging review of employment law and social security provisions in relation to part-time work. This appears unlikely unless there is a change of government.

Race discrimination

The Race Relations Act 1976 is almost identical to the Sex Discrimination Act 1975 in all respects but, importantly, it lacks the potential effects of EC law. The Commission for Racial Equality (CRE) has similar powers to the EOC (see Appendix 5, Section A5.1), and further details of its activities can be found in its Annual Report (CRE, 1994).

In *Showboat Entertainment Centre* v *Owens* [1984] IRLR 7 EAT, a white manager (Mr Owens) was dismissed for disobeying an order not to admit black people to the centre. The EAT found that the employer had unlawfully discriminated against Mr Owens on grounds of race, on the basis of someone else's race; Mr Owens had suffered less favourable treatment on racial grounds. This outcome was only possible because of a difference in wording between the Race Relations Act and the Sex Discrimination Act. Under Sex Discrimination law the claimant has to have suffered discrimination himself or herself. A licensee who put up a notice at his public house saying 'Sorry, no travellers' in response to the setting up of a gypsy camp nearby was found to have discriminated indirectly against gypsies as a racial group in *Commission for Racial Equality* v *Dutton* [1989] IRLR 8 CA. In *Dhatt* v *McDonald's Hamburgers* [1991] IRLR 130 CA, there was no discrimination in a refusal to employ a non-national without a work permit.

The number of complaints for race discrimination is slightly smaller than for sex discrimination, and these have remained relatively static (926 in 1991–2; 1,070 in 1992–3). The chance of the applicant being successful at industrial tribunal (as a proportion of all complaints) is only 6 per cent, although the median award has increased from £1,749 (1990–1) to £3,333 (1992–3) (Employment Department, 1993). As for sex discrimination, the upper limit for damages for injury to feelings of £11,000 has been removed by the Race Relations (Remedies) Act 1994.

The pattern of race discrimination cases differs from that of those concerned with sex discrimination. Race cases are most likely to emanate from firms in other services (median size 75 employees) which recognize trade unions (Employment Department, 1994).

Legal loopholes

Employers may discriminate against other identifiable groups in the labour market (see Appendix 5, Section A5.1), such as homosexuals and young people, although in many cases this is not unlawful. The current government opposes legislation in principle, preferring a voluntary approach, and therefore progress to legislation can only be made via a private member's bill.

Many employers and other interested parties concur that the extant legislative arrangements relating to the disabled are outdated and need to be changed (Smith *et al.*, 1991; Barnes, 1992; Doyle, 1994). The most recent attempt (1994) to legislate against disability (the Civil Rights (Disabled Persons) Bill) was thwarted by the government, which estimated that forcing companies to provide facilities for employees and customers who are disabled would cost £18 billion. However, the government has since announced a proposal to introduce a Disability Rights Bill in the 1994–5 parliamentary session that would include provisions to improve disabled people's employment (see Appendix 5, Section A5.1). Efforts to secure legislation that would make ageism unlawful failed in 1989 and 1991 (see Lucas, 1993, 1995) although there are plans for another private member's bill in the 1994–5 parliamentary session.

Both disability and ageism look set to remain on the public policy agenda as potential areas for anti-discrimination and equality legislation, although neither area is likely to gain much impetus from developments in European social policy. Ironically, disabled workers look set to lose out on jobs through the decision of the UK government to comply strictly with an EC Directive on competition. This would mean that disabled workers would no longer get 'a second chance to bid for government contracts at a lower price if their bid is substantially higher than commercial sector private bids' (Wintour, 1994).

Earnshaw (1990) has suggested that discrimination against rehabilitated criminal offenders is still rife in spite of the existence of the Rehabilitation of Offenders Act 1974, because the Act has no enforcement mechanism (see Appendix 5, Section A5.1). If an employee is dismissed because of a spent conviction, which he or she had no duty to disclose, there is no apparent remedy, unless the employee has two years' service. In *Hendry and another* v *Scottish Liberal Club* [1977] IRLR 5 IT, the dismissal of the club manager, who had a spent conviction for possession of cannabis, was held to be unfair. There is arguably a case for adding dismissal for a spent conviction to the list of inadmissible reasons for dismissal or, as Earnshaw suggests, amending the Rehabilitation of Offenders Act to match the Sex Discrimination Act and Race Relations Act more closely.

These non-protected groups are also considered later in relation to the voluntary policies that employers may adopt in order to pursue equal opportunities in the workplace, and in Chapter 11 where there is a health, safety and welfare implication.

Equal pay

The Equal Pay Act 1970 (as amended by the Sex Discrimination Act 1975) was aimed at removing and preventing discrimination between men and women as regards terms and conditions of employment, but only where a contractual relationship exists. There was a closing of the earnings gap early on: women's pay rose from 63 per cent of men's pay in 1970 to 75.5 per cent in 1977, but then stuck at around 75 per cent

(Smith *et al.*, 1993, p. 211). For full-time workers in all industries and services, women's hourly pay was 72 per cent of men's hourly pay in 1993 (see Table 5.6). In two former wages council sectors in hotels and catering the gap between men and women was smaller; women's pay was around 80 per cent of men's pay in 1993 (see also Table 5.6).

While some explanation for the pay gap is related to the segregation of women and men occupationally (see Chapter 3), whether the difference in pay is *justifiable* is another matter. Even within the same broad occupational groups, such as chefs/cooks, waiting staff and bar staff, men earn significantly more than women (see Table 3.4). Full-time female waiting staff earned 81 per cent of their male counterparts' hourly pay, while female chefs and cooks fared rather better, earning 87 per cent of their male counterparts' hourly pay. These figures do not reflect other important differences in the number of hours worked, such that men's total earnings are higher (see Table 3.6) or how men and women may be graded differently within the same occupation. Most of all, these differences do not reflect the less favourable position of female part-timers (44 per cent of all jobs), who as a general rule earn less per hour than their full-time female counterparts (see Tables 3.4 and 5.6).

As a general rule, the pay gap is smallest among young workers because workers newest to the labour force are least segregated on-the-job. In occupations where women's earnings are closest to men's, both sexes' pay tends to be low (Reskin and Padavic, 1994, pp. 107–9). A smaller pay gap may also be caused by men's pay falling, rather than because women's pay has risen. Given the relatively high proportion of young workers in the industry and the incidence of low pay, these factors may help to explain the smaller gap between men's and women's pay in some catering employment compared to the economy as a whole.

However, EC law has had a profound effect on equal pay at a more general level. Under Article 119 of the Treaty of Rome, each member state shall 'ensure and subsequently maintain the principle that men and women shall receive equal pay for equal work'. 'In a series of landmark decisions, the ECJ has established that Article 119 (and its supporting directives ...) is wider than domestic law in several important respects, not least of which is that the definition of "pay" under Community law is significantly wider than under domestic law' (Smith *et al.*, 1993, p. 184). First, the European Commission can bring enforcement proceedings in the ECJ if domestic law is found wanting. According to Smith *et al.* (1993) the most important instance is the decision of *Commission of the European Communities* v *United Kingdom of Great Britain and Northern Ireland* 61/81 [1982] IRLR 333 ECJ. As a result the Equal Pay Act 1970 had to be amended (Equal Pay (Amendment) Regulations 1983) to allow women to claim equal pay for work of equal value, even where no job evaluation scheme was in force. Second, Article 119 is directly applicable following *Defrenne* v *Sabena (No. 2)* Case 43/75 [1976] ECR 455 ECJ. This means that claims for equal pay can be made under the Equal Pay Act 1970 and under Article 119 (for more details, see Smith *et al.*, pp. 184–9, 211–30; and also *Mediguard Services Ltd* v *Thame* [1994] IRLR 504 EAT, mentioned below).

Equality clause

Every woman's contract is deemed to include an equality clause, but the difficulties of determining its meaning are usefully illustrated from *Hayward* v *Cammell Laird Shipbuilders Ltd* [1988] IRLR 257 HL, [1987] IRLR 186 CA, [1986] 287 EAT and [1984] 463 IT. This case was one of the first under the equal value provisions to be tested in the courts, and when she had won her case, the courts had to decide what Miss Hayward was entitled to. She (a cook) claimed the same basic pay as her male comparators (thermal insulators, painters and engineers), but her employers objected, claiming she received a better overall 'package', including sick pay. The decision of the Court of Appeal to uphold the 'package' approach was overturned by the House of Lords, which allowed equality of basic pay.

In approaching equal pay claims for 'like work', the courts have taken a wide view. In *Capper Pass Ltd* v *Lawton* [1976] IRLR 366 EAT, a female cook who prepared 10 to 20 lunches for directors was held to be on like work with two male assistant chefs who helped provide many more meals over a longer period of the day in the works canteen, particularly as similar levels of skill and knowledge were required. For claims to be made under the 'work rated as equivalent' provision, jobs must have been given equal value (in terms of demands made relating to skill, effort and decision-making) under a job evaluation scheme. As noted in Chapter 5 (see Table 5.2), job evaluation schemes are in the minority in hotels and catering and, therefore, the opportunity for workers to make claims under this provision are inevitably limited.

Smith *et al.* (1993) state that the equal value provisions are subject to an 'exceptionally complicated, not to say tortuous, procedure' (p. 217). The average time to resolve claims is 17 months, and problems come from tribunals being brought into the realms of assessing 'value'. While the procedure remains judicial, tribunals are assisted by an 'expert' on-job evaluator appointed by ACAS from a nominated panel (see Smith *et al.*, 1993, pp. 216–21).

Genuine material differences and factors

'Even if there is a prima facie case for equality, the applicant may still fail if the employer can show, on a balance of probabilities, that the variation in terms or conditions is genuinely due to a material factor other than sex' (Smith *et al.*, 1993, p. 221). Factors that could justify a difference in pay could include longer service and higher output (see Smith *et al.*, 1993, pp. 221–7).

Enderby v *Frenchay Health Authority and Secretary of State for Health* 127/92 [1993] IRLR 591 ECJ, [1992] 15 CA and [1991] IRLR 44 EAT concerned an equal value claim across separate bargaining groups with the same employer. The judgement stated that in circumstances where predominantly female employees doing job A earn less than predominantly male employees doing job B which is of equal value, it is for the employer to justify the difference objectively; employees do not need to show any deliberate discrimination. The whole difference must be justified and it is not sufficient justification that there are bargaining structures for different groups including job A and job B.

A major weakness of the legislation also remains the fact that claims are only valid between different sexes, thus preventing a part-time female room maid claiming equal pay with a full-time female room maid.

Remedies

An industrial tribunal can make a declaration of an equality clause, and make an award of arrears of pay or damages subject to a maximum period of two years before the date on which proceedings were brought. Smith *et al.* (1993, p. 227) suggest that this upper limit is now in doubt following *Marshall* v *Southampton and South-West*

Hampshire Area Health Authority (Teaching) (No. 2) 271/91 [1993] 445 ECJ, discussed above.

Equal pay claims have declined substantially from a figure of 508 (1990–1) to 240 (1992–3). Only 9 per cent of all claims succeeded at industrial tribunal in 1992–3 (Employment Department, 1993).

'Pay' under Article 119

The complex issue of pay is inextricably linked to issues related to death, retirement, pensions and sex discrimination, as noted in the following citation. 'While the differential state of pensionable age is outside the scope of Article 119, retirement benefits (indeed retirement ages) which are based on the state pensionable age are not, which means that the exclusion in the domestic legislation for provisions relating to death or retirement does not comply with EC law' (Smith *et al.*, 1993, p. 228).

Smith *et al.* (1993, pp. 228–30) suggest that this complex 'pay' picture centres on three main points, which are summarized and updated by recent decisions. First, sex-based differences in retirement benefits are unlawful. This was established in *Garland* v *British Rail Engineering Ltd* [1982] IRLR 111 ECJ and [1982] IRLR 257 HL, whereby differences in travel concessions to retired males and females were held to be unlawful. Second, following *Marshall* v *Southampton and South-West Hampshire Area Health Authority (Teaching)* 152/84 [1986] IRLR 140 ECJ and [1983] IRLR 237 EAT, the Sex Discrimination Act 1975 and the Equal Pay Act 1970 were amended to extend unlawful discrimination against a woman in relation to retirement, such that it became unlawful compulsorily to retire a woman at an earlier age than a man in the same position. This applies to both the public and the private sectors.

The third point arises because the above cases did not appear to affect different pension ages: 65 for men and 60 for women. Indeed, EC law allowed member states to fix their own pension ages – a clear example of subsidiarity. Therefore, the *Marshall* case appeared to create an anomaly to the extent of requiring equal retirement ages for men and women, but still allowing for pensions to be payable at different ages. This area became even more confused by *Barber* v *Guardian Royal Exchange Assurance Group* 262/88 [1990] IRLR 240 ECJ. Mr Barber was made redundant at age 52, and while a woman of the same age would have got a pension at age 50, he had to wait until he was 55. The ECJ ruled that benefits paid out under a contracted-out private occupational scheme fell within pay under Article 119. The effect of this decision was to require the equalization of pensionable ages under private occupational pension schemes with effect from 17 May 1990.

The *Barber* case raised a number of complex issues, principally how far the decision was to be retrospective. Most points have since been resolved but the practical implications of the judgements remain to be seen. First, in *Ten Oever* v *Stichting Bedrijfspensioenfonds voor het Glasenwassers en Schoonmaakbedrijf* 109/91 [1993] IRLR 601 ECJ, equality in pension rights applies only to so much of a pension as relates to or is attributable to service from 17 May 1990 unless a claim was made before that date. From *Moroni* v *Firma Collo GmbH*, C-110/91 [1994] IRLR 130 ECJ and *Neath* v *Hugh Steeper Ltd*, C-152/91 [1994] IRLR 91 ECJ, it has been established that Article 119 applies equally to all private occupational pension schemes (contributory and non-contributory). The same principles (including the limit on retrospectivity) apply to survivors' benefits, and survivors can bring benefits claims under Article 119 (*Ten Oever*). Bridging pensions payable to men between age 60 and 65 are not pay, and a woman of the same age who does not receive such payments, because she does not qualify for a state pension because she opted for the lower rate married woman's National Insurance contribution, has no claim under Article 119 (*Birds Eye Walls Ltd* v *Roberts*, C-132/91 [1994] IRLR 29 ECJ). It is probable that sex-based actuarial calculations of reductions in pension for early payment are also lawful (the *Neath* case).

The judgements of the ECJ in six pensions cases given on 28 September 1994 have further clarified outstanding questions raised by the *Barber* case. These include whether equal treatment can be achieved only by rounding up benefits, whether a pension fund can be sued directly and whether statutory public sector pension schemes fall within Article 119. The cases are *Bestuur van het Algemeen Burgerlijk Pensioenfonds* v *Beune*, C-7/93 [1994] ECJ, *Coloroll Pension Trustees Ltd* v *Russell and others*, C-200/91 [1994] ECJ, *Fisscher* v *Voorhuis Hengelo BV and another*, C-128/93 [1994] ECJ, *Smith and others* v *Avdel Systems Ltd*, C-408/92 [1994] ECJ, *Van 0den Akker and others* v *Stichting Shell Pensioenfonds*, C-28/93 [1994] ECJ, and *Vroege* v *NCIV Instituut voor Volkshuisvesting BV and another*, C-57/93 [1994] ECJ. The issues are complex but the two most widely reported aspects relate to the consequences of equalizing pension ages and the rights of part-time workers *vis-à-vis* pension schemes (for more details, see Incomes Data Services, 1994).

The consequences of equalizing pension schemes are now placed on three dimensions. First, there is no justification to raise women's pension age prior to 17 May 1990. Second, between that date and the date when pension ages are equalized, both sexes are entitled to the most beneficial treatment afforded. This means that men are entitled to have their pensions calculated on the basis that they have the lower retirement age for women. Third, there is nothing to stop the raising of women's pension age for future service. Women would still be able to retire at 60 but this would constitute early retirement on a reduced pension. Thus employers look set to be faced with the practical job of coping with three different regimes for each employee.

As regards part-timers' pension rights, where the exclusion of part-timers from schemes has disproportionately affected women, and this could not be objectively justified, then such an exclusion is unlawful. The question then becomes: how far back in time can those unreasonably excluded claim? Claims cannot go back beyond the date on which the ECJ first declared that Article 119 had direct effect (8 April 1976), but national time limits may be applied in certain circumstances. At a more practical level, in contributory schemes employees must pay the appropriate retrospective contributions in order to benefit from this ruling. In non-contributory schemes it is expected that there will be litigation over the issue of time limits because of the potentially high costs to employers.

Further relevant points are that public sector pensions do fall within Article 119 and trustees and pension administrators are bound by the principles of equality contained in Article 119. Legislation to equalize the state pension ages of men and women at 65 by the year 2020, to increase equality in pension schemes and tighten up their management is planned in the 1994–5 parliamentary session.

Implementation of Schedule 5 of the Social Security Act 1989 (implementing EC Directive 86/378), postponed pending the outcome of the *Coloroll* case, was implemented in June 1994. As a result requirements of equal treatment in occupational pensions, including detailed rules as to pension accrual and contributions during maternity leave and the prohibition of indirect discrimination, are now in force.

Additionally the EAT has affirmed an industrial tribunal decision that compensation for unfair dismissal falls within the definition of 'pay' under Article 119 in

Mediguard Services Ltd v *Thame* [1994] IRLR 504 EAT. Thus a worker working fewer than 16 hours a week with less than five years' service was entitled to claim compensation by relying directly on Article 119.

Summary propositions

- Anti-discrimination and equal pay legislation, as a foundation for achieving greater equality, has had some limited success.
- Eradication of deeper seated differences as regards the unequal treatment of men and women in employment cannot be achieved by the law alone.
- Although EC law is having a marked effect on equality issues in relation to the broader aspects of 'pay', including pensions, this looks unlikely to apply to basic pay.

Discussion points

- Present a case for or against extending anti-discrimination legislation to other areas (specify which areas).
- How can anti-discrimination legislation be made more effective?
- Is there a difference between eliminating discrimination and promoting equality of opportunity? Defend your argument.
- Is the achievement of equal pay an achievable goal? What would be the effects of achieving this goal on the organization, management and workforce?
- As a manager, what steps would you take to implement an equal pay policy in your organization?

10.4 PREGNANCY AND MATERNITY

Although statutory provisions relating to pregnancy and maternity have been in force in the UK since the mid-1970s, cases decided in the ECJ and recent legislative changes effected by TURERA 1993 designed to implement the protection of women at work Directive (92/85) have made important changes to this whole area. In broad terms, more women are now protected from dismissal on grounds of pregnancy and childbirth and are eligible to take a period of paid maternity leave. Pregnant women also have other rights; for example to take time off for antenatal care (see Appendix 5, Section A5.3).

As yet there are no provisions on statutory paternity leave. In September 1994 the UK government vetoed the proposed Directive to give either parent three months off work without pay after birth or adoption (this would not have affected mothers' rights to leave and absence, discussed below). A new parental leave Directive is to be proposed under the Maastricht protocol, which would apply in the other member states but not in the UK.

Dismissal

The significant change brought about by TURERA 1993 (Section 24) is the removal of the two years' service qualification with regard to dismissals for pregnancy and maternity, the grounds for which have been made more explicit (see Appendix 5, Section A5.3). This is not to say that women with less than two years' service had no legal protection, because they could make claims under the Sex Discrimination Act. However, the main difficulty was finding a comparable male, a point that was resolved, to an extent and not entirely satisfactorily, in *Hayes* v *Malleable Working Men's Club and Institute* [1985] IRLR 367 EAT. The comparable male was deemed to be one suffering from a temporary disability, which had the effect of treating pregnancy as an illness.

This notion of a comparable male was rejected in two important ECJ cases in 1991. In *Dekker* v *Stichting Vormingscentrum Voor Jong Volwassen (VJV-Centrum) Plus* 179/88 [1991] IRLR 27 ECJ, refusal to engage for a reason connected with pregnancy (specifically the financial burden on the employer of maternity leave) constituted direct sex discrimination in contravention of the equal treatment Directive 76/207. In *Hertz* v *Aldi Marked K/S (sub nom Handels-og Kontorfunktionaerernes Forbund i Danmark* v *Dansk Arbejdsgiverforening)* 199/88 [1991] IRLR 31 ECJ, dismissal for pregnancy was also direct discrimination but protection for pregnancy or maternity ends on return from maternity leave. The effect of these cases was to throw into confusion the status of UK law *vis-à-vis* EC law.

Some of this confusion has subsequently been resolved as a result of *Webb* v *EMO Air Cargo (UK) Ltd*, C-32/93 [1994] IRLR 482 ECJ, although the full consequences of this ruling are not yet clear. Mrs Webb, recruited on an indefinite contract to cover for an another employee absent on maternity leave, was dismissed when she was found to be pregnant. Although her claim that her dismissal was unlawful under the Sex Discrimination Act did not succeed at industrial tribunal, the EAT or the Court of Appeal, the decision in the House of Lords ([1993] IRLR 27) made a ruling that made further progress possible. The House of Lords ruled that dismissal for pregnancy was automatically unfair but, where the reason for adverse treatment related to the consequences of pregnancy, the comparable male test applied. Mrs Webb's claim fell into the latter category and, therefore, failed.

However, the House of Lords accepted that the matter was unclear under EC law (as per *Dekker* and *Hertz*) and, before giving final judgement, referred to the question of whether unavailability for work rather than pregnancy would be the real reason for dismissal. The result was that the ECJ ruled out the situation of comparing the situation with that of a man. Pregnancy was not comparable with any pathological condition and even less so with unavailability for work on non-medical grounds, both of which might justify the dismissal of a woman without discrimination on grounds of sex. No exceptions or derogations were possible from the prohibition on dismissing a woman from the beginning of her pregnancy to the end of her maternity leave (exceptions could only be unconnected to her condition). Mrs Webb had been unlawfully dismissed on grounds of sex.

Although the *Webb* case undoubtedly represents a landmark ruling, its outcome in terms of UK domestic legislation is as yet unclear. The case must now be referred back to the House of Lords, where it will be decided if the Sex Discrimination Act can be construed consistently with European law as interpreted by the ECJ. If the House of Lords finds it cannot interpret domestic law consistently, then Mrs Webb will get no compensation because as a private sector worker she cannot rely directly on the equal treatment Directive. She will have to take a '*Frankovich*' claim (see Appendix 3) against the government on the grounds that she has suffered losses as a result of the government's failure to implement European law in the UK. However, the comparable male argument is no longer tenable, at least with regard to pregnancy. At a more practical level employers can circumvent some of the problems raised in covering for temporary absence through the use of short-term fixed contracts.

Leave and absence

One of the tensions about the concept of maternity leave is striking a balance between reasonable employee rights and employer inconvenience (Smith *et al.*, 1993, p. 232). The 'great British compromise' has been to put employees through a series of hurdles that have to be jumped in order to meet the statutory requirements, which have, up until now, made for complexities in this area (for leading cases, see Smith *et al.*, 1993, pp. 236–45; for a study of employers' and employees' experiences of maternity rights in Britain, see McRae, 1991). However, an employer has no recourse against an employee who says she will return to work and does not do so. The complex picture has been made even more difficult as a result of the way in which the UK has chosen to implement the Directive on pregnant women. The opportunity to rationalize the whole area of leave and absence has been missed, and employers have been left to cope with two types of leave and levels of maternity pay from October 1994, which are summarized in Appendix 5, Section A5.3.

The terminology used here, and in Appendix 5, reflects that used by Smith *et al.* (1993). The *new maternity leave of 14 weeks*, which is a universal right among all women, is so called because the contract of employment (except remuneration) continues through the period of leave. The *old maternity absence of 40 weeks*, which is subject to two years' service, does not necessarily allow for the contract to continue (except for the first 14 weeks with the exception of remuneration). Thus there is no right to return under maternity leave, rendered redundant by continuity of the contract, but this right remains necessary under maternity absence in the absence of a contract.

A recent pregnancy/maternity case, *Hilton International Hotels (UK) Ltd* v *Kaissi* [1994] IRLR 270 EAT, has raised some interesting issues. A pregnant employee (chambermaid) was off sick with a back injury and received sick pay until after the birth of her child. Although her contract continued, she was required to give three weeks' notice of her intention to return to work at the end of her period of maternity leave. She failed to do this and her employers refused to take her back. She made claims for unfair dismissal and sex discrimination. Although an industrial tribunal found she had been unfairly dismissed, this was not on grounds of sex or pregnancy. Following a cross-appeal to the EAT, the unfair dismissal has been upheld, but the case has been remitted to industrial tribunal to determine if the dismissal (for sickness) was pregnancy related, and whether the matter is automatically unfair or amounts to sex discrimination (see Suter, 1994).

The new provisions clearly mark a substantial expansion in employee rights, and it is going to be very interesting to learn what strategies employers will attempt to follow in order to minimize the impact of these rights, which need to be viewed in the context of some of the leading ECJ judgements on sex discrimination and pay, discussed above. One obvious move would be to employ more women beyond child-bearing age, with the consequence that many more 'mature' workers will no longer find themselves subject to ageism policies (see below).

Summary proposition

- The extension of paid maternity leave to all women represents a substantial change for employers and female employees.

Discussion point

- Analyse and evaluate the implications of the new maternity leave regulations for organizations, managers and workers.

10.5 MANAGING EQUALITY AND WORKPLACE DIVERSITY

Equal opportunities: policy and practice

The preceding part of the chapter has shown that the law, particularly EC law that affects women, important though it may be, is not enough on its own to bring greater equality in the workplace (for further details of European initiatives on equality, see Collins, 1992, pp. 99–127). Demographic factors, such as a gradually ageing workforce and a declining birth rate, and the increasing participation of women in the labour market are more pragmatic reasons why employers may need to revise their employment strategies (see Lucas and Jeffries, 1991; Lyon and Mogendorff, 1991; Lucas, 1993, 1995). Such developments have led to Opportunity 2000, a self-financing campaign launched in 1991, with the support of the Prime Minister, as a result of a campaign to encourage British industry to take full advantage of the economic potential of women (see Collins, 1992, pp. 88–96). Other initiatives exist with regard to particular groups, including those to promote the cause of the disabled (see Collins, 1992, pp. 40–50), to prevent age discrimination (pp. 51–5), and to overcome discrimination on grounds of sexuality (pp. 55–61).

The issue of ethnic minorities is important to employers in some of Britain's large cities, particularly London, because these are where ethnic minorities tend to be clustered. People from ethnic minorities are also disproportionately concentrated in the younger age groups. Thus 25 per cent of new entrants to the labour market will be from ethnic minorities in areas such as central London, where hotels are a major competitor for labour; 'more people from ethnic minorities were likely to be recruited or promoted by employers who took action targeted specifically at the barriers to equality of opportunity in their firms' (Jameson and Hamylton, 1992, p. 21).

Guidance to employers on how they should comply with anti-discrimination legislation and be proactive through the adoption of equal opportunities policies and monitoring systems, are provided by the EOC (1985a, b) and the CRE (1984, 1988). In practice, 'good' employers will extend equal opportunities beyond just sex- and race-related issues to include other areas, such as disability, sexual orientation and age (see, for example, Collins, 1992, pp. 69–70).

Collins (1992, p. 63) has identified seven fundamental steps (identified below in italics) involved in translating government policy into everyday organizational practice (see also Beck and Steel, 1989; Torrington and Hall, 1991, pp. 362–4). The *law* needs to be reflected in a *policy* that is brought to employees' attention through *awareness-raising and training*. Managers and workers need to *incorporate good practice* into their activities, the policy needs to be *monitored* and its effects *evaluated*, and *improvements* need to be made where necessary.

Although issuing a policy does at least focus everyone's attention on a particular issue and make public the organization's intentions, the main problem with policies, as noted in Chapter 2, is that they can be rendered ineffective by the actions of line managers (Brewster and Richbell, 1982). Price (1994) found that although equal opportunities policies were more likely to be found in larger high quality hotels and restaurants, they often did not meet the full requirements or spirit of the legislation.

Collins (1992, pp. 70–7) offers more detailed guidance on successful policy implementation. Policies are useless without good practice and 'teeth'; for example, disciplining managers who discriminate unfairly. A more searching approach to devising more proactive equality policies is discussed in Chapter 11 with regard to HIV/AIDS (Adam-Smith *et al.*, 1992). This approach may offer some lessons for those seeking to construct more effective equal opportunities policies elsewhere.

One of the more controversial aspects of equal opportunities policies relates to monitoring the sex and racial mix of employees and job applicants. This is usually done by asking the employee or candidate to complete a simple form which seeks basic details relating to the sex and racial or ethnic group of the individual. While the identity of the person completing the form remains anonymous, the organization can identify the proportion of successful candidates by sex or race. If, for example, a business with an equal opportunities policy was located in a locality with a high ethnic minority population, yet it employed few ethnic minority staff and was tending to select white candidates for jobs, then the senior manager responsible for that policy might need to ask a few questions. The 1991 Census of Population was the first to include a race question. If data from the Census are linked to an analysis of local labour market data, firms will be in a better position to identify if they are achieving genuine equality of opportunity (Coombes and Hubbock, 1992).

Typical equality measures include targeting 'alternative' labour sources, such as the unemployed, women returners, ethnic minorities, disabled people and older workers, and introducing associated measures and benefits, such as term-time contracts, career breaks, training and induction (for example, Lucas and Jeffries, 1991; Luckock, 1991; Lyon and Mogendorff, 1991; Lucas, 1993, 1995). More tangible progress is likely to be achieved within workplaces if targets are set. For example, Littlewoods made a commitment to ensure that *all* disadvantaged groups would be fairly represented at *all* levels of the company in five years (Beck and Steel, 1989, p. 51). This initiative was different from most because it aimed to be pervasive rather than selective. Proactive employers need to change attitudes and reconstruct jobs. Another major barrier to the progress of many women is the dearth of affordable childcare facilities (Bischoff, 1990; Berridge, 1991). This is a gender-related issue that has moved out of the private domain into the public domain (Bischoff, 1990), and the case for it can be justified by demographic trends and economic, political and social/psychological factors (Berridge, 1991, pp. 10–11).

The problem with a lot of equal opportunities initiatives, and much of the attendant commentary in the literature, is that they tend to reduce the issue of equal opportunities to something that can be resolved relatively simply and smoothly. But they really fail to address the deep-seated societal prejudice that engenders sex, and other, segregation in the workplace, noted at the beginning of the chapter. A similar 'naivety' is associated with customer care programmes (Wynne, 1993).

Some of the more searching literature on equal opportunities policies relates to women and suggests that such policies are positively damaging. On the one hand, 'the notion of rights is removed and replaced by an anodyne equal opportunities policy awaiting implementation by a human resources manager' (Buswell and Jenkins, 1994, p. 8). On the other hand, the policy becomes an instrument of public patriarchy. Policies divide and individualize women in the workplace. By being controlled through control of time (that is, flexible working), women are subordinated in the labour process (Buswell and Jenkins, 1994). Whereas many men dominated women within marriage (private patriarchy), now they dominate women anonymously outside it (public patriarchy) (Hochschild, 1989; see also Collinson and Hearn, 1994, pp. 6–8). A study of women with young children entering the labour market found that the women themselves placed a high premium on training and development, but the gender-based social construction of organizations was likely to make this difficult to achieve (Healy and Kraithman, 1991, p. 27).

Progress in hotels and catering?

The reality of achieving a greater extent of equal opportunities for women in hotels and catering still remains remote, particularly at managerial level, largely because men are more powerful and make the rules and, as Price (1994) has noted, because policies may be defective. McKenna and Larmour's (1984) suggestion that colleges should consider giving women special training and that organizations should recruit more women is too simplistic. Guerrier (1986) noted how men formally excluded women by the way jobs and working hours were constructed, and informally through the use of the 'old boy' network in recruitment, which was circumscribed by prejudice and stereotypical thinking.

Hicks (1990) showed how the shared values and norms of (mostly) male managers were so unanimous that they excluded women from positions of power. Women's part in hotels was perceived in a narrow way, and women were often offered less inspiration and encouragement than men because it was thought that they were only temporary. Additionally, the managerial trait of 'being there', referred to in Chapter 2, is a noticeable way of demonstrating commitment. Presumptions about women's 'other responsibilities' – that is, in relation to the home – would militate against the encouragement of women towards duty management positions. Hicks uncovered common stereotypical views that female managers did not share the same goals as male managers, thus providing male managers with justification for their actions.

The proposition that equal opportunities are more likely to be achieved in areas of the labour market where formal qualifications are a prerequisite to entry and promotion (Crompton and Sanderson, 1989) is not supported by Purcell's (1993) evidence. In a study of hospitality students graduating between 1985 and 1987, it was found that women were less likely to be recruited, developed and rewarded as career staff with senior management potential. She has suggested that in industries where women predominate in lower level jobs, 'it is particularly difficult for professional women to gain credibility as committed management colleagues and career staff' (Purcell, 1993, p. 138).

With regard to race, a similarly bleak picture has been found in the hotel industry. The CRE undertook a general formal investigation into recruitment and selection between 1987 and 1989 in the 20 largest hotel groups, with particular reference to managerial, supervisory and skilled jobs (CRE, 1991; Jameson and Hamylton, 1992). At the beginning of the investigation, most groups had equal opportunities statements in the staff handbook but no company kept records of the ethnic origins of staff or job applicants. A subsequent monitoring exercise found that 16 per cent of all staff (8 per cent male, 8 per cent female) were from ethnic minorities. There were no ethnic minority general managers but among cleaners, waiters and porters (53 per cent of all jobs) there were '70 per cent of all ethnic minority male employees (compared to 43 per cent of all white males) and 76 per cent of all ethnic minority female employees (compared to 55 per cent of all white females)' (Jameson and Hamylton, 1992, p. 26). The CRE made recommendations for action at group and unit level (see Jameson and Hamylton, 1992, pp. 25–6), but there appears to have been no reported follow-up progress.

Summary propositions

- Equal opportunities policies are, perhaps, one of the most difficult areas of employment policy to frame and implement successfully.
- Employers may need to ensure that women are not being perceived as receiving unfair advantage; good practice must be available for all.
- Hotels and catering have particular problems with regard to equality issues, but progress has not been satisfactory.

Discussion points

- Why should hotels and catering have such a poor record of managing equality issues when it employs above average proportions of women and ethnic minorities?
- What practical steps can managers take to broaden access to greater equality of opportunity in the workplace? What impact are these measures likely to have on members of the existing workforce?

REFERENCES

Adam-Smith, D., Goss, D., Sinclair, A., Rees, G. and Meudell, K. (1992) AIDS and employment: diagnosis and prognosis. *Employee Relations*, **14** (3), 29–40.

Bagguley, P. (1987) *Flexibility, Restructuring and Gender: Changing Employment in Britain's Hotels.* Lancaster Regionalism Group, University of Lancaster.

Barnes, C. (1992) Disability and employment. *Personnel Review*, **21** (6), 55–73.

Beck, J. and Steel, M. (1989) *Beyond the Great Divide: Introducing Equality into the Company.* London: Pitman.

Berridge, J. (1991) Childcare provisions – the perk whose time is yet to come. *Employee Relations*, **13** (3), 10–16.

Bischoff, J. (1990) Corporate-sponsored child care: news from abroad. *Employee Relations*, **12** (1), 13–16.

Bourn, C. (1992) *Sex Discrimination Law – a Review*. London: Institute of Employment Rights.

Brewster, C. and Richbell, S. (1982) Getting managers to implement personnel policies. *Personnel Management*, December, 34–7.

Buswell, C. and Jenkins, S. (1994) Equal opportunities policies, employment and patriarchy. *Gender, Work and Organization*, **1** (2), 83–93.

Cockburn, C. (1983) *Brothers: Male Dominance and Technical Change*. London: Pluto Press.

Cockburn, C. (1989) The long and short agenda. *Industrial Relations Journal*, **20** (3), 213–25.

Collins, H. (1992) *The Equal Opportunities Handbook: a Guide to Law and Best Practice in Europe*. Oxford: Blackwell.

Collinson, D. and Hearn, J. (1994) Naming men as men: implications for work, organization and management. *Gender, Work and Organization*, **1** (1), 2–22.

Commission for Racial Equality (1984) *Race Relations Code of Practice: for the Elimination of Racial Discrimination in Employment and the Promotion of Equality of Opportunity*. London: Commission for Racial Equality.

Commission for Racial Equality (1988) *Monitoring an Equal Opportunity Policy: a Guide for Employers*. London: Commission for Racial Equality.

Commission for Racial Equality (1991) *Working in Hotels*. London: Commission for Racial Equality.

Commission for Racial Equality (1994) *Annual Report 1993*. London: Commission for Racial Equality.

Coombes, M. and Hubbock, J. (1992) Monitoring equal employment opportunity in the workplace: the crucial role of the 1991 census. *Ethnic and Racial Studies*, **15** (2), 193–213.

Cox, S. (1993) Equal opportunities. In M. Gold (ed.), *The Social Dimension: Employment Policy in the European Community*, pp. 41–63. London: Macmillan.

Crompton, R. and Sanderson, K. (1990) *Gendered Jobs and Social Change*. London: Unwin Hyman.

Deakin, S. (1994) Part time employment, qualifying thresholds and economic justification. *Industrial Law Journal*, **23** (2), 151–5.

Doyle, B. (1994) *New Directions Towards Disabled Workers' Rights*. London: Institute of Employment Rights.

Earnshaw, J. (1990) Criminal convictions: a bar to equality of employment. *Employee Relations*, **12** (3), 23–6.

Earnshaw, J. (1994) Sex discrimination in the workplace – UK legal round up. *Gender, Work and Organization*, **1** (2), 110–19.

Employment Department (1993) Industrial and employment appeal tribunal statistics 1991–92 and 1992–93. *Employment Gazette*, November, 527–31.

Employment Department (1994) The 1992 survey of industrial tribunal applications. *Employment Gazette*, January, 21–8.

Employment Department (1995) Statutory rights for part-time workers. *Employment Gazette*, February, 43.

Equal Opportunities Commission (1985a) *A Model Equal Opportunity Policy*. London: Equal Opportunities Commission.

Equal Opportunities Commission (1985b) *Equal Opportunities: a Guide for Employers to the Sex Discrimination Act 1975*. London: HMSO.

Equal Opportunities Commission (1994) *Annual Report 1993*. London: Equal Opportunities Commission.

Guerrier, Y. (1986) Hotel manager: an unsuitable job for a woman. *The Service Industries Journal*, **6** (2), 227–40.

Healy, G. and Kraithman, D. (1991) The other side of the equation – the demands of women on re-entering the labour market. *Employee Relations*, **13** (3), 17–28.

Hicks, L. (1990) Excluded women: how can this happen in the hotel world? *The Service Industries Journal*, **10** (2), 349–63.

Hochschild, A. (1983) *The Managed Heart*. Berkeley: University of California Press.

Hochschild, A. (1989) *The Second Shift*. London: Viking.

Incomes Data Services (1994) Post-Barber confusion over? *IDS Brief 527*, October, 2–6.

Jameson, S.M. and Hamylton, K. (1992) The CRE's investigation into the UK hotel industry. *International Journal of Contemporary Hospitality Management*, **4** (2), 21–6.

Kitching, J. (1994) Employers' work–force construction policies in the small service sector enterprise. In J. Atkinson and D. Storey (eds.), *Employment, the Small Firm and the Labour Market*, pp. 103–46. London: Routledge.

Lashley, C. (1984) Why women don't become chefs. MA thesis, University of Warwick.

Lucas, R.E. (1993) Ageism and the UK hospitality industry. *Employee Relations*, **15** (2), 33–41.

Lucas, R.E. (1994) Industrial relations theory, discourse and practice – are hotels and catering merely a case of oversight? Paper presented to the British Universities Industrial Relations Association Conference, Worcester College, Oxford, July.

Lucas, R.E. (1995) Some age-related issues in hotel and catering employment. *The Service Industries Journal*, **5**(2), 234–50.

Lucas, R.E. and Jeffries, L.P. (1991) The 'demographic timebomb' and how some hospitality employers are responding to the challenge. *International Journal of Hospitality Management*, **10** (4), 323–37.

Luckock, S. (1991) Flexible working practices for women returners. *International Journal of Contemporary Hospitality Management*, **3** (3), 4–9.

Lyon, P. and Mogendorff, D. (1991) Grey labour for fast food? *International Journal of Hospitality Management*, **10** (1), 25–34.

McKenna, M. and Larmour, R. (1984) Women in hotel and catering management in the UK. *International Journal of Hospitality Management*, **3** (3), 107–22.

McRae, S. (1991) *Maternity Rights in Britain: the Experience of Women and Employers*. London: Policy Studies Institute.

Peters, R. (1992) *Essential Law for Hotel and Catering Students*. London: Hodder and Stoughton.

Peters, T.J. and Waterman, S. (1983) *In Search of Excellence*. New York: Harper and Row.

Price, L. (1994) Poor personnel practice in the hotel and catering industry: does it matter? *Human Resource Management Journal*, **4** (4), 44–62.

Purcell, K. (1993) Equal opportunities in the hospitality industry: custom and credentials. *International Journal of Hospitality Management*, **12** (2), 127–40.

Reskin, B. and Padavic, I. (1994) *Women and Men at Work*. Thousand Oaks, California: Pine Forge Press.

Rideout, R. (1989) *Rideout's Principles of Labour Law*, 5th edition. London: Sweet and Maxwell.

Robinson, O. and Wallace, J. (1983) Employment trends in the hotel and catering industry in Great Britain. *The Service Industries Journal*, **3** (3), 260–78.

Robinson, O. and Wallace, J. (1984) Earnings in the hotel and catering industry in Great Britain. *The Service Industries Journal*, **4** (2), 143–60.

Smith, B., Povall, M. and Floyd, M. (1991) *Managing Disability at Work – Improving Practice in Organizations*. London: Jessica Kingsley Publishers and The Rehabilitation Resource Centre.

Smith, I.T., Wood, Sir J.C. and Thomas, G. (1993) *Industrial Law*, 5th edition. London: Butterworths.

Suter, E. (1994) Chambermaid unfairly dismissed after leave. *Personnel Management Plus*, **5** (7), p. 30.

Torrington, D. and Hall, L. (1991) *Personnel Management: a New Approach*, 2nd edition. London: Prentice Hall.

Whyte, W.F. (1948) *Human Relations in the Restaurant Industry*. New York: McGraw-Hill.

Wintour, P. (1994) Portillo hits disabled jobs with EC rule. *The Guardian*, 12 August, p. 3.

Wynne, J. (1993) Power relationships and empowerment in hotels. *Employee Relations*, **15** (2), 42–50.

ELEVEN

Managing Health, Safety and Welfare

The purposes of this chapter are to:

- review the importance of health, safety and welfare legislation in the employment relationship in hotels and catering;
- consider the effect of EC legislation on British health, safety and welfare law and practice;
- discuss contemporary workplace issues in occupational health and welfare.

11.1 INTRODUCTION

The area of health, safety and welfare covers matters as diverse as the occupational health of the workforce, the safety of machinery, processes and substances, and the physical and emotional well-being of workers through the provision of canteen facilities, rest breaks and satisfying, relatively stress-free, work. There has always been something of a tension between employers seeking to maximize production and service levels, increase efficiency, productivity and output, and the need to provide employee protection against hazards and risks in the workplace. Indeed, welfare has a 'soft' image, and many personnel professionals have sought to distance themselves from it, preferring the 'harder' approach implied by HRM.

British and EC law has provided an important framework to guide employers and employees towards more effective health and safety practice in the workplace. However, the state of health, safety and welfare in hotels and catering remains far from satisfactory. The industry is perceived to be a stressful place in which to work, where working conditions may be arduous and dangerous (for example, Byrne, 1986; Gabriel, 1988; Wood, 1992). Accident and injury rates are high and there are other associated negative outcomes, such as low productivity and high labour turnover. Health and safety issues are less likely to be dealt with on a joint employer–employee basis. Work may be less than satisfying and training and development opportunities relatively limited. Welfare benefits, such as the level of paid holidays, access to sick pay and pension schemes, are likely to be less generous than elsewhere.

This chapter first considers the British health and safety legislative framework and how this relates to what is actually happening in hotels and catering. Data from the WIRS3 and other sources are also discussed. Second, the effects of current and proposed EC measures on health, safety and welfare in the industry are assessed. Finally, some contemporary issues related to occupational health and employee welfare are discussed, such as absenteeism, sick pay, stress, HIV/AIDS and smoking.

11.2 HEALTH AND SAFETY LAW : THE BRITISH FRAMEWORK

According to Smith *et al.* (1993, p. 624), 'There is no aspect of industrial law more important than those rules, which are largely statutory, that attempt to secure the health and safety of workers at their place of work.' Such legislation dates from the beginning of the nineteenth century, but it developed in a piecemeal fashion, largely on a selective industry basis. It was not until the mid-1970s that the Health and Safety at Work Act (HASAWA) 1974 provided an umbrella and a framework to extend health and safety law to workers in every type of workplace (for a summary of the main features of the Act, see Appendix 6, Section A6.1). Hotel and catering workers (and others in 'new' industries and workplaces) had enjoyed relatively limited legal protection; for example, some were covered as 'shop workers' under the Shops Act 1950 with regard to matters such as days off and rest breaks (see Appendix 6, Section A6.7). Many catering workers in the catering services sector, such as those working in hospitals, had been denied statutory health and safety protection.

The HASAWA 1974 came about as a result of the Robens Committee on Safety and Health at Work, which envisaged three levels of control: legislation, regulations and codes of practice (statutory or approved). The Act imposed new duties on employers; these duties and their relationship with the previous law on common law negligence and breach of statutory duty are complex and best understood from reading Smith *et al.* (1993, pp. 604–23, 630–3). The Health and Safety Commission (HSC) and the Health and Safety Executive (HSE) were set up to oversee, administrate and enforce the law.

Measuring success in health and safety is difficult, and there has always been a tendency to measure relative failure, such as examining accident and injury statistics. However, this is as flawed as measuring the state of industrial relations by the number of days lost due to industrial action, partly because of the way in which data are recorded, and partly because it detracts from other less noticeable but equally valid negative behaviours and outcomes like absenteeism, labour turnover and stress. Even so, accidents dropped in the late 1970s, but increased in the 1980s, suggesting that the initial impetus provided by the new law soon evaporated (a similar situation occurred in relation to the Equal Pay Act 1970 – see Chapter 10). It is estimated that 80 per cent of accidents are preventable (Drummond, 1994). A useful analysis and evaluation of how safety at work is undermined by individualism, free market policies and deregulation is given by Moore (1991).

James (1993a, p. 24) believes that HASAWA was seriously flawed in structure and operation, for reasons that will be discussed in due course. The motto of the legislation was 'safety is everybody's responsibility', and James argues that the Committee overestimated the degree of commonality of interest between employers and workers, so that insufficient reliance was placed on external regulation. Drawing on the work of Dawson *et al.* (1988), James (1993a) argues that three main influences are important in assessing the role of the legislation: the relevance of specialists in an organization, pressure from employee representatives and the activities of the inspectorate. Suffice it to say that most hotel and catering workers may have to rely rather more on the last influence than workers in other industries, particularly in unionized workplaces where safety representatives and committees function. These points and evidence from the WIRS3 are now considered in broad terms in relation to the state of health and safety in hotels and catering.

The state of health and safety in hotels and catering

Management organization

In spite of the importance attached to health and safety rules, at least by lawyers (Smith *et al.* 1993), Ralston's study (1989, p. 241) of 55 hotel and catering organizations found that health and safety was ranked in the middle of all personnel activities in terms of the amount of time spent on it, its importance and whether the activity had increased in importance over the previous few years. The activity was fifth (of 14 activities) in terms of the amount of discretion that the respondents had over it. Only 6 per cent of the respondents saw welfare as being a central activity of the personnel function.

Firms employing fewer than five employees are exempt from the requirement to prepare and revise a written safety policy which must specify the organization and arrangements for carrying out the policy. The requirements for health and safety policy may lead to no more than woolly, inadequate policies (James, 1993a, p. 13), a point also noted with regard to equal opportunities (see Chapter 10). Although management still bears the brunt of responsibility for accidents, rather than workers, the development of decentralized management structures enables senior managers to evade health and safety responsibility by delegating such responsibility to managers further down the managerial hierarchy.

Ralston's (1989) study in the hotel and catering industry also found that a considerable number of individuals with health and safety responsibilities did not come under the personnel function: their role was largely advisory and administrative rather than concerned with formulating policy or safe systems of work. James (1993a, p. 23) also suggests that management responsibility could be reinforced by making regulations under the Companies Act 1985 'to require companies to include in directors' reports information about their health and safety activities and standards'.

As noted earlier from WIRS3 data, the majority of workplaces (88 per cent) in hotels and catering had formal health and safety procedures in common with workplaces across all sectors (Table 7.1). Management in a higher proportion of workplaces compared to elsewhere also claimed to provide a lot of information before implementing changes relating to safety and occupational health arrangements (see Table 8.8). As Table 11.1 shows, the tendency for hotel and catering managers to deal with health and safety matters in a more authoritarian way is considerably more marked than in other sectors. The extent of 'some employee consultation' is considerably less marked in hotels and catering compared to elsewhere, occurring in only one-fifth of workplaces. This is not inconsistent with observations about authoritarian management that have featured throughout the text, and the notable absence of trade unions.

Table 11.1 *How health and safety matters are dealt with in establishments.*

			PSS (by no. of employees)				HCI (by no. of employees)			
	AIS	SI	All	25–49	50–99	100+	All	25–49	50–99	100+
Unweighted base	2061	1299	702	174	153	375	61	21	18	22
Weighted base	2000	1462	886	518	218	151	99	59	26	14
JC[a] for health and safety only (%)	23	21	18	14	17	36	4	6	[b]	1
JC[a] for health and safety etc. (%)	9	9	8	7	8	11	10	7	13	18
Workforce representatives/ no committee (%)	24	26	17	15	20	17	7	–	21	7
Management in consultation with employees (%)	5	5	5	6	5	3	1	–	–	5
Management alone/no consultation (%)	37	37	51	57	48	33	75	82	65	63
Some employee consultation (%)	61	61	48	41	51	67	21	12	35	31

Table excludes missing cases.
[a]JC = joint committee. [b]Fewer than 0.5 per cent.
Any slight discrepancies between figures are due to the rounding of decimal points.
Source: WIRS3 (1990).

Poor performance indicators

Although sickness, absence and injury statistics arguably provide only limited measures of the success or failure of health and safety at work, they cannot be ignored. Data from the WIRS3 on these points in hotels and catering are not encouraging. Table 11.2 shows that the mean number of employees sick or absent in hotels and catering was lower than elsewhere, although there was a higher proportion of hotel and catering workplaces with recent incidences of sickness or absence. This suggests that short-term absence may be more marked in hotels and catering than elsewhere, although it should also be remembered that such workers are less likely to enjoy the benefits of a sick pay scheme (see below) and, therefore, the need to attend work may be more marked than elsewhere. Table 11.3 shows an injury rate in hotels and catering that is three times higher than the norm in the economy as whole, and that is even more disturbingly high in the larger workplaces where more systematic arrangements could be expected to exist.

Table 11.2 *Proportion of employees sick or absent.*

	AIS	SI	PSS (by no. of employees) All	PSS 25–49	PSS 50–99	PSS 100+	HCI (by no. of employees) All	HCI 25–49	HCI 50–99	HCI 100+
Unweighted base	2061	1299	702	174	153	375	61	21	18	22
Weighted base	2000	1462	886	518	218	151	99	59	26	14
Last week (%)	34	31	31	31	36	26	46	47	49	36
Last month (%)	32	33	34	37	28	28	23	25	15	32
Mean number of employees sick or absent	6.7	7.1	6.3	6.7	5.2	6.5	3.7	3.0	3.9	6.3

Table excludes missing cases.
Any slight discrepancies between figures are due to the rounding of decimal points.
Source: WIRS3 (1990).

Table 11.3 *Any employees sustaining major injury in past 12 months.*

	AIS	SI	PSS (by no. of employees) All	PSS 25–49	PSS 50–99	PSS 100+	HCI (by no. of employees) All	HCI 25–49	HCI 50–99	HCI 100+
Unweighted base	2061	1299	702	174	153	375	61	21	18	22
Weighted base	2000	1462	886	518	218	151	99	59	26	14
Yes (%)	20	15	15	12	15	25	17	11	29	21
Injuries per 1000	7.1	7.9	7.7	5.9	3.8	10.4	23.3	11.9	14.2	43.2

Table excludes missing cases.
Any slight discrepancies between figures are due to the rounding of decimal points.
Source: WIRS3 (1990).

The HSE has observed a rising trend in accidents since the introduction of the Reporting of Injuries, Diseases and Dangerous Occurrences Regulations (RIDDOR) 1985 (see Appendix 6, Section A6.2). In 1991–2 there were 2,295 major injuries reported (Jones, 1992) and in 1992–3 the figure rose to 2,635 (Drummond, 1994). An HSC report has also identified accidents in the catering industry as being particularly problematic among young workers, and in relation to slips, trips and falls in kitchens and serveries (HSC, 1990). The kitchen slicer is responsible for two-thirds of all machine accidents in catering. There were 40,000 accidents in catering premises in 1991–2, with a marked increase in incidence occurring around the Christmas period when firms were busier and more likely to employ temporary staff (Jones, 1992). These injury statistics may well be symptomatic of an even deeper health and safety malaise that would seem to require a rethink in terms of the law and its application. Workers sustaining industrial injury are entitled to seek compensation for damages and to receive state benefits (see Smith *et al.*, 1993, pp. 586–623).

Workforce involvement

Safety representatives and committees, in terms of HASAWA 1974, are limited to trade union appointees. HASAWA 1974 Section 2 (5), providing for safety representatives to be elected from among employees, was repealed by the Employment Protection Act 1975 so as not to undermine the development of workplace trade union organization. Even if that provision were to be reintroduced, its success would be dependent on strong autonomous worker organization, which is largely absent in hotels and catering.

As Table 11.1 shows, joint committees for health and safety alone are almost entirely absent from hotels and catering. The WIRS3 also shows that there are very few workplaces with employee representatives on the committee dealing with health and safety that are chosen by trade unions or staff associations, which is hardly surprising given low trade union density (Table 4.1), and looks unlikely to change given the limited attempts of trade unions to recruit members in hotels and catering (Table 4.2). Ralston's study (1989, p. 299) found very few safety committees in the commercial services sector, but they were present in all public sector establishments. Where committees existed, their value was recognized. A useful study of worker participation in health and safety from a European perspective is given by Walters (1990).

Enforcement

Although James (1993a, p. 20) has argued that the Robens Committee placed too much emphasis on self-regulation, and that there have been too few inspectors to enforce the law, it is probable that inspection has been more effective than in the now defunct wages council system. James (1993a, p. 21) suggests that fixed workplaces can expect an inspection once every three years. Estimates for catering establishments employing 50 or more staff are slightly better – one visit every 12 months to two years (Jones, 1992, pp. 36–7) – but this seems inadequate given the injury rates in the industry noted in Table 11.3. By contrast, in 1989 an hotel in one of the former hotel and catering wages council sectors could expect to have received a visit only once every 18 years (Lucas, 1991, p. 283), although the wages inspectorate was being deliberately run down at the time (Radiven and Lucas, 1994).

The preferred approach to health and safety enforcement is through the issue of improvement or prohibition notices; prosecutions are a matter of last resort. James (1993a) argues that the penalties on managers and employers found guilty of health and safety offences should be increased. In 1993 the maximum penalty applicable to a person found guilty of an offence on summary conviction was £5,000 for the generality of offences, and £20,000 and/or six months' imprisonment for breaches of Sections 2 to 6 of HASAWA 1974 or contraventions of improvement and prohibition notices.

Many of James's criticisms remain to be answered, although as the result of a recent review of health and safety legislation, the HSC has announced in a White Paper (1994) that existing regulations will be simplified and modernized. The Commission found little support for any major deregulation as signalled by the government or for exempting small firms and the self-employed from the law. The effects of the review are likely to lead to a 40 per cent reduction in legislation and the removal of almost all pre-1974 legislation.

Summary proposition

- Authoritarian management and lack of workforce involvement may contribute to the poor state of health and safety practice in hotels and catering.

Discussion points

- Is the poor state of health and safety practice inevitable or could it be managed more effectively?
- How far would more rigorous enforcement mechanisms improve the poor state of health and safety?
- What can be done to reduce the high rate of short-term absence and injury in the industry?

11.3 EUROPEAN DEVELOPMENTS

The role of the EC in health and safety matters increased dramatically after the Single European Act 1986 introduced a new article (Article 118A) into the Treaty of Rome. This gave the European Commission for the first time direct authority to take legislative action on health and safety matters and allowed for such proposals to be adopted by qualified majority voting. This has been used in two ways.

First, Article 118A has been used to progress Directives on 'mainstream' health and safety issues, which have not been subject to much controversy among the member states. An early measure was the Control of Substances Hazardous to Health Regulations (COSHH) 1988 (see Appendix 6, Section A6.3). The most important measures include the Framework Directive on measures to improve the health and safety of workers, and 'daughter' Directives dealing with matters such as display screen equipment, manual handling of loads and minimum workplace requirements, which have been introduced into British law by six sets of Regulations from 1 January 1993 (HSE, 1992a, b, c, d, e, f – all summarized in Appendix 6, Section A6.3).

The second way in which Article 118A has been used is to progress more controversial Directives on pregnant workers, working time and young persons. The UK government's opposition, both to the principle that issues such as working time and pregnancy are health and safety matters and to much of the substance of the proposals, has been a notable feature of the discussions that have led up to the adoption of each proposal. The upshot is that each proposal has been watered down considerably, and there are substantial derogations, in particular from the working time and young persons' Directives. These two Directives are discussed as ancillary health and safety.

The ways in which these EC developments have affected mainstream and ancillary British health and safety law are now discussed in more detail, with the exception of pregnancy, which has already been discussed in Chapter 10.

Mainstream health and safety

Although new regulations implementing the Framework and daughter Directives (see Appendix 6, Section A6.3) are not all yet fully operational, there is a broad consensus that the changes are, and will be, for the better (James, 1993a; Smith *et al.*, 1993). They mark the start of a 'comprehensive spring cleaning of the English law in this area that is long overdue' (Smith *et al.*, 1993, p. 626). The Framework Directive is based on a similar approach to HASAWA 1974 but it is more detailed in terms of its specific duties on employers, particularly with regard to Sections 2 to 6 of HASAWA 1974. Smith *et al.* (1993, p. 645) foresee that it

> will be difficult to resist breaches of regulations if there has not been a considerable amount of forethought, careful consideration of risks and how to protect against them. There must be repeated inspection, information and training. Responsibilities overlap so that both employer and employee may have a concurrent responsibility.

However, James (1993a, pp. 25–8) has expressed some doubt as to whether the spirit of some aspects of the Directives has been fully written into the letter of the new regulations (HSE 1992a, b, c, d, e, f). He argues that the HSC has adopted a minimalist approach to implementation, and identifies four areas where the Management of Health and Safety at Work Regulations may not comply with the Framework Directive. One is in relation to employee involvement and three concern the duties regarding employer cooperation and coordination: the obligations imposed on employers concerning the development of protective and preventative services and the provision of information to workers.

The position of safety representatives and committees is strengthened by the requirement that representatives should not be placed at any disadvantage by their activities. The need for emergency procedures to allow workers to stop work in certain circumstances is provided under the Management of Health and Safety Regulations and under TURERA 1993 (Section 28 and Schedule 5), such that a worker who is dismissed or suffers a detriment for exercising this right can complain to an industrial tribunal (see Appendix 4, Section A4.6). These health and safety responsibilities and rights apply to any worker, as well as safety representatives or committees that are trade union appointed under the Safety Representative and Safety Committees Regulations 1976. The 1976 Regulations have also been extended, such that employers must provide appointed representatives with reasonable facilities and assistance for all functions, not just inspections. There are now more stringent requirements for employers to consult with representatives in 'good time' over specific matters.

Thus, for the first time, non-union representatives have specific statutory rights in relation to health and safety, which should help to strengthen workers' rights in hotels and catering. On a broader level, this whole area begs the question of whether the union-only channels for consultation as per the newly extended 1976 Regulations are sustainable, particularly in light of the decisions reached in *Commission of the European Communities* v *United Kingdom of Great Britain and Northern Ireland*, C-382/92 [1994] IRLR 392 ECJ and *Commission of the European Communities* v *United Kingdom of Great Britain and Northern Ireland*, C-383/92 [1994] IRLR 412 ECJ about consultation in relation to collective redundancies and transfer of undertakings legislation (see Chapter 9).

Finally, the success of the new regulations will owe much to the enforcement process. British enforcement mechanisms are a great deal more advanced than those in some of the member states and for this reason, and others, James (1993a, pp. 31–4) implies that Britain cannot expect the same kind of European intervention by the European Commission which was so instrumental in changing equal pay legislation. The HSC has stated that it will seek to promote awareness and act as a source of advice in the early stages, rather than seek rigorous compliance with the new regulations.

Ancillary health and safety

Working time Directive

Even before the working time Directive was adopted in November 1993, it had become increasingly obvious that its proposals had been diluted significantly, and that its effects on hotels and catering had been overstated (Lucas, 1993, pp. 97–8). Many of its provisions are not wholly different from similar extant provisions contained in the Shops Act 1950, which apply to hotel and catering workers defined as 'shop workers' – those working in restaurants and public areas of hotels (see Appendix 6, Section A6.7). As the summary of the Directive's provisions shows (Appendix 6, Section A6.4), there are substantial derogations that mean that some provisions, such as the maximum 48-hour week, do not become effective until 2003, while others may never become effective at all.

The Sunday Trading Act 1994 gives workers the right not to work on Sundays (see Appendix 6, Section A6.8). However, it is not clear how these provisions relate to provisions about Sunday working contained in the Shops Act 1950 (see Appendix 6, Section A6.7).

Since 1980, declining proportions of hotel and catering workers have been working in excess of 48 hours. This practice is most marked among full-time males employed in public houses and clubs (19.3 per cent in 1993), although they represent a relatively small proportion of all hotel and catering employees (see Table 3.1). A working week in excess of 48 hours for women is extremely rare. Even so, employers may be able to derogate from the maximum 48-hour week under the 'particular activities' arrangements.

Many of the other requirements, such as minimum hours of rest and rest breaks, seem to be aspects further to encourage employers to continue to create even more part-time jobs of fewer hours; for example, through the use of young labour in full-time education. In other words, it would seem to be relatively easy for employers simply to circumvent such obligations by continuing with workforce construction policies that have characterized much of the industry since the 1960s. Such a prospect makes the argument that the Social Charter will 'cost jobs' in the present government's terms look thin, although it is likely to discourage the creation of more 'traditional' full-time jobs. Others believe that the Directive will have major implications for the British workplace (see, for example, Bercusson, 1994a, b), but the author remains rather more circumspect about such predictions with regard to hotels and catering (see Lucas, 1993).

The link between working hours and health and safety is not accepted by the government, whose case is supported by Harrington (1993) from a review of relevant literature, although the arguments are not wholly convincing. Harrington (1993, p. 34) maintains that 'The EC Explanatory Memorandum supporting the Working

Hours Directive was found to be seriously inadequate. The references cited would not be considered to provide any basis for justifying the proposals outlined in the Directive'. The bulk of the literature in this area is rooted in shiftwork, not weekly working hours or rest days. Harrington rightly identifies the need for further research on the effects of shift length, rotating hours and variable start times on health and performance.

Harrington finds some evidence of a relationship between abnormal hours and negative outcomes such as quantitative and qualitative sleep loss, fatigue, stress and cardiovascular and gastro-intestinal disease. Shiftwork carries a poorer performance and safety record. Yet the thrust of his arguments and the basis for his case centre on points like 'There is no scientific basis for setting Sunday as a rest day nor for setting holiday periods' (Harrington, 1993, p. 37), which may be true, but seem to be rather thin arguments, given the other more relevant supporting evidence he cites.

There is a further issue that Harrington does not address, which could have particular implications for many hotel and catering employees. How many workers have two or more jobs? If the requirements of separate jobs are examined, these workers may not appear to be subjected to working unduly long or unsocial hours. But when the jobs are added together, this would give an even greater continuity of work effort, which must undoubtedly increase the risk of stress, fatigue and accidents.

If the UK government's challenge to the Directive on the grounds that it does not fall within Article 118A fails, there will be relatively little time to implement its provisions, which will have to be implemented by November 1996.

Protection of young people at work Directive

This Directive was eventually adopted in June 1994 (see Appendix 6, Section A6.5), with the UK government having secured a number of renewable opt-outs, such that the full impact of the Directive may never fall on young workers in industries such as hotels and catering. Currently the Children and Young Persons Act 1933 (as amended) regulates the employment of young people aged 18 and below (see Appendix 6, Section A6.6).

Summary propositions

- Recent mainstream health and safety measures impose clearer duties and obligations on employers and workers.
- Future ancillary health and safety measures may have relatively little impact on employment practice in hotels and catering.

Discussion points

- Analyse and evaluate the impact of recent mainstream health and safety measures on managers and workers in hotels and catering.
- Analyse and evaluate the potential impact of ancillary health and safety measures on managers and workers in hotels and catering.

- How could the workforce become more actively involved in the management of health and safety issues?

11.4 CONTEMPORARY ISSUES IN OCCUPATIONAL HEALTH AND WELFARE

Organizations attempting to pursue good employment practices need to address both the physical and emotional well-being of employees (Torrington and Hall, 1991, pp. 329–52). As noted above, many of the basic physical provisions have been prompted by health and safety legislation with regard to matters such as systems of work, the use and handling of articles and substances, limitations on hours of work and the provision of rest breaks. 'Good' employers may go beyond these minimum requirements by providing emergency treatment beyond first aid, medical and dental facilities on-site, regular medical examinations for workers and generous paid holiday entitlements. Because of growing concern about the effects of passive smoking (HSE, 1988), no-smoking policies are becoming more widespread (see Jones and Kleiner, 1990; Painter, 1990). Additionally, those subscribing to the Human Relations School of thought (Elton Mayo) will pursue job satisfaction policies in order to increase productivity. Thus the physical side of occupational health and welfare has both tangible and less tangible aspects.

The emotional well-being of employees is even more difficult to deal with because aspects such as stress are even more intangible than matters such as physical injury. Stress at work has not received as much recognition and attention in the UK as in the USA and Japan (McHugh, 1993), but it is recognized as a serious threat to an individual's physical and psychological well-being. An employee (John Walker, a social worker) has successfully sued his employer (Northumberland County Council) as responsible for his nervous breakdown. The court ruled that the employer was 'liable for damages because of its unreasonable failure to provide a safe system of work – thereby putting stress on the same footing as accidents or industrial diseases' (Dyer, 1994).

In a study of managerial job stress in the hotel industry, Brymer *et al.* (1991) examined the stressors (causes) in the work domain that generate strain (outcomes) in managers. Not only did managers perceive their jobs to be stressful, they also seemed unable to develop satisfactory strategies to cope with both the stressors and the strains. The use of alcohol or drugs was related to increased strain. The clear implication is that organizations are better placed to make positive interventions; for example, by increasing employees' control over their work.

Many employees are unable to draw a dividing line between home and work. Marital, family and financial problems inevitably pervade the employment relationship, as does the issue of the work itself. The adverse effects of these problems may contribute to negative behaviours and outcomes, including low productivity, high turnover and low morale, issues that have been discussed in more detail in earlier chapters. However, establishing cause and effect is problematic, with the result that many managers adopt the 'ostrich' approach to managing these issues. They simply bury their heads in the sand and ignore the problems completely.

In practice, there is a large element of overlap between the physical and emotional aspects of occupational health and welfare. HIV/AIDS may be a very limited physical threat to people in the workplace, but it is an issue that provokes an emotional, often ill-informed, irrational response from many people, which could create employee relations problems. The removal of physical stress from the work process may lead to emotional stress if the employee feels unable to cope or has not been given the appropriate training and support. Empowerment initiatives may also create stresses caused by customers that workers may be able to cope with in positive or negative ways (Wynne, 1993).

These key issues are now examined under the headings of absence (including stress, sickness and sick pay) and welfare benefits (including holidays and pensions). The final section of the chapter considers HIV/AIDS.

Absence

Bona fide employee absence clearly arises from circumstances of illness or injury, although the term absenteeism tends to be used to denote situations where the absent person is, perhaps, not 'really' ill. This is not to say that one occasional Monday absentee is not justifiably absent for genuine 'stress-related' reasons (see McHugh, 1993), but the person will usually have to fabricate a more generally acceptable reason for the absence. Another occasional Monday absentee may be absent because of Sunday football match commitments, and he or she will also have to manufacture an acceptable reason for that absence. Rules are there to be manipulated, and may have to be manipulated on the employee's part to provide an acceptable reason for deviating from the normal duty to attend for work. Employers may experience difficulty in, and therefore avoid, differentiating between the 'worthy' and the 'unworthy'. These are examples of short-term absence, which need to be differentiated from cases of long-term absence (see ACAS, 1987, pp. 41–3). In short, the circumstances of absence are complex and may be difficult to determine.

McHugh (1993) estimates that 60 per cent of absence is caused by stress-related illness. Some 100 million days are lost per annum, which, when compared to the number of days lost through industrial action (649,000 in 1993; Employment Department, 1994, p. 202), puts the matter rather more firmly at the centre of contemporary employee relations problems than is addressed in most of the literature. Put bluntly, many people cannot face the prospect of going into work, a trend that may well be on the increase. Stress may also be manifested in poor job performance, high levels of absenteeism and high labour turnover (see also Brymer *et al.*, 1991). Other means of individual behavioural response, or coping strategies, include smoking, alcohol or substance abuse and eating disorders. Stress remains one of the great intangibles in the management of health and welfare in Britain and, more particularly, in the extent to which it impinges on the management of the employment relationship.

One of the main reasons why absence or absenteeism is important to organizations is its costs (Huczynski and Fitzpatrick, 1989a, b). For every £1 that the employee loses, the cost to the company is twice as much. In 1987 the CBI estimated that the cost of non-attendance and sickness in Britain came to £5 billion. This had more than doubled by 1993, when its estimated cost had risen to £11 billion. Average sickness absence was eight days, equivalent to 3.5 per cent of working time (Industrial Relations Services, 1994).

Huczynski and Fitzpatrick (1989b) differentiate between direct (including sick pay and replacement staff) and indirect costs (including recruitment and training costs, extra management time in coping with absence issues and reduced productivity). The management-driven prescription is that absence should be addressed as an opportunity for gaining productivity improvements, not treated as a problem. It is interesting how notions of employee welfare tend to be forgotten in discussions on this issue.

One of the most significant direct employment costs in circumstances of absence is sick pay, a matter that has been brought to a situation of prominence by the requirement from April 1994 that employers fund the total cost of statutory sick pay (SSP) (see Appendix 6, Section A6.9). Although some small employers are given exemptions, this change is potentially more costly to employers than the EC-driven requirements for maternity pay during maternity leave discussed in Chapter 10 (see Appendix 5, Section A5.3). Employers must fund SSP to a maximum of 28 weeks compared to a normal maximum of 14 weeks' pay during maternity leave, and perhaps more importantly, SSP applies to both men and women. Precise comparisons are hampered by the need to account for exclusions of some part-timers from entitlement to SSP. The change requiring employers to fund the total cost of SSP is likely to prompt a tightening up of absence that is cost-driven. This development has been noted in a CBI survey, which found that 77 per cent of employers were strengthening their sickness controls, 35 per cent were tightening their eligibility rules and 15 per cent might consider reducing occupational sick pay benefits to compensate for the loss of SSP rebates (Industrial Relations Services, 1994).

Employers may operate more advantageous sick pay schemes, although most evidence from hotels and catering shows that this is not likely to be the case, particularly in smaller firms (for example, Lucas, 1991, p. 281). This scenario is also confirmed by the WIRS3, shown in Table 11.4. When this information is viewed in relation to figures contained in Table 11.2, it is interesting to note that there is a tendency for smaller workplaces without sick pay schemes to show a higher than average propensity for short-term sickness or absence.

Table 11.4 *Availability of sick pay and occupational pension schemes to any employees at establishment.*

			PSS (by no. of employees)				HCI (by no. of employees)			
	AIS	SI	All	25–49	50–99	100+	All	25–49	50–99	100+
Unweighted base	1644	932	482	83	109	290	50	12	17	21
Weighted base	1355	890	529	263	155	111	70	33	24	13
Sick pay[a] (%)	75	71	80	75	80	91	64	47	76	84
Pension (%)	86	87	83	77	87	94	62	41	82	80

[a]Over and above statutory requirements.
Source: WIRS3 (1990).

Welfare benefits

Many welfare or 'fringe' benefits, such as paid holidays, sick pay and sports and social clubs, have their origins in the schools of philanthropy and paternalism that took root in the earlier part of the twentieth century. Even now, these items cannot be said to constitute a major aspect of remuneration for many workers in hotels and catering, although such extra 'benefits' are more widely available to workers in larger organizations (Lucas, 1991) and in larger workplaces (see Table 11.4). The classification of what constitutes such a benefit in mid-1990s Britain is, perhaps, changed. Private medical insurance may now be more widely coveted, and measures of employee involvement (see Chapter 8), such as the provision of a regular news sheet or newsletter, may be seen by employers as means to promote employee well-being. From her study of 55 hotel and catering establishments Ralston noted that, 'Given the size of establishments and, more importantly, the size of the organizations to which they belong, the provision of welfare is very poor' (Ralston, 1989, p. 301). The public sector was more welfare conscious for historical cultural reasons.

The issue of pensions is complex and will not be discussed here in any detail (for a discussion on pensions as 'pay' see Chapter 10; for a discussion on pensions more generally see Torrington and Hall, 1991, pp. 607–28). It is probable that large proportions of hotel and catering employees, particularly those working part-time and those employed in smaller firms, are not covered by an occupational pension scheme. This point has certainly been established in the hotel industry: only 9 per cent of small hotels (25 rooms or fewer) offered employees a pension scheme, although 93 per cent of hotel groups had pension schemes (Lucas, 1991).

Data from the WIRS3 shown in Table 11.4 confirm that hotel and catering workplaces are less likely to have occupational pension schemes, and that the likelihood of provision reduces in smaller workplaces. Ralston (1989) found that contributory pensions were available to operatives and administrative staff in around half the firms surveyed, whereas two-thirds of firms had contributory pension schemes for managers. Private medical insurance was available to managers in around one-third of firms, but was rarely available to operatives and administrative staff.

Although a minimum of four weeks' paid holiday may become a universal entitlement under the working time Directive by the end of the 1990s, many hotel workers appeared not to be receiving this level of holidays in 1989, particularly in small units (Lucas, 1991, p. 281). This level of entitlement seems to be the norm in the hotel groups, many of whom also give additional service leave. Ralston (1989) found that a majority of the firms in her survey gave extra service holidays.

Other ways in which employers can provide for employees' welfare include providing a counselling service, allowing compassionate leave for circumstances such as close family bereavements, allowing time off for moving house, setting up pre-retirement courses, introducing positive health care programmes and introducing alcohol and drug abuse policies.

Summary proposition

- Stress and absence are important issues that may give rise to problems in the management of the employment relationship.

Discussion points

- Outline an occupational health and welfare policy for an hotel and catering organization of your choice. Explain how this will improve the management of health and safety and will benefit the health and welfare of the workforce.
- Argue the case for and against employers providing good welfare benefits and assess the effects of this on the organization and the workforce.

11.5 HIV/AIDS

The issue of HIV/AIDS featured briefly in Chapter 10 as one potential area for inclusion in equal opportunities policies, but is considered in more detail here because of its perception as a health and safety issue. With regard to HIV/AIDS, Adam-Smith *et al.* (1992, p. 31) argue that 'ignorance is damaging employment relationships'. Since 90 per cent of those contracting AIDS will be in employment, employers must encourage open discussion and promote non-discrimination policies (see also Department of Employment/Health and Safety Executive, 1990). Interestingly, Adam-Smith and Goss (1993) note that the spread of HIV/AIDS may have an accentuated effect in hotels and catering, for reasons that will be expanded on below.

Adam-Smith *et al.* (1992) suggest that there are three legal implications associated with HIV/AIDS. Discrimination could be potentially unlawful under both the Sex Discrimination Act 1975 and the Race Relations Act 1976. Following an Equal Opportunities Commission investigation, Dan Air was required to change its recruitment of only women to cabin crew because this was direct discrimination against men. The airline argued that 30 per cent of applicants were male homosexuals, and that they constituted a health and safety risk under the HASAWA 1974, but this defence was not accepted. Similarly, race discrimination could occur if an employer refused to engage, or dismissed, someone from a country of origin with a high incidence of AIDS.

With regard to the laws of confidentiality, an employer cannot disclose the fact that an employee has AIDS, except in very exceptional cases. Adam-Smith *et al.* (1992) suggest that disclosure could be justified to someone like a first-aider, although the risk of infection is negligible. With regard to dismissal, apart from the potential sex and race implications mentioned above, the dismissal of someone with full-blown AIDS should be treated in accordance with normal long-term sickness procedures (see ACAS, 1987, pp. 42–4). At a more practical employee relations level an employee may refuse to work with someone who has HIV/AIDS, but to dismiss the HIV/AIDS carrier would probably be unfair dismissal. On the other hand, if a person with the disease is known to be employed in a restaurant, and this leads to loss of customers, 'commercial considerations' may make the dismissal of the carrier fair.

Adam-Smith *et al.* (1992, pp. 35–7) suggest that employers can be more successful in changing attitudes and removing prejudice if they construct a humanistic HIV/AIDS policy that is low on conditionality and exclusion. Such a policy would provide for no discrimination in recruitment, and state that there would be no requirement for employees or job applicants to be tested for HIV. Defensive policies which are high on conditionality and exclusion seek to present AIDS as potentially dangerous, and are more likely to lead to discrimination, less likely to lead an individual to disclose his or her condition and more likely to impede education and awareness programmes.

Hotels and catering may be perceived as having a particular AIDS risk for four reasons (Adam-Smith and Goss, 1993). First, the industry employs high proportions of young workers who are more closely associated with so-called risk behaviour (sex and drugs). Such employment is more likely to be concentrated in resorts and inner cities where such practices are encountered. Second, there is a reported tendency for the industry to employ high concentrations of male homosexuals. Third, the nature of some work may expose employees to a very small risk of contact with HIV, such as cleaning in areas where body fluids are spilt or dealing with violence with blood-to-blood contact in public houses and nightclubs. However, the *risks are exceptionally small.* Finally there is also a perception that HIV can be transmitted through food preparation. The great problem about all of these issues is the disparity between an objective assessment of risks by the experts, and the exaggerated perceptions of risk of those working in the industry.

Gaps in these two sets of values are likely to create workplace problems and here Adam-Smith and Goss (1993) propose that employers must 'assess the particular pattern of risk assessment which is dominant and locate this within the social/organizational context' (p. 28) in order to understand how employees will respond to AIDS as a workplace issue. Their pilot study in hotels and catering found that employers did not really think that HIV/AIDS was an issue. However, managers and students demonstrated a clear awareness of AIDS as a workplace issue, although this was likely to be based on incomplete information, which overstated the problem.

Summary proposition

- HIV/AIDS is more of a perceived risk and workplace issue in industries such as hotels and catering, although the *risk is exceptionally small.*

Discussion points

- Prepare an HIV/AIDS policy and explain how you would go about implementing it.
- Workers refuse to work with a colleague who is found to be HIV positive. As their manager, how would you handle this situation?

REFERENCES

Adam-Smith, D., Goss, D., Sinclair, A., Rees, G. and Meudell, K. (1992) AIDS and employment: diagnosis and prognosis. *Employee Relations*, **14** (3), 29–40.

Adam-Smith, D. and Goss, D. (1993) HIV/AIDS and hotel and catering employment: some implications of perceived risk. *Employee Relations*, **15** (2), 25–32.

Advisory, Conciliation and Arbitration Service (1987) *Discipline at Work: the ACAS Advisory Handbook*. London: HMSO.

Bercusson, B. (1994a) *Working Time in Britain: towards a European Model. Part I: the European Union Directive*. London: The Institute of Employment Rights.

Bercusson, B. (1994b) *Working Time in Britain: towards a European Model. Part II: Collective Bargaining in Europe and the UK*. London: The Institute of Employment Rights.

Brymer, R.A., Perrewe, P. L. and Johns, T.R. (1991) Managerial job stress in the hotel industry. *International Journal of Hospitality Management*, **10** (1), 47–58.

Byrne, D. (ed.) (1986) *Waiting for Change*. London: Low Pay Unit.

Dawson, S., Willman, P. , Bamford, M. and Clinton, A. (1988) *Safety at Work: the Limits of Self-regulation*. Cambridge: Cambridge University Press.

Department of Employment, Health and Safety Executive (1990) *AIDS and the Workplace*. London: HMSO.

Drummond, G. (1994) Danger at work. *Caterer and Hotelkeeper*, 14 April, 38–40.

Dyer, C. (1994) Damages for overwork. *The Guardian*, 17 November, 1.

Employment Department (1994) Labour disputes in 1993. *Employment Gazette*, June, 199–209.

Gabriel, Y. (1988) *Working Lives in Catering*. London: Routledge & Kegan Paul.

Harrington, J.M. (1993) *The Health and Safety Aspects of Working Hours – a Critical Review of the Literature.* Birmingham: University of Birmingham Institute of Occupational Health.

Health and Safety Commission (1990) *Accidents in Service Industries: 1988/89 Health and Safety Statistics for Premises Inspected by Local Authorities*. London: Health and Safety Commission.

Health and Safety Executive (1988) *Passive Smoking at Work*. London: HSE.

Health and Safety Executive (1992a) *Display Screen Equipment at Work: Health and Safety (Display Screen Equipment) Regulations 1992, Guidance on Regulations L 26*. London: HMSO.

Health and Safety Executive (1992b) *Management of Health and Safety at Work: Management of Health and Safety at Work Regulations 1992, Approved Code of Practice L 21*. London: HMSO.

Health and Safety Executive (1992c) *Manual Handling: Manual Handling Operations Regulations 1992, Guidance on Regulations L 23*. London: HMSO.

Health and Safety Executive (1992d) *Personal Protective Equipment at Work: Personal Protective Equipment at Work Regulations 1992, Guidance on Regulations L 25*. London: HMSO.

Health and Safety Executive (1992e) *Work Equipment at Work: Provision and Use of Work Equipment Regulations 1992, Guidance on Regulations L 22*. London: HMSO.

Health and Safety Executive (1992f) *Workplace Health, Safety and Welfare: Workplace (Health, Safety and Welfare) Regulations 1992, Guidance on Regulations L 24*. London: HMSO.

Huczynski, A. and Fitzpatrick, M. (1989a) *Managing Employee Absence for a Competitive Edge*. London: Pitman.

Huczynski, A. and Fitzpatrick, M. (1989b) End of the mystery – calculating the true cost of employee absence. *Employee Relations*, **11** (6), 12–15.

Industrial Relations Services (1994) Average sickness absence is eight days, says CBI. *Employment Trends*, 564, July, 5.

James, P. (1993a) *The European Community: a Positive Force for UK Health and Safety Law?* London: The Institute of Employment Rights.

James, P. (1993b) Occupational health and safety. In M. Gold (ed.), *The Social Dimension – Employment Policy in the European Community*, pp. 135–52. Basingstoke: Macmillan.

Jones, T.H. and Kleiner, B.H. (1990) Smoking and the work environment. *Employee Relations*, **12** (6), 29–31.

Jones, W. (1992) Health and efficiency. *Caterer and Hotelkeeper*, 17 December, 36–7.

Lucas, R.E. (1991) Remuneration practice in a wages council sector: some empirical observations in hotels. *Industrial Relations Journal*, **22** (4), 273–85.

Lucas, R.E. (1993) The Social Charter – opportunity or threat to employment practice in the UK hospitality industry? *International Journal of Hospitality Management*, **12** (1), 89–100.

McHugh, M. (1993) Stress at work: do managers really count the costs? *Employee Relations*, **15** (1), 18–32.

Moore, R. (1991) *The Price of Safety: the Market, Workers' Rights and the Law*. London: The Institute of Employment Rights.

Painter, R. (1990) Smoking policies: the legal implications. *Employee Relations*, **12** (4), 17–21.

Radiven, N.A. and Lucas, R.E. (1994) Wages councils – did they matter and will they be missed? Paper presented to the Third CHME Research Conference, Napier University, April.

Ralston, R. (1989) The changing nature of personnel management in the hotel and catering industry. MSc thesis, University of Manchester.

Smith, I.T., Wood, Sir J.C. and Thomas, G. (1993) *Industrial Law*, 5th edition. London: Butterworths.

Torrington, D. and Hall, L. (1991) *Personnel Management: a New Approach*, 2nd edition. London: Prentice Hall.

Walters, D.R. (1990) *Worker Participation in Health and Safety*. London: The Institute of Employment Rights.

Wood, R.C. (1992) *Working in Hotels and Catering*. London: Routledge.

Wynne, J. (1993) Power relationships and empowerment in hotels. *Employee Relations*, **15** (2), 42–50.

TWELVE

Conclusions and Future Issues

The purposes of this short chapter are to:

- draw together some of the themes of earlier chapters;
- highlight key issues.

In the pursuit of an overriding theme, all the main chapters are self-standing and include conclusions about the subject matter addressed. However, there is some merit in drawing together key issues from the subject matter that has been reviewed, particularly from the previously unpublished WIRS3 data. The WIRS3 data are to be further developed in a separate journal article, so more discussion will follow.

This chapter is deliberately brief because, even after the process of writing the book, the author still holds the view that a dearth of empirical work makes for difficulties in attempting to propose theoretical perspectives on the employment relationship in hotels and catering. Even if Dunlop's (1958) systems approach is used to organize facts about employee relations in hotels and catering (this proved to be far too complex to depict satisfactorily), it is still not possible to state with any degree of certainty how this complex mix of variables reacts to generate the web of rules. More fundamental explanation of issues in the employment relationship thus awaits more searching analysis. The issues identified below encapsulate some of these difficulties.

Much of what has been observed about the state of employee relations in the hotel and catering industry from the first systematic and in-depth analysis of this topic does support many of the stereotypical assumptions that are held about the industry. Hotels and catering embodies poor employee relations practices that seemingly lead to 'bad' employee relations outcomes. This statement raises an important question: are employee relations in the industry badly managed or are these outcomes a deliberate result of managerial 'strategies'? It is possible to give partial answers to this question in the following way.

From the WIRS3 data discussed in Chapter 2, it can be seen that hotels and catering has a greater extent of 'specialist' managers than other sectors in the economy.

This finding was unexpected. First, although the level of 'strong' personnel specialists is similar to that found in other broad sectors in the economy, this overall difference is mainly explained because there are more line managers who spend 25 per cent or more of their time on personnel ('weak' specialists). But a greater proportion of hotel and catering managers also claimed to have personnel qualifications and had assistance to deal with personnel matters than managers elsewhere. Therefore, it is not unreasonable to claim that the hotel and catering industry has more personnel 'specialists' than elsewhere.

Even so, in conventional terms the end product is a consistently poor set of employee relations outcomes including: high labour turnover; skills shortages; high rates of absenteeism, dismissals and accidents; less employee involvement; and low pay. From this it can be posited that these 'specialists' are not doing a 'good job', an unpopular and controversial claim that has been made at a more general level by Fernie *et al.* (1994) from their analysis of the complete WIRS3 data.

Does this proposition go far enough, or is it not necessary to qualify these remarks by asking a supplementary question: what is good and what is poor? If a business is driven by technological features and market/budgetary constraints that require high flexibility, management strategy may seek deliberately to perpetuate high labour turnover and high rates of dismissals, which in themselves militate against the concept of employee involvement. In other words, some of the outcomes mentioned above are not 'poor' but 'good' in terms of a dominant frame of reference that appears to drive managerial actions in much of this industry (see also the discussion on managerial style in Chapter 2). It may well be the case that managers do not perceive such outcomes to be a manifestation of 'poor' employee relations. Managers might believe, in the absence of trade unions, strikes and disputes, that employee relations in hotels and catering are actually 'good'. Conversely, it might well be the case that organizations do not intend such poor outcomes to occur. If that is the case, some of the fault may lie in the way in which the personnel function is operationalized, such that those charged with the responsibility are unable to do a 'good' job.

The second area where there is difficulty of judging what is 'good' and what is 'bad' lies in pay levels and how pay is determined; these are issues of real substance at the heart of the employment relationship. Hotels and catering has always been a low-paid industry. Employers and workers have not organized formally, with the consequent effect that there have been no voluntary collective bargaining arrangements at industry level. The former wages councils, tantamount to state-sponsored quasi-collective bargaining arrangements, did not lead to any marked improvement in pay levels. Very few voluntary collective bargaining arrangements have existed at company level. The WIRS3 confirms all these factors: high proportions of low-paid workers, low trade union density, a high incidence of unilateral decision-making and little discussion or consultation with regard to determining pay.

What is also striking about findings from the WIRS3 is the fact that performance appraisal is more widely used in hotels and catering, particularly in relation to determining pay. Individual employee performance was a more significant factor influencing pay increases than elsewhere. These are employment practices that are supposedly relatively 'sophisticated', but they do not lead to a 'better deal' for hotel and catering workers. Whatever the pay practice in this industry, authoritarian management is highly instrumental in achieving an end product of low pay.

As noted above, the WIRS3 data will be reviewed more comprehensively elsewhere. Here is not the place to reflect further on their contribution to the understanding of employee relations in hotels and catering, other than to reaffirm their importance, because they have undoubtedly opened up a number of new avenues that merit further exploration.

The extent to which organizations in the hotel and catering industry fit Guest and Hoque's 'bad' and 'ugly' typology (Guest and Hoque, 1994) or Sisson's bleak environments (Sisson, 1993) that are driven by strategies of cost and control (Braverman, 1974), externalization (Gospel, 1992) or low individualism (Purcell, 1987) remains to answered. Employee relations in hotels and catering is undoubtedly interesting, but explaining it is always going to prove something of a challenge.

REFERENCES

Braverman, H. (1974) *Labor and Monopoly Capital*. New York: Monthly Review Press.

Dunlop, J. (1958) *Industrial Relations Systems*. New York: Henry Holt and Company.

Fernie, S., Metcalf, D and Woodland, S. (1994) *Does HRM Boost Employee-Management Relations?* Centre for Economic Performance Working Paper 546. London: London School of Economics.

Gospel, H.F. (1992) *Markets, Firms and the Management of Labour in Modern Britain*. Cambridge: Cambridge University Press.

Guest, D. and Hoque, K. (1994) The good, the bad and the ugly: employment relations in new non-union workplaces. *Human Resource Management Journal*, **5** (1), 1–14.

Purcell, J. (1987) Mapping management styles in employee relations. *Journal of Management Studies*, **24** (5), 533-48.

Sisson, K. (1993) In search of HRM. *British Journal of Industrial Relations*, **31** (2), 201-10.

APPENDICES

Appendix ONE

Workplace Industrial Relations Survey 1990 (WIRS3)

The WIRS 1990, a structured sample survey of British workplaces employing 25 or more employees, and the third of its kind (the first two were in 1980 and 1984), was designed to:

- contribute to the debate on the reform of British industrial relations;
- aid the better understanding of the processes that underlie the employment relationship and;
- provide evidence on the way industrial and employment relations change over time.

The survey was sponsored by the Employment Department, the Economic and Social Research Council, the Policy Studies Institute (with funds from the Leverhulme Trust) and the Advisory, Conciliation and Arbitration Service.

The overall findings of the WIRS3 have been published elsewhere (Millward *et al.*, 1992; Millward, 1994) but it has been possible to obtain a separate breakdown of data for the hotel and catering industry. This marks an important breakthrough in terms of furthering knowledge about, and understanding, employee relations in hotels and catering. First, such data constitute by far the most comprehensive account of employee relations management practices, procedures and processes in the industry that has ever been obtained. Second, the data on hotels and catering can be compared with what has been established nationally and, perhaps more logically, with all service industries and private sector services. Such comparisons should make a useful contribution to the debate about contemporary employee relations in Britain.

However, invaluable though the WIRS3 data undoubtedly are, they cannot be said to be fully representative of the hotel and catering industry by size of establishment or by industry sector. Virtually all industries have a highly skewed distribution of establishments by size, with small establishments greatly outnumbering larger ones. This is even more extreme in hotels and catering. In fact hotels and catering is one of the very few industries where size distribution is so skewed towards smaller units that the majority of *employees* work in units of fewer than 25 employees (see Table 1.1).

It is conceivable that the WIRS series could be extended to include some smaller workplaces at a future date. Proportionate change in the balance of industry sectors and establishment size could be expected to be represented in the results for hotels and catering. However, it is unrealistic to suggest that a future WIRS could or would cover units as small as one employee, because very small units would dominate the sample but represent only a minority of employees in employment.

A further effect of the sampling method related to workplace size is that nearly three-quarters of responding workplaces in the WIRS3 survey are hotels and restaurants, yet in terms of the industry as a whole these two sectors probably account for around half of all workplaces.

Given the relatively small number of hotel and catering respondents (see Table A1.1), and the even smaller numbers of respondents from the other constituent parts of the industry – public houses and bars, clubs, canteens (contract catering) and other accommodation – analysis on an industry sector basis is not really meaningful.

A more instructive approach is to make comparisons on the basis of establishment size – 25 to 49 employees, 50 to 99 employees and 100 employees or more – and such an analysis informs much of the discussion throughout the main body of the text.

The tables at the end of this appendix show the main overall characteristics of the WIRS3 responding sample, which form the basis for the more specific analyses discussed throughout the book (see Notes to Tables below).

The inclusion of the WIRS3 data would not have been possible without the help of Neil Millward of the Policy Studies Institute, who provided invaluable assistance in producing the data. The provision of funds for this work by the Economic and Social Research Council's Business Resources Board is gratefully acknowledged. Neil Millward also offered helpful, incisive and constructive comments on earlier drafts of the analysis.

NOTES TO TABLES

AIS (all industries and services) denotes all SIC Divisions 1 to 9.

SI (all service industries) denotes all SIC Divisions 6 to 9.

PSS (private sector services) denotes SIC Divisions 6 to 9, not owned by the state or a public sector body.

HCI (hotels and catering) denotes SIC Class 66, which comprises restaurants and takeaways (661), public houses and bars (662), clubs (663), canteens or contract catering (664), hotels (665) and other accommodation (667).

The responding sample was weighted to compensate for the higher probability of larger workplaces being selected than smaller workplaces. The effect of this has been to increase the number of hotel and catering workplaces in the weighted results. The weighted base numbers for the individual categories (numbers of employees) within PSS and HCI may not add up to the total weighted base number (All) owing to the rounding of decimal points.

Column percentages do not all sum to 100 owing to the rounding of decimal points.

REFERENCES

Millward, N. (1994) *The New Industrial Revolutions?* London: Policy Studies Institute.

Millward, N., Stevens, M., Smart, D. and Hawes, W.R. (1992) *Workplace Industrial Relations in Transition*. Aldershot: Dartmouth Publishing Company Limited.

Table A1.1 *Total employees at establishment.*

	AIS	SI	PSS	HCI
Unweighted base	2061	1299	702	61
Weighted base	2000	1462	886	99
By number (%)				
25–49	53	58	58	60
50–99	25	23	25	26
100–199	13	12	12	13
200–499	7	5	4	1
500–999	2	1	1	[a]
1000+	1	1	[a]	–
Mean	102	94	78	62
Base employees in millions	15.568 (100%)	10.469 (67%)	5.260 (34%)	0.471 (3%)

[a]Fewer than 0.5 per cent.
Source: WIRS3 (1990).

Table A1.2 *Formal status of establishment.*

			PSS (by no. of employees)				HCI (by no. of employees)			
	AIS	SI	All	25–49	50–99	100+	All	25–49	50–99	100+
Unweighted base	2061	1299	702	174	153	375	61	21	18	22
Weighted base	2000	1462	886	518	218	151	99	59	26	14
Limited company or PLC (%)	60	48	80	75	83	91	78	66	95	97
Partnership/self-proprietorship (%)	5	6	10	14	5	3	17	29	–	–
Other (%)	35	46	10	11	12	6	5	6	5	3

Source: WIRS3 (1990).

Table A1.3 *Type of establishment.*

	AIS	SI	PSS (by no. of employees) All	25–49	50–99	100+	HCI (by no. of employees) All	25–49	50–99	100+
Unweighted base	2061	1299	702	174	153	375	61	21	18	22
Weighted base	2000	1462	886	518	218	151	99	59	26	14
Single independent establishment (%)	21	15	23	29	17	12	23	32	16	–
One of several establishments (%)	79	85	77	71	83	88	77	68	84	100
Of which head office	4	5	7	4	6	16	1	–	–	3
Other administrative office	5	5	4	2	6	5	–	–	–	–
Working establishments	91	90	89	93	88	79	99	100	100	97

Source: WIRS3 (1990).

Table A1.4 *Workforce composition.*

	AIS	SI	PSS (by no. of employees) All	25–49	50–99	100+	HCI (by no. of employees) All	25–49	50–99	100+
Unweighted base	2061	1299	702	174	153	375	61	21	18	22
Weighted base	2000	1462	886	518	218	151	99	59	26	14
Per cent female										
High (>70)	25	32	25	26	26	18	31	34	33	16
Medium (31–70)	32	35	38	37	38	44	54	46	66	62
Low (0–30)	35	24	29	32	26	23	2	3	–	3
Per cent manual										
High (>70)	38	30	34	32	38	37	87	82	100	83
Medium (31–70)	26	24	21	24	19	17	8	11	[a]	7
Low (0–30)	36	47	44	44	43	46	5	7	–	9
Per cent part-time										
High (>70)	22	30	26	25	31	25	57	58	65	34
Medium (31–70)	35	39	37	42	30	27	27	26	29	28
Low (0–30)	42	30	35	32	35	44	14	16	5	25
Per cent ethnic minorities (>10)	7	8	9	10	7	8	16	13	9	36

[a]Fewer than 0.5 per cent.
Source: WIRS3 (1990).

Table A1.5 *Whether particular occupational groups present.*

	AIS	SI	PSS (by no. of employees) All	25–49	50–99	100+	HCI (by no. of employees) All	25–49	50–99	100+
Unweighted base	2061	1299	702	174	153	375	61	21	18	22
Weighted base	2000	1462	886	518	218	151	99	59	26	14
Manual workers (%)										
Unskilled	69	64	63	61	68	62	68	60	84	76
Semi-skilled	50	40	43	39	47	53	69	65	80	67
Skilled	54	42	48	46	52	53	82	81	85	79
Any manual	82	77	76	73	83	76	92	88	100	92
Non-manual workers (%)										
Clerical/ administrative	89	86	88	87	89	91	68	58	80	91
Supervisor	61	57	60	56	59	73	51	42	64	65
Junior technical/ professional	53	53	48	45	50	58	16	12	11	41
Senior technical/ professional	55	52	45	41	49	56	16	10	11	52
Managers	87	85	89	88	89	89	87	80	100	95
Any non-manager	95	96	96	98	95	93	93	89	100	95

Source: WIRS3 (1990).

Table A1.6 *Percentage of part-timers in workforce.*

	AIS	SI	PSS (by no. of employees) All	25–49	50–99	100+	HCI (by no. of employees) All	25–49	50–99	100+
Unweighted base	2061	1299	702	174	153	375	61	21	18	22
Weighted base	2000	1462	886	518	218	151	99	59	26	14
Numbers (%)										
0	16	12	15	16	8	17	3	4	[a]	1
1–19	47	38	43	42	44	45	23	23	21	29
20–49	20	25	18	20	18	12	24	22	28	24
50+	16	22	21	20	26	22	48	50	50	33
Missing (%)	1	1	2	1	4	3	2	–	–	13
Aggregate[b] (%)	18	24	25	25	27	23	41	44	43	34

[a]Fewer than 0.5 per cent.
[b]Excluding missing cases.
Source: WIRS3 (1990).

Table A1.7 *Percentage of females (full- and part-time) in workforce.*

			PSS (by no. of employees)				HCI (by no. of employees)			
	AIS	SI	All	25–49	50–99	100+	All	25–49	50–99	100+
Unweighted base	2061	1299	702	174	153	375	61	21	18	22
Weighted base	2000	1462	886	518	218	151	99	59	26	14
Numbers (%)										
0	1	[a]	[a]	[a]	–	1	2	3	–	–
1–19	25	15	18	19	18	18	[a]	–	–	3
20–49	21	20	26	25	26	25	37	29	52	40
50+	46	56	48	50	49	44	48	50	48	37
Missing (%)	8	9	8	5	10	15	13	17	[a]	19
Aggregate[b] (%)	44	52	50	49	51	49	62	65	61	59

[a]Fewer than 0.5 per cent.
[b]Excluding missing cases.
Source: WIRS3 (1990).

Table A1.8 *Percentage of manuals in workforce.*

			PSS (by no. of employees)				HCI (by no. of employees)			
	AIS	SI	All	25–49	50–99	100+	All	25–49	50–99	100+
Unweighted base	2061	1299	702	174	153	375	61	21	18	22
Weighted base	2000	1462	886	518	218	151	99	59	26	14
Numbers (%)										
0	16	20	23	26	15	22	4	7	–	3
1–19	13	18	16	13	19	22	1	–	–	6
20–49	18	22	15	14	20	10	3	6	[a]	–
50+	52	40	48	48	46	47	87	88	99	92
Aggregate (%)	46	36	42	42	45	40	78	78	84	75

[a]Fewer than 0.5 per cent.
Source: WIRS3 (1990).

Table A1.9 *Percentage of manuals in workforce who are skilled.*

			PSS (by no. of employees)				HCI (by no. of employees)			
	AIS	SI	All	25–49	50–99	100+	All	25–49	50–99	100+
Unweighted base	1692	998	535	121	123	291	55	18	17	20
Weighted base	1627	1119	672	380	179	113	91	52	26	13
Numbers (%)										
0	35	46	37	38	38	33	10	7	15	14
1–19	27	25	26	23	26	36	57	61	33	69
20–49	13	10	12	11	13	12	15	9	27	14
50+	25	18	24	28	23	21	19	24	16	3
Aggregate (%)	28	22	26	32	30	21	23	32	23	15

Source: WIRS3 (1990).

Table A1.10 *Ethnic minority groups in workforce in establishment.*

			PSS (by no. of employees)				HCI (by no. of employees)			
	AIS	SI	All	25–49	50–99	100+	All	25–49	50–99	100+
Unweighted base	2061	1299	702	174	153	375	61	21	18	22
Weighted base	2000	1462	886	518	218	151	99	59	26	14
None (%)	57	57	54	63	50	27	44	48	46	21
West Indian (%)	22	21	19	13	22	36	19	6	35	47
Indian/Pakistani/ Bangladeshi (%)	25	24	24	17	25	45	19	15	18	40
East African Asian (%)	7	7	7	4	8	18	5	2	–	25
Other Black African (%)	11	11	11	8	9	20	20	9	36	38
Malay/Chinese/ Filipino/Far East (%)	10	11	12	7	13	28	26	26	25	28
Other Asian (%)	8	8	9	8	6	16	13	13	5	24

Table excludes missing cases.
Source: WIRS3 (1990).

Table A1.11 *Proportion of total workforce in establishment from ethnic groups.*

			PSS (by no. of employees)				HCI (by no. of employees)			
	AIS	SI	All	25–49	50–99	100+	All	25–49	50–99	100+
Unweighted base	2061	1299	702	174	153	375	61	21	18	22
Weighted base	2000	1462	886	518	218	151	99	59	26	14
Numbers (%)										
None	61	63	62	70	59	38	46	48	46	34
<5%	24	22	21	15	25	39	30	29	34	29
5–10%	7	7	7	4	9	13	8	9	11	1
11–19%	2	2	2	2	1	4	2	–	–	14
20%+	5	5	6	7	5	4	14	13	9	23

Table excludes missing cases.
Source: WIRS3 (1990).

Appendix TWO

The Author's Perspective

It is generally the convention for authors to 'declare their hand' and explain their own perspective or approach to managing employee relations. Much of the evidence on which this book has been based is strongly suggestive of a relatively poor state of employee relations in hotels and catering, from which it follows that some improvement needs to be made. Thus the author's perspective should help to justify and explain what has been deemed to be appropriate or 'good' practice with regard to the more effective management of the employment relationship in the hotel and catering industry.

In seeking ways to improve employee relations, some caution must be exercised because the theoretical understanding of what really happens in the hotel and catering industry needs much more development. In other words, should a 'conventional' approach be taken and the industry damned for high labour turnover, or should an 'alternative' approach be taken that supports this feature as an inevitable and necessary fact of life?

As a general rule, the author rejects the 'totality' of the unitary perspective on the grounds that it represents, at best, a naive view. To suggest that employers 'hold all the cards' is a comforting theoretical concern, and true in an economic sense, but it is not a practical reality. Deviance, as implied in the unitary approach, is actually a good deal more subtle, unstructured and informal, and manifests itself in a wide variety of ways other than through unionization. Nevertheless, the author believes that managers should 'manage' and that there are circumstances when managers need to act unilaterally.

Having said this, an HRM approach, as a way forward for 'people' management, has undoubted appeal, but it does not necessarily fall within the unitary perspective. During a ten-year career in industrial relations and personnel management, it never occurred to the author to think that what was being done was not integrated with the wider objectives of the organization as a whole, or one of its constituent parts, although the nature of that integration may not have been entirely 'rational' or explicit. This experience spanned mainly unionized businesses, but even in the 1970s performance-related pay systems were operational and organizations attempted to secure employee involvement and commitment through a variety of means.

Therefore, accepting Rose's (1988) point that conflict and cooperation are always found together, it is taken as read that conflict and cooperation are inherent dynamic forces in a pluralist employment relationship that needs to be managed. Consent, which forms a basis for resolving conflict and achieving cooperation, is also present to varying extents (see Marchington, 1992). Conflict may be hidden and this should be recognized (see, for example, Kolb and Bartunek, 1992; Analoui and Kakabadse, 1993). Cooperation is based on trust between individuals, which is engendered at the level of the workplace rather than through elaborate organizational mechanisms.

The broad framework that the author envisages involves a state that acts as the guardian of the national interest, within which bargaining about rules and their application takes place at a variety of levels and between a variety of 'interested' groups and individuals. The most important level is the workplace because that is where most of the action centring on control, conflict and cooperation occurs. The state would legislate to confer rights on organizations and individuals in relation to issues of employment protection and bargaining. The state would also 'sponsor' appropriate institutional arrangements to underpin this approach. The state might also need to intervene more directly in matters of the national interest, but this would be the exception rather than the rule, on the assumption that the 'system' of rights and institutions was adequately constituted for the majority of circumstances.

To some, this approach may be reminiscent of 'consensus' pluralism, but this is not entirely so in that the role of the state is more positive and is in some ways much more in keeping with the approaches taken by Britain's European partners and implied in much of the social dimension of European integration. As an example, in the author's 'ideal world' there would be some kind of Statutory National Minimum Wage, and other conditions of employment would also be regulated by mechanisms appropriate to particular industries (see, for example, Lucas, 1989). This approach also acknowledges the role of the individual, but not in a free market sense.

Many of the means to secure these ends require considerable refinement because they break with some important traditions in British industrial relations, but then most of the hotel and catering industry has always been, and still remains, isolated from the mainstream of British industrial relations. As an example, it is difficult to know how to reconcile individualism and collectivism satisfactorily within a conceptual framework that gives the state clear responsibilities and places the onus of 'action' on the individual workplace. Therefore specific definitions of employee relations or theoretical perspectives on employee relations are not proposed in this book on the grounds that there is insufficient empirical evidence to do so. Only when such evidence becomes available, through further research, will it be possible to make significant advances in the theory and practice of employee relations in the hotel and catering industry.

As a final point, the state of affairs that has been conceptualized is, in the final analysis, radical because it envisages a revolution of social, economic and political proportions that would elevate the 'disadvantaged' majority, particularly women, to a position of greater formal influence and status. This it not to deny that many of the disadvantaged already have power. But what is missing from virtually all analysis and the development of perspectives on employee relations is any recognition that there may be a gender dimension involved.

REFERENCES

Analoui, F. and Kakabadse, A. (1993) Industrial conflict and its expressions. *Employee Relations*, **15** (1), 46–62.

Kolb, D.M. and Bartunek, J.M. (eds), (1992) *Hidden Conflict in Organizations*. Newbury Park, California: Sage Publications.

Lucas, R.E. (1989) Minimum wages – straitjacket or framework for the hospitality industry into the 1990s? *International Journal of Hospitality Management*, **8** (3), 197–214.

Marchington, M. (1992) Managing labour relations in a competitive environment. In A. Sturdy, D. Knights and H. Willmott (eds), *Skill and Consent in the Labour Process*, pp. 149–84. London: Routledge.

Millward, N. (1994) *The New Industrial Relations?* London: Policy Studies Institute.

Millward, N., Stevens, M., Smart, D. and Hawes, W.R. (1992) *Workplace Industrial Relations in Transition*. Aldershot: Dartmouth Publishing Company Ltd.

Rose, M. (1988) *Industrial Behaviour*, 2nd edition. Harmondsworth: Penguin.

Appendix THREE

Outline Summary of the Effect of European Community Law

The effect of Community law in the UK is defined in the European Communities Act 1972, which had to be enacted because the UK has no written constitution. However, the issue of whether national law or EC legislation has primacy or supremacy remains extremely complicated (see Hepple, 1991) and confused (Aikin, 1992). Detailed discussion of this area falls outside the main purpose of this text, but an outline summary is given below. For a good description of the history, developments and legal principles involved, see Steiner (1994).

Hepple (1991, p. 1061) summarizes six general principles but two of these have been affected by subsequent decisions made by the European Court of Justice (ECJ), the ultimate decision-making court with regard to Community law.

If a state fails to conform to Community law the European Commission takes up the matter and the ECJ may take enforcement proceedings, as in the recently heard cases on collective redundancies and the transfer of undertakings: *Commission of the European Communities* v *United Kingdom of Great Britain and Northern Ireland*, C-382/92 [1994] IRLR 392 ECJ and *Commission of the European Communities* v *United Kingdom of Great Britain and Northern Ireland*, C-383/92 [1994] IRLR 412 ECJ. Individual cases may also be referred to the ECJ by a UK court; for example, the referral by the House of Lords of *Webb* v *EMO Air Cargo (UK) Ltd*, C-32/93 [1994] IRLR 482 ECJ.

The principles summarized by Hepple (1991, p. 1061) are outlined and developed below.

EC Directives which are 'sufficiently precise' have direct effect, making them enforceable by individuals against organs of the state. Individuals may not take action against private persons or bodies. As a result of *Frankovich* v *Italian Republic* 6/90 [1992] IRLR 84 ECJ, individuals may take action against the state for failure to implement a Directive. The government may thus be liable to compensate those who suffer loss where rights to compensation from a private sector employer are excluded. The principle in *Frankovich* may allow some claims against the government by workers harmed by non-implementation of the acquired rights Directive. This is further highlighted by the decision reached in *Commission of the European Communities* v

United Kingdom of Great Britain and Northern Ireland, C-382/92 [1994] IRLR 392 ECJ.

Community obligations may be put into effect by Order in Council or regulations under the European Communities Act 1972. Six sets of health and safety regulations introduced in 1992 (see Chapter 11) fall within this principle.

A UK statute passed before an EC Directive will not be interpreted in the light of that Directive even if the proposals for that Directive were being discussed at the time the UK statute was passed. However, in the case of *Marleasing SA* v *La Commercial Internacional de Alimentacion SA* [1992] 1 CMLR 305 ECJ, the ECJ decided that the pre-existing national law had to be interpreted in the light of a subsequent Directive. As noted by Aikin (1992, p. 56), 'Clearly this can only apply where the pre-existing law is ambiguous – but it is not clear how "ambiguous" the law needs to be for *Marleasing* to apply'.

The words of a UK statute passed in order to implement Community obligations which have 'direct' effect will apparently be construed in accordance with those obligations, provided the words are 'reasonably capable of such construction'.

The words of a UK statute passed in order to implement Community obligations which have 'indirect' effect will apparently be construed in accordance with those obligations.

A UK statute passed subsequently to an EC Directive, other than for the purposes of implementing the Directive, will be construed in accordance with the directly effective provisions of the Directive.

REFERENCES

Aikin, O. (1992) A matter of precedence. *Personnel Management*, April, 56–7.

Hepple, B. (1991) Institutions and sources of labour law: European and international standards. In *Encyclopedia of Employment Law Volume 1*, pp. 1052–61. London: Sweet and Maxwell.

Steiner, J. (1994) *Textbook on EC Law*. London: Blackstone Press.

Appendix FOUR

Workplace Employment Law

A4.1 WRITTEN STATEMENT OF TERMS AND CONDITIONS OF EMPLOYMENT

Contracts of Employment Act 1963 (as amended by Contracts of Employment Act 1972, Employment Protection Act 1975 (EPCA 1978), Employment Acts 1980, 1982, 1988 and 1989, TURERA 1993 and EPPTER 1995).

- To be issued to all new employees (EPPTER 1995) within two months of commencing employment to employees working eight hours a week.
- Employees with at least one month's service who then leave are still entitled to be given particulars.

The principal statement must include (the few exceptions where reference to another document is permitted are marked*):

1 Names of employer and employee.
2 The date when the employment began.
3 The date on which the employee's period of continuous (statutory) employment began (taking into account any employment with a previous employer which counts towards that period).
4 Scale or rate of remuneration or method of calculating remuneration (including overtime, and bonus/commission if this is a contractual arrangement).
5 Intervals at which remuneration is paid (weekly, monthly etc.).
6 Any terms relating to hours of work, including normal hours (this would include compulsory overtime).
7 Any terms relating to holiday entitlement (including pay and public holidays. Information must be sufficient to enable entitlement, including accrued entitlement, on leaving to be 'precisely calculated'.
8 Any terms and conditions due to sickness or injury including, but not limited to, sick pay*.

9 Pension and pension schemes except in relation to certain pension schemes set up by Act of Parliament*.
10 Length of notice that employee must give or receive to terminate the contract (see Statutory Notice)*.
11 Job title or a brief description of the work for which the employee is employed (care is needed since the latter may amount to a contractual job description which is less flexible or amenable to change; TURERA 1993).
12 Expected duration of non-permanent employment, and the expiry date of a fixed-term contract (the former needs handling with care to avoid creating an enforceable guarantee).
13 Place of work or for itinerant employees, the employer's address (TURERA 1993) (inclusion of a statement of place of work without more detail may entail a risk of curtailing implied mobility clauses; an express mobility clause is prudent).
14 Any collective agreement directly affecting terms and conditions, including the parties, if the employer is not one (TURERA 1993).
15 For employees required to work outside the UK for more than a month, the duration of overseas work, currency in which salary will be paid during absence, any additional remuneration to be paid to the employee or additional benefits for overseas service and any conditions in relation to return to the UK (TURERA 1993).
16 A note specifying disciplinary rules (not health and safety) referring to a reasonably accessible document containing them*.
17 A note specifying by description or otherwise a person to whom the employee can appeal against a disciplinary decision, and to whom he or she can apply when raising a grievance, and in either case, the manner of application*.
18 A note explaining any further stages of grievance and disciplinary procedures or referring to an accessible document containing them*.

Notes

- Firms employing fewer than 20 employees are exempted from the provision of disciplinary rules and procedural details (Employment Act 1989)).
- Employers must inform employees in writing about changes, no more than one month after their introduction or, if the change results from the employee being required to work outside the UK for a period of more than one month, at the time when he or she leaves the UK to commence work, if that is earlier.
- Change of employer requires the issue of a new, full written statement. If the identity of the employer remains unchanged (the name may have altered), or continuity is preserved, notification is required within one month. Changed identity or new name and the dates on which the employee's continuous employment began must be given.

A4.2 PAY

Itemized pay statement

Employment Protection Act 1975 (now EPCA 1978) amended by TURERA 1993 and EPPTER 1995.

Employers must issue an itemized pay statement to all employees specifying:

- gross wages or salary;
- amounts of fixed deductions and the purposes for which they are made (e.g. savings schemes) – the aggregate amount can be shown if the employee is given a standing statement of fixed deductions, to be renewed annually;
- amounts of variable deductions and the purposes for which they are made (e.g. income tax and national insurance);
- net wages or salary;
- where different amounts of the net wage or salary are paid in different ways, the amount and method of payment of each part payment (e.g. weekly wage, monthly bonus).

Guarantee payment

Employment Protection Act 1975 (now EPCA 1978 amended by EPPTER 1995) applies to all employees with one month's service if pay is lost through short-time working or lay-offs, providing:

- a full day's work is lost;
- an offer of alternative work has not been unreasonably refused;
- lost time does not result from a trade dispute involving any employee of the employer or an associated employer;
- limited to £14.10 per day for up to five workless days in any three-month period (i.e. £70.50 in any three-month period), or the equivalent number of days normally worked by that employee;
- excludes temporary employees taken on for up to three months.

Insolvency payment

Employment Protection Act 1975 (now EPCA 1978) and Insolvency Act 1986.

- Employment Department may administer certain payments from the Redundancy Fund owing to employees by insolvent employers.
- The limit on a week's pay for recoverable debts is £205.
- Upper limits applicable to debts recoverable are basic award (£6,150), pay arrears (£1,640), holiday pay (£1,230) and statutory notice pay (£2,460).

Deductions from pay

Wages Act 1986.

- If required or permitted by a statutory or contractual provision.
- The employee has already given written permission for the deduction to be made.
- Deductions or making good cash shortages or stock discrepancies are limited to 10 per cent of the gross amount of the wages payable on any pay day (including any deductions owing to alleged dishonesty) – employers must notify the

employee in writing of the full amount owed (those who work in retail employment only).

Holiday pay

The working time Directive provides for three weeks' paid annual holiday, rising to four weeks after a specified period after the Directive is implemented.
See also:

- Equal pay (Section A5.2).
- Maternity pay (Section A5.3).
- Redundancy pay (Section A4.7).
- Sick pay (Section A6.9).
- Medical suspension (Hotel and Catering Training Company, 1990, p. 49).
- Time off with pay (Section A4.3).

A4.3 TIME OFF WORK

'Reasonable' time off with pay

Trade union duties

Employment Protection Act 1975 (now EPCA 1978) (amended by Employment Act 1989) now TULRCA 1992 (amended by EPPTER 1995).

- For officials (shop stewards) of independent, recognized trade unions to carry out industrial relations duties between employer and employees and attend TUC or union approved industrial relations training (EPCA 1978).
- Time off is limited to duties concerned with matters in respect of which the employer recognizes the union, or in relation to functions that are outside the terms of recognition, but which the employer has agreed the union may perform (Employment Act 1989).
- Time off or training must be relevant to those duties (Employment Act 1989).

For further guidance, see ACAS Code of Practice (No. 3) *Time off for Trade Union Duties and Activities* (revised 1991) (ACAS, 1991).
See also:

- Antenatal care and maternity leave (Section A5.3).
- Safety representative duties (Sections A4.6, A6.1 and A6.3).
- To look for work or rearrange training if redundant (Section A4.7).
- Sick pay (Section A6.9).

Without pay

Employment Protection Act 1975 (now EPCA 1978) (amended by Employment Act 1989) now TULRCA 1992 (amended by EPPTER 1995).

Trade union activities

For members to take part in union's activities (not industrial action). For further guidance, see ACAS Code of Practice (No. 3) *Time off for Trade Union Duties and Activities* (revised 1991) (ACAS, 1991).

Public duties

Employment Protection Act 1975 (now EPCA 1978).

- Employees who are JPs or members of other specified public bodies should be allowed 'reasonable' time off to carry out their duties, subject to business needs.
- Duties include membership of a local authority, any statutory tribunal, a health authority and the governing body of an educational establishment maintained by a local education authority.
- Time off already allowed under other time off provisions can be taken into account in determining what is 'reasonable'.

A4.4 TRANSFER OF AN UNDERTAKING

Transfer of Undertakings (Protection of Employment) Regulations 1981 (TUPE) amended by TURERA 1993.

- No hours/service qualifications.
- Employees employed by the old employer at the time of transfer automatically become employees of the new employer.
- The new employer takes over the employment liabilities of the old employer (except criminal liabilities and occupational pension rights).
- Where there are recognized independent trade unions, representatives of the employees must be informed of the transfer.[a]
- Trade union representatives to be consulted with a view to seeking agreement about proposed employer measures in relation to the transfer.
- Maximum penalty for failure to consult is four weeks' pay and this cannot be offset against other emoluments.
- Dismissal for a reason connected with the transfer will be unfair (some other substantial reason) unless the dismissal was necessary for economic, technical or organizational reasons, and the employer acted reasonably in dismissing for such a reason.
- 'Reasonable' dismissal may constitute redundancy.
- Applies to any undertaking (or part), whether or not a commercial venture (TURERA 1993).
- Transfer effected in whatever manner, including cases where no property passes between transferor and transferee (TURERA 1993).

Note

[a] Subject to decisions reached in *Commission of the European Communities* v *United Kingdom of Great Britain and Northern Ireland*, C-382/92 [1994] IRLR 392 ECJ and *Commission of the European Communities* v *United Kingdom of Great Britain and Northern Ireland*, C-383/92 [1994] IRLR 412 ECJ.

See also EPCA 1978 for circumstances when employment may remain continuous even though there is a change in employer.

A4.5 STATUTORY NOTICE

Employment Protection Act 1975 (now EPCA 1978) amended by EPPTER 1995.

- All employees to receive: one week after one month's continuous service; two weeks after two years' continuous service; one week for each additional year's service to a maximum of 12 weeks after 12 years' service;
- Employees to give one week's notice after one month's service;
- Employees with a contract for three months or less are not entitled to a week's notice unless they work longer than three months (to be made clear at time of engagement).

A4.6 DISMISSAL

Complex history and not consolidated as one area, now EPCA 1978, TULRCA 1992 and TURERA 1993 (amended by EPPTER 1995).

- When employment is terminated, with or without notice.
- An employer refuses to renew a fixed-term contract.
- An employee leaves because of the employer's conduct.
- An employer refuses to allow a woman to return to work after confinement.

Written reasons for dismissal

All employees (EPPTER 1995) with two years' continuous service.

Unfair dismissal

- The most frequently tested individual employment right at industrial tribunal.
- Subject to an employee (no hours qualification – EPPTER 1995) having two years' continuous service and not being over the normal retirement age (unless

otherwise specified, this will be taken to be 65). In certain circumstances without these qualifications, the dismissal is automatically unfair (see below).
- Claims to be lodged within three months of the effective date of dismissal.
- The ex-employee has to show that a dismissal occurred (see constructive dismissal below).
- The employer must show that the reason for dismissal was valid in terms of capability, conduct, illegality, redundancy or another substantial reason (as specified in the EPCA 1978).
- The tribunal must decide whether the 'employer acted reasonably or unreasonably in treating it as a sufficient reason for dismissing the employee; and that question shall be determined in accordance with equity and the substantial merits of the case.' The size and the administrative resources of the employer's undertaking must also be taken into account in reaching a decision.
- Tribunal may order re-engagement/reinstatement (an employer's refusal to comply may lead to the tribunal making an additional award) but compensation is the norm.
- Compensation is based on three elements. *Basic award* is similar to redundancy pay. Based on a week's pay of £205 (unchanged since 1 April 1992) subject to a maximum of £6,150 (see Smith *et al.*, 1993, pp. 389–90). *Compensatory award* is meant to constitute realistic recompense for pecuniary loss to a maximum of £11,000[a] (see Smith *et al.*, 1993, pp. 390–8) except in sex and race cases.[b] *Additional award* is for an employer's refusal to re-engage/re-instate (between 13 and 26 weeks' pay) or because of sex or race discrimination (between 26 and 52 weeks' pay). Based on a week's pay of £205 subject to a maximum of £10,660 (see Smith *et al.*, 1993, pp. 385–9).

Notes

[a] Although the unfair dismissal compensatory award is still subject to £11,000, where a re-employment order is not complied with (or only partially) the limit on the compensatory award may be lifted to ensure that the employee receives full compensation for arrears of pay or benefits.

[b] Maximum limits removed by the Sex Discrimination and Equal Pay (Remedial) Regulations 1993 and Race Relations (Remedies) Act 1994.

Automatically unfair dismissal (inadmissible reasons)

No hours/service qualification.

Trade unions[a]

- Being a member of a trade union or proposing to become one.
- Taking part, or proposing to take part, in the activities of an independent trade union at an appropriate time (outside working hours or agreed by the employer for such activity).
- Not being a member of any trade union (or particular trade union), or refusing or proposing to refuse to become or remain a member (see also Section A5.1).

Pregnancy

Automatically unfair to dismiss for reasons connected with pregnancy/taking maternity leave from date employer informed of pregnancy to date woman returns from leave (see Section A5.3).

Health and safety[a]

- For carrying out any health and safety duties or activities designated by the employer or as a recognized safety representative or safety committee member.
- For raising concerns about safety issues in the absence of a safety representative.
- For leaving the workplace or taking protective measures in circumstances of what the employee reasonably believed to be serious and imminent danger.

Assertion of statutory rights

Automatically unfair to dismiss (but not to take action short of dismissal) for assertion of statutory rights under the EPCA 1978, the Wages Act 1986 or TULRCA 1992.

Refusal to work on Sundays

Shop workers (not Sunday only workers) (see Sunday Trading Act 1994, and Section A6.8).

See also Transfer of undertakings in Section A4.4.

Note: Dismissal for union or health and safety reasons

[a] There are differences in the compensation arrangements relating to dismissal on grounds of trade union membership, trade union activities, non-membership of a trade union and health and safety reasons. The maximum *basic award* (£6,150) is subject to a minimum (£2,700). The *compensatory award* remains subject to a maximum of £11,000. Where a reinstatement/re-engagement order is sought, but not made, there is a maximum *special award* of £26,800 subject to a minimum of £13,400. Where an order is made and complied with, there is no maximum (the employee receives all losses from dismissal to re-employment). Where an order is made and not complied with the special award is up to 156 weeks' pay, subject to a minimum of £20,100 but with no maximum. Where a reinstatement/re-engagement order is not sought no special award will be made.

Constructive dismissal

- Where an employee terminates the contract by resignation because the employer's behaviour is tantamount to breach or repudiation.
- Breaches can be express or implied, and can derive from one significant breach or a series of breaches.

Wrongful dismissal

- Where an employee is not given due notice (or wages in lieu), or has not received other entitlements, such as holiday or sick pay, that are part of the employment contract.
- Gross misconduct on the part of the employee repudiates any claim to notice and other entitlements.

Some other useful cases

Trust Houses Forte Leisure Ltd v *Aquilar* [1976] IRLR 251 EAT (dismissal for misconduct/crime).
Trust Houses Forte Hotels Ltd v *Murphy* [1977] IRLR 186 EAT (tribunal's approach to 'range of reasonable responses' test/misconduct/theft).
Taylorplan Catering (Scotland) Ltd v *McInally* [1980] IRLR 53 EAT (incapability).
Bouchaala v *Trust House Forte Hotels Ltd* [1980] IRLR 382 EAT (reasons for dismissal).
Les Ambassadeurs Club v *Bainda* [1982] IRLR 5 EAT (compensation, basic award, compensatory award, contributory fault).
O'Kelly and others v *Trusthouse Forte plc* [1983] IRLR 369 CA (meaning of 'employee', remission of appeal back to industrial tribunals).
Hotson v *Wisbech Conservative Club* [1984] IRLR 422 EAT (dismissal, dishonesty, tribunal procedure).
Trusthouse Forte (Catering) Ltd v *Adonis* [1984] IRLR 382 EAT (written particulars, legal effect, breach of contract, disciplinary rules, gross misconduct).
Babar Indian Restaurant v *Rawat* [1985] IRLR 57 EAT (no reason for dismissal shown, no evidence, redundancy/selection).
Moyes v *Hylton Castle Working Men's Social Club and Institute Ltd* [1986] IRLR 482 EAT (dismissal, reasonableness in the circumstances, fair and effective disciplinary procedure, compensation, contributory fault).
McGrath v *Rank Leisure Ltd* [1985] IRLR 323 EAT (reason for dismissal, transfer of undertaking, redundancy).
BSC Sports and Social Club v *Morgan* [1987] IRLR 391 EAT (termination, wrongful dismissal, conduct and capability, summary dismissal).
Aberdeen Steak Houses Group plc v *Ibrahim* [1988] IRLR 420 EAT (formality of industrial tribunal proceedings).
Ford v *Stakis Hotels and Inns Ltd* [1988] IRLR 46 EAT (exclusions and qualifications, claim in time).
Gateway Hotels Ltd v *Stewart and others* [1988] IRLR 287 EAT (reason for dismissal, transfer of undertaking).
Roadchef Ltd v *Hastings* [1988] IRLR 142 EAT (compensation, deductions).
Society of Licensed Victuallers v *Chamberlain* [1989] IRLR 421 EAT (terms of employment, hours of work, continuity of employment).
Hilton International Hotels (UK) Ltd v *Protopapa* [1990] IRLR 316 EAT (constructive dismissal).
Hewcastle Catering Ltd v *Ahmed and another* [1991] IRLR 473 CA (meaning of 'employee', illegal contract).

Soros and Soros v *Davison and Davison* [1994] IRLR 264 EAT (breach of confidence after dismissal/compensation).
Hilton International Hotels (UK) Ltd v *Faraji* [1994] IRLR EAT 267 (unavailability for work/compensation).
Burgess v *Bass Taverns Ltd* [1994] IRLB 499 (constructive dismissal/trade union activities).

A4.7 REDUNDANCY

Complex history and not consolidated as one area (see below), but mainly in EPCA 1978 (as amended), TULRCA 1992 and TURERA 1993 (amended by EPPTER 1995).

- Where a business has, or is about to be closed down.
- The business in moving to another location.
- 'Work of a particular kind' is no longer needed by the business, or is not needed by the business at that particular place.
- Definition of redundancy enlarged for consultation purposes only (TURERA 1993).

Redundancy pay

Redundancy Payments Act 1965 (as amended by the EPCA 1978, the Employment Act 1989 and EPPTER 1995).

- Subject to an employee (no hours qualification – EPPTER 1995) having two years' continuous service over the age of 18.
- Employee must be aged over 18 or below 65, or the company's normal, non-discriminatory, retirement age, if lower.
- Payment must be claimed by the employee within six months of the termination date.
- Payment varies according to age and length of service, subject to a limit of 30 weeks' pay up to the statutory maximum level of a week's pay (companies may operate a more generous scheme of redundancy compensation on top of statutory requirements).
- Maximum statutory payment £6,150 (based on a week's pay of £205, 1 April 1992).
- Employers must now fund the total amount of statutory redundancy pay (Employment Act 1989).

Alternative employment

- All employees (EPPTER 1995) facing redundancy should be given reasonable time off with pay (subject to a maximum of two-fifths of a week's pay) to look for work or make arrangements for training.
- No redundancy payment is necessary if the employee accepts an alternative offer of work, made before the old contract expires and taking effect within four weeks of the previous employment ending.

- The four-week trial may be extended, but the agreement must be in writing.
- An employee may refuse an alternative job offer if he or she considers the alternative 'unreasonable', and make a claim to an industrial tribunal that the employer is unreasonably withholding the redundancy payment in the circumstances.

Dismissal and discrimination

- Unfair selection for redundancy enables an employee to claim unfair dismissal (subject to meeting unfair dismissal qualifications).
- Dismissal during a takeover or merger may be redundancy (see Transfer of undertakings).
- Automatically unfair to select a woman for redundancy on grounds of pregnancy/maternity (see Section A5.3).
- Automatically unfair to make a shop worker who refuses to work on a Sunday redundant (see Section A6.8).
- Selection of part-time employees could constitute unlawful indirect discrimination against females.

Consultation

- Consultation is good employee relations practice.
- It covers proposed dismissal for a reason (or number of reasons) not related to the individual concerned (TURERA 1993).
- Statutory presumption that dismissals are for redundancy as now defined unless shown otherwise (TURERA 1993).
- It is compulsory where there are recognized independent trade unions, even though employees affected are not union members.
- Consultation with the union should begin at the earliest opportunity, and within a specified time before the first dismissal takes effect. If 10 to 99 employees may be dismissed over a period of 30 days or less, at least 30 days in advance; if 100 or more employees may be dismissed over a period of 90 days or less, at least 90 days in advance.
- Consultation must be undertaken with a view to reaching agreement with union representatives (TURERA 1993).
- Consultation must extend to consultation about ways of avoiding dismissals, reducing the numbers to be dismissed and mitigating the consequences of the dismissals (TURERA 1993).
- Trade union representatives must be given details of the proposed redundancy, the numbers and descriptions of those to be dismissed, the proposed method of selection and dismissal, and (TURERA 1933) the method proposed to calculate any redundancy payments in excess of the statutory minimum.
- Employers must take note of, reply to and if possible adopt any representations that are made about the proposals.
- Failure to consult the unions, or take note of their proposals or reply to them, entitles the unions to apply to an industrial tribunal for a 'protective award' which requires the employer to continue paying the employees for a specific period.

- Protective awards (maximum four weeks' pay) are additional and cannot be offset against other emoluments (TURERA 1993).

Notification

- *All employers*, even where there is no trade union involvement;
- Department of Employment to be notified (on form HR1) of redundancy proposals within the same time period that applies to consultation with trade unions (see above).

Note

With regard to both consultation and notification, where decisions leading to redundancy are taken by a parent company (inside or outside the UK), the failure to provide information by the parent company will be no defence to failure to consult properly or in good time, or failure to give the appropriate information or to give the necessary information to the Department of Employment (TURERA 1993).

A4.8 COLLECTIVE LAW: SUMMARY OF THE MAIN CHANGES TO BRITISH EMPLOYMENT LEGISLATION SINCE 1980

The legal base has a complex history but the main provisions are contained in TULRCA 1992 and TURERA 1993. This summary is based largely on information contained in Hendy (1993, pp. 50–68).

Trade union autonomy

Constitutional matters

- All members of the principal executive committee of every trade union must be elected by postal ballot at least every five years. Ballots on amalgamations and mergers must also be fully postal. Ballots must be carried out by a body independent of the union, which may include the scrutineer.
- Unions must to maintain a register of members' names and addresses. This is subject to inspection by the scrutineer in executive elections and political fund ballots.
- Requirements are imposed on the circulation of election addresses.
- State approved scrutineers are required for elections, mergers, amalgamations and political funds.
- Candidates are not to be 'unreasonably excluded' from standing for election. A Code of Practice may be published on the regulation of elections.

Trade union funds and activities

- Courts have increased powers over the use of union funds.
- Indemnification by a union in respect of members' or officials' fines is precluded.

- Unions are required to reveal 'accounting records' to member of accountant.
- A statement of the salaries of senior union officials is to be included in the trade union's annual return to the Certification Officer, who now has increased investigatory powers.
- Check-off union subscriptions must be authorized in writing by the member within the past three years so that the employer can lawfully deduct.
- Having or maintaining a separate fund for political activities is now subject to a ballot.

Trade union membership

- Unreasonable exclusion and expulsion from a union in a 'closed shop' (union membership agreement) can be challenged. Prohibition on exclusion or expulsion from membership only applies on specifically permitted grounds.[a]
- 'Unjustifiably disciplined' members can seek compensation. 'Unjustifiable' charges extended to include joining another union and refusing to join the check-off.[a]
- Members can take a grievance against a union to Court.
- Employers can withhold benefits from employees on grounds of trade union membership.

Note

[a] At industrial tribunal, awards for unreasonable exclusion/expulsion from a union and unjustifiable discipline by a union may not exceed the combined limit of the basic and compensatory awards for unfair dismissal (£17,150). Awards at the EAT are subject to a £5,000 minimum, although the £17,150 maximum still applies.

Trade union and employer relations

Collective bargaining

- The extension of the benefits of collectively bargained conditions throughout a trade or locality has been removed.
- Statutory recognition procedures have been repealed.
- Clauses in commercial contracts relating to the use of union labour only or union recognition are void.

The right to strike and take industrial action

- Number of torts for which immunity is given have been reduced. Definition of a trade dispute is now more limited.
- Pickets have considerably more limited protection to function. Code of Practice on Picketing limits the number of pickets to six.
- All secondary action is unlawful.
- Industrial action in support of a closed shop, pressing for a commercial contract concerning union only labour or recognition, and supporting strikers is now unprotected.
- Secret ballot required before strike action is governed by the Code of Practice on Industrial Action Balloting. Ballot requirements are now more stringent.

- Trade union immunities from civil actions in tort in relation to industrial activities have been removed.
- Full-time officials are responsible for unofficial industrial action, unless action is repudiated 'as soon as reasonably practicable'.
- Any individual can seek an injunction against a union if industrial action has prevented or delayed the supply of goods or services. The Commissioner for the Protection against Unlawful Industrial Action can finance individuals to sue unions.
- Strikers on unofficial strike are excluded from all unfair dismissal protection.
- All strikers are excluded from social security benefits.

The closed shop (union membership agreement)

- Dismissal for not being a member of a union is automatically unfair whether there is a union membership agreement or not, and subject to a huge 'special award'.
- An individual cannot be refused employment because he or she is not a union member or is not prepared to join a union (employment cannot be refused if the individual *is* a union member).
- Where there is a union membership agreement, a person who is unreasonably refused admission to or unreasonably expelled from the relevant union(s) can take a case to industrial tribunal.
- All industrial action in support of a closed shop is unlawful.

Closed shops or union membership agreements are not unlawful, but have become more difficult to maintain.

Anti-union litigation

- The number of potential plaintiffs has increased and litigation has been made easier.
- A receiver and sequestrator have considerable powers relating to substantial fines and/or the sequestration of union assets.
- Interlocutory injunctions are relatively easy to obtain.
- The Commissioner for the Rights of Trade Union Members provides legal advice and finance for members to litigate against unions.

REFERENCES

ACAS (1991) *Code of Practice No. 3, Time off for Trade Union Duties and Activities*. London: HMSO.

Hendy, J. (1993) *A Law unto Themselves. Conservative Employment Laws: a National and International Assessment*, 3rd edition. London: The Institute of Employment Rights.

Hotel and Catering Training Company (1990) *Employee Relations in the Hotel and Catering Industry*, 7th edition. London: HCTC.

Smith, I.T., Wood, Sir J.C. and Thomas, G. (1993) *Industrial Law*, 5th edition. London: Butterworths.

All cases mentioned in this Appendix are noted in the list of cases.

Appendix FIVE

Equalizing the Employment Relationship

A5.1 DISCRIMINATION

Sex and marital status

Sex Discrimination Act 1975 (amended 1986).

- Applies to persons seeking employment, in employment (no hours or service qualification) and in the provision of goods, services and accommodation (customers).
- Right not to be discriminated against directly and indirectly unless sex etc. is a genuine occupational qualification (discrimination against single people not unlawful).
- Prohibits discrimination at all stages of employment including selection, promotion, transfer, training and other benefits, and dismissal.
- Disparity in actual terms and conditions once engaged falls under Equal Pay Act 1970.
- Applies to all firms (no size qualification).
- Code of Practice for the Elimination of Discrimination on the Grounds of Sex and Marriage and the Promotion of Equality of Opportunity (two parts).
- Equal Opportunities Commission monitors legislation and provides guidance. EOC may set up formal investigations, issue and enforce non-discrimination notices, take action against discriminatory advertisements and persistent discrimination, and help individuals bring claims against employers.

Race, colour, nationality, ethnic or national origins

Race Relations Act 1976.

- Applies to persons seeking employment, in employment (no hours or service qualification) and in the provision of goods, services and accommodation (customers).

- Right not to be discriminated against directly or indirectly on grounds of race etc. unless race etc. is a genuine occupational qualification.
- Prohibits discrimination at all stages of employment, including selection, promotion, transfer, training and other benefits, and dismissal.
- Code of Practice for the Elimination of Racial Discrimination and the Promotion of Equality of Opportunity in Employment (four parts).
- Commission for Racial Equality monitors legislation and provides guidance. CRE may set up formal investigations, issue and enforce non-discrimination notices, take action against discriminatory advertisements and persistent discrimination, and help individuals bring claims against employers.

Ex-offenders

Rehabilitation of Offenders Act 1974.

- Enables ex-offenders to 'live down' their past and make a new start.
- For any offender not convicted during a specified rehabilitation period, the conviction becomes spent.
- Only relatively short sentences can become spent in this way.
- Mention of a spent conviction in a reference could lead to a charge of defamation of character by the ex-offender.

Disablement

- No anti-discrimination legislation but firms employing 20 or more staff have a duty to employ a quota of 3 per cent registered disabled people (Disabled Persons Employment Acts 1944 and 1958).
- Code of Practice on the Employment of Disabled People 1984.
- Every director's report of companies employing more than 250 people must contain a statement concerning the disabled.
- Disablement Advisory Service offers practical guidance and help.

Note

Proposed measures for disabled people announced on 24 November 1994 include abolition of the quota system and a new right to non-discrimination at work in firms employing more than 20 people.

Trade unions

- Unlawful to refuse to employ someone on grounds of trade union membership, or on the grounds that they are not trade union members – direct discrimination only (Employment Act 1990 (now TULRCA 1992)).
- Employees are protected from dismissal, or action short of dismissal, on grounds of union membership or taking part in trade union activities, or on the opposite ground of non-membership of a trade union (Employment Protection Act 1975 (now TULRCA 1992)).

Sexual preference

- No anti-discrimination legislation.
- Refusal to employ male homosexuals but not female homosexuals because of HIV/AIDS possibilities would constitute direct discrimination on grounds of sex.

Age

- No anti-discrimination legislation.
- Right to equality in retirement age.

Pregnancy/maternity

- Relationship of UK law and EC law subject to complex litigation in the ECJ.
- *Prima facie* sex discrimination (*Dekker* v *Stichting Vormingscentrum Voor Jong Volwassenen (VJV – Centrum) Plus* 179/88 [1991] IRLR 27 ECJ; *Hertz* v *Aldi Marked K/S (sub nom Handels-og Kontorfunktionaerernes Forbund i Danmark* v *Dansk Arbejdsgiveforening)* 179/88 [1991] IRLR 31 ECJ).
- The test of whether a dismissal of a pregnant employee is sex discrimination in UK law – how a male employee with a comparable need for period of absence would be treated – now seemingly defunct following *Webb* v *EMO Air Cargo Ltd*, C-32/93 [1994] IRLR 482 ECJ.

Sunday work

Sunday Trading Act 1994 (see Section A6.8).

Right not to suffer any other detriment for refusing to work on Sundays. 'Detriment' is not defined but could include the denial of overtime, promotion or training opportunities.

A5.2 EQUAL PAY

Equal pay

Equal Pay Act 1970 (amended by Equal Pay (Amendment) Regulations 1983).

- Applies to employees regardless of size of firm and hours worked.
- Claims can only be made to industrial tribunal if the complainant has been in employment within the six months preceding the date of the reference.
- Has the effect of writing an equality clause into a woman's (man's) contract of employment.
- Is designed to eliminate sex discrimination from pay and employment conditions.
- A woman (man) has the right to equal pay and other contractual employment terms when she (he) is doing the same or broadly similar work or work rated as

equivalent under a job evaluation scheme as a man (woman) (employed by the same or an associated employer) unless there is a material difference (not sex based) between their cases.
- A woman (man) has the right to equal pay and other contractual employment terms when she (he) is doing work of equal value to a man (woman) (employed by the same or an associated employer) (1983 amendment).
- Relationship of UK law and EC law subject to complex litigation in the ECJ (see Chapter 10).

A5.3 MATERNITY PROVISIONS

Antenatal care

Employment Act 1980.

- No hours or service qualification.
- Pregnant employees have the right not be unreasonably refused paid time off to attend for antenatal care on the advice of a doctor, midwife or health visitor.
- Except for the first appointment, the employee may be asked to produce proof of her appointment and a certificate of pregnancy.
- If she produces due evidence and is unreasonably refused permission to attend, she can complain to an industrial tribunal, which can order the employer to pay her the sum she would have received had she been granted the time off.

Maternity leave and absence

There are now differential arrangements based on existing UK legislation (maternity absence) and the European social dimension (maternity leave).

Maternity leave

TURERA 1993 Sections 23 and 24 (Directive on the Protection of Pregnant Women).

- No minimum hours or service qualification.
- Fourteen weeks' paid maternity leave (protection from dismissal where the reason for dismissal is connected with pregnancy for a further four weeks if covered by a medical certificate).
- Subject to certification of pregnancy and 21 days notice (or as soon as is reasonably practicable).
- Begins no earlier than the eleventh week before the expected week of childbirth.
- Can return at any time during the 14 weeks provided seven days' notice is given.
- No notice is required for return at the end of the 14-week period.
- At least two weeks of leave must be taken.
- All contractual rights continue (except remuneration).
- Automatically unfair to dismiss for pregnancy/maternity (see Dismissal: pregnancy or childbirth).

- Entitled to written reasons for dismissal without asking (penalty two weeks' pay).

Maternity absence

EPCA 1978 (amended by Employment Act 1980), now TURERA 1993 Schedule 2 and EPPTER 1995.

- Subject to having two years' continuous service (no hours qualification).
- Right to return to work at any time up until the end of 29 weeks beginning with the week in which the baby is born.
- Employers employing five or fewer employees are exempt from allowing a woman to return to work if it is 'not reasonably practicable' to do so, or suitable alternative work cannot be offered.
- Employee must be employed (not necessarily at work) up to 11 weeks before the expected week of childbirth.
- Employee must inform employer that she will be absent from work due to pregnancy, and that she intends to return to work, at least 21 days before absence begins (or as soon as is reasonably practicable).
- Not earlier than 21 days before the end of the maternity leave period, if the employer requests written confirmation of intention to return, confirmation must be provided within 14 days or the right to return is forfeited.
- Intention to exercise the right of return must be made in writing at least three weeks in advance, or the right to return is forfeited.
- Return can be postponed for up to four weeks by either party (medical/business grounds).
- Contractual rights are not automatically maintained during the period of absence, except remuneration during the first 14 weeks.
- Automatically unfair to dismiss for pregnancy/maternity (see Dismissal: pregnancy or childbirth).
- Entitled to written reasons for dismissal without asking (penalty two week's pay).
- Reinstatement in original job not necessary: a 'suitable alternative' job may be offered as long as terms and conditions are not 'substantially less favourable' than those of the original job.

Note

Employees with two years' service may take *both* maternity leave and maternity absence. In some cases employees working in small organizations (five or fewer employees) may not be able to take maternity absence.

Suspension

TURERA 1993 (Section 25 and Schedule 3).

- A woman has the right to be suspended from work with pay on maternity grounds where continued employment would be unlawful or contrary to a Code of Practice.

- A woman has the right to suitable alternative work.
- If there is no suitable alternative work, the woman has the right to be suspended with pay.

Dismissal: pregnancy or childbirth

- No hours or service qualification.
- Automatically unfair: pregnancy or any related reason (e.g. miscarriage); dismissal occurs within maternity leave period after the birth; taking maternity leave; where medical certificate provided at end of maternity leave period and the contract is terminated within four weeks; if suspended on maternity grounds (see Suspension); made redundant during maternity leave and not offered alternative employment.

Maternity pay

With 26 weeks' service[a] plus earnings of more than £57 per week plus employed at fifteenth week before the expected week of childbirth:

- Ninety per cent earnings for six weeks (Statutory Maternity Pay (SMP)) plus £52.50 for 12 weeks.
- Employer's rebate for SMP reduced to 92 per cent, except for small employers.

Fewer than 26 weeks' service;[a] Maternity allowance from Department of Social Security if eligible (18 weeks at £52.50).

Note

[a] Service as at qualifying week. All figures effective to 5 April 1995.

Rest places

For pregnant women and nursing mothers (see Workplace (Health and Safety and Welfare) Regulations 1992 section A6.3).

Appendix SIX

Health and Safety Law

A6.1 HEALTH AND SAFETY AT WORK ACT 1974

Employers

- Every employer has the duty, so far as is reasonably practicable, to ensure the health, safety and welfare of all employees (Section 2 (1)).
- Provision and maintenance of safe and risk-free plant and systems of work.
- Ensuring the safety etc. in the use, handling, storage and transport of articles and substances.
- Provision of information, instructions, training and supervision.
- Maintenance of a safe workplace and the provision of risk-free means of entry and exit.
- Provision and maintenance of a safe and risk-free working environment with adequate facilities and arrangements (Section 2 (2)).
- Duty on employers extends to virtually all workers and workplaces, including non-employees and the self-employed (Section 3).
- Where safety representatives have been appointed by recognized trade unions, employers have a duty to consult them, and if requested to do so, to establish a safety committee (Section 2 (4), (6), (7) and Safety Representative and Safety Committees Regulations 1976, extended by Code of Practice and Management of Health and Safety Regulations).
- Employers to use the best practical means to prevent, remove or render harmless emissions of noxious or offensive substances into the atmosphere (Section 5).
- Employers to prepare, and revise as necessary, a written safety policy that brings matters to the attention of all employees (Section 2 (3)).

Beyond employers

Duties on occupiers and controllers of premises, and the manufacturers, suppliers, importers and designers of articles and substances for use at work (Sections 4 and 6).

Employees

- Duty to take reasonable care for own health and safety, and that of others who may be affected by his acts or omissions.
- To cooperate with others regarding any duty imposed under the Act (Section 7).

The Health and Safety Commission

The HSC is under a duty to:

- assist and encourage all the purposes of the general provisions of the Act;
- encourage research and its publication and safety training;
- ensure wide dissemination of advice and information particularly to those practically involved in this field;
- prepare and propose regulations (the power to make regulations lies with the Secretary of State for Employment);
- approve and issue suitable Codes of Practice.

The Health and Safety Executive

The HSE is under a duty to:

- make adequate arrangements for the administration and enforcement of the Act (the Secretary of State can transfer the duty of enforcement to local authorities).
- Enforcement in hotels and catering is by environmental health officers.
- Inspectors have quite wide-ranging powers, including the power of entry at any time where there is danger and to issue improvement and prohibition notices.
- The HSE can instigate prosecutions (civil and criminal) through the inspectorate or the Director of Public Prosecutions.

A6.2 REPORTING OF INJURIES, DISEASES AND DANGEROUS OCCURRENCES REGULATIONS (RIDDOR) 1985

- Major accidents (causing death or defined injuries) and dangerous occurrences must be notified by telephone and written reports on standard form must be sent within seven days.
- Industrially linked diseases must also be reported on the standard form.

- Minor accidents leading to incapacity for work for more than three consecutive days must be reported on the standard form within seven days.

A6.3 HEALTH AND SAFETY REGULATIONS AND CODES (FROM EC DIRECTIVES)

Control of Substances Hazardous to Health Regulations (COSHH) 1988

- Applies to all workplaces where substances hazardous to health are produced, stored, used or manufactured.
- Employers are obliged to adopt principles of good occupational hygiene in dealing with these substances (backed by four Approved Codes of Practice and HSE guidance).
- Identifies substances which are prohibited for certain purposes.
- Where such listed substances are used there has to be an assessment of risks to health.
- There must be prevention of exposure to the substance or the necessary, appropriate control.
- Appropriate control measures have to be used and their use supervised along with their examination, maintenance and testing.
- The health of employees concerned has to be monitored and they have to be given information as to what steps to take if there is exposure to a harmful substance.

Six sets of regulations, all implementing EC Directives, were made under the Health and Safety at Work Act 1974 and came into force on 1 January 1993. All place duties on employers, employees and the self-employed (HSE, 1992a, b, c, d, e, f).

Management of Health and Safety at Work Regulations 1992

Regulations and Code of Practice (see HSE, 1992b).

- Assessment must be undertaken of the risks associated with all work activities, and results recorded where five or more employees are employed.
- Adequate arrangements, which must be recorded where five or more employees are employed, must exist for securing health and safety.
- Provides for health surveillance measures to be operated in certain cases.
- Employers must designate one or more competent persons (this may be an external adviser) with responsibility for health and safety.
- Procedures, including the designation of competent persons to supervise evacuation, must be established to deal with serious and imminent danger.
- Employees must be given adequate safety information and training.
- Non-employees working on the employer's premises must be given information about hazards.
- Employers sharing a workplace must cooperate on safety.
- The duty to consult safety representatives is extended.
- Employees must report all hazards potentially affecting them or arising from their work.

Workplace (Health, Safety and Welfare) Regulations 1992

Regulations and Code of Practice (see HSE, 1992f).

- Apply to new workplaces from 1 January 1993 and existing workplaces from 1 January 1996.
- Replaces and updates the law on the construction of the workplace and minimum facilities and amenities, including the Offices, Shops and Railway Premises Act 1963.
- Basic duty of employers is to maintain the place of work, equipment, devices and systems. They have to be cleaned and maintained in an efficient state, working order and good repair.
- Covers matters such as ventilation, temperature in indoor workplaces, lighting, cleanliness, working space, workstations, floor and traffic routes.
- Deals with the prevention of falls and falling objects, and dangers/substances likely to cause scalds or burns, loading and unloading of vehicles.
- Covers escalators, moving walkways, sanitary conveniences, drinking water, facilities for clothing, for rest and for eating meals.

Provision and Use of Work Equipment Regulations 1992

Regulations and Guidance (see HSE, 1992e).

- Replaces and updates existing regulations.
- To ensure the provision of safe work equipment and its safe usage.
- Existing equipment is exempt from some specific requirements until 1 January 1997.
- All work equipment must be suitable, properly maintained and used only for the purpose for which it is suitable, and by individuals capable of using it safely. Any necessary information and training must be given to such individuals (effective 1 January 1993).
- Equipment must be such as to prevent a range of hazards, and/or control measures must be taken to protect employees from risks associated with the hazards (effective from 1 January 1997).

Health and Safety (Display Screen Equipment) Regulations 1992

Regulations and Guidance (see HSE, 1992a).

- Applies to users (extensive guidance given), who regularly use VDU equipment to a significant extent, and their equipment.
- Applies to laptops and homeworkers.
- All post-1992 VDU workstations must meet ergonomic requirements for hardware, software, lighting, noise, work surface and chair.
- Existing workstations must comply by 31 December 1996.
- All workstations must be analysed for risks to health and safety of users.
- Users' work must be planned to minimize health hazards, for example, the use of short breaks.
- Users must be informed and trained about hazards and how to avoid them; for example, bad posture.

- Users are entitled to a free eye test on request (and must be informed of their right to request it) and corrective appliances needed for VDU work must be supplied free. Future eye tests must be provided at intervals specified by the ophthalmologist.

Manual Handling Operations Regulations 1992

Regulations and Guidance (see HSE, 1992c).

- First time such operations have been covered by detailed regulations.
- Put forward largely as guidance to cover infinitely wide range of actions a worker might be called upon to undertake.
- Framed in the form of a flow chart of questions to be asked and considered.
- Require the avoidance of unnecessary handling operations.
- Each task to be assessed and a way devised to perform it so as to reduce risk to a minimum.
- Precise information on the weight and characteristics of loads to be given when necessary.

Personal Protective Equipment at Work Regulations 1992

Regulations and Guidance (see HSE, 1992d).

- Provision is a last resort to be used only if there is no way of removing the danger.
- Covers a wide range of clothing extending to adverse weather protection (excludes ordinary clothing, uniforms and personal protective equipment required under certain existing Regulations, such as noise protectors and hard hats for construction workers).
- Equipment must be assessed prior to supply, properly maintained and accommodated.
- Users must receive adequate information and training.
- Employers must take reasonable steps to ensure the equipment is properly used.
- Employees must use and look after items properly and report losses or defects.

Management of Health and Safety (Amendment) Regulations 1994

- Effective 1 December 1994.

Special measures for:

- Pregnant women, those who have recently given birth and those who are breastfeeding.
- Risks must be assessed.
- Work to be adjusted to remove risk or women transferred to safer work.
- Suspension on full terms and conditions if no safe work available.
- May refuse night work on medical grounds.

A6.4 WORKING TIME DIRECTIVE

Subject to a UK challenge that the Directive's legal base (Article 118A) is invalid. Judgement from the ECJ is not expected before late 1995. If the challenge fails, the Directive will have to be implemented by all member states by November 1996.

For all workers

- Maximum average 48-hour week (including overtime) normally over a reference period of four months. The reference period may rise to six months if the situation is covered by a derogation, or to 12 months if the situation is subject to a collective agreement.
- Eleven consecutive hours of rest in 24 hours.
- Thirty-five consecutive hours of rest a week, in principle including Sundays. These can be averaged out over two weeks. There are some exceptions, including a minimum of 24 hours of rest a week on technical or work organization grounds.
- Rest break whenever the working day exceeds six hours.
- Four weeks' paid annual leave. No payment in lieu except on termination of the contract. Three weeks' paid leave for the first three years (to 1999).
- Work organization to take account of health and safety and workers' needs.

Night workers

Night work to be a seven-hour period set by member states and to include midnight to 05.00. Night workers are those working regularly over three hours during this period. The following rights are in addition to those above.

- Night shift not to exceed an average of eight hours in a 24 hours, over an agreed period. The reference period for averaging to be determined after consultation, or by agreement.
- An absolute limit of eight hours applies where special hazards or strain are involved (to be determined nationally by law or agreement).
- Health and safety protection and provision as for day workers.
- Free medical checks at start of work, and at regular intervals thereafter.
- Right to transfer to day work on medical grounds.
- Introduction of night work to be notified, if required, to relevant authorities.

Derogations

- From all provisions for certain types of worker: managers, transport workers, work at sea and family or religious workers.
- To all limits (excluding the maximum working week) in law, collective agreements and agreements between two sides of industry for particular activities, including where continuity of service or production is required and shift work. Compensatory rest must be given, or exceptionally, appropriate protection.
- Exemptions to daily and weekly rest periods, rest breaks and night work by collective agreements or agreements between the two sides of industry, at national, regional or a lower level.

Exceptions

Voluntary opt-out from 48-hour week will be allowed for seven years after implementation (to 2003). The worker must have freely consented to work longer hours and the employer must keep records to be made available for inspection by the competent authorities.

A6.5 PROTECTION OF YOUNG PEOPLE AT WORK DIRECTIVE

All those under 18

- Employers must carry out risk assessments before they start work and inform the young workers and their parents/guardians of possible risk involved.
- Exposure ban for certain work activities (some exceptions for supervised vocational training).

For those aged 15 to 18 and no longer in full-time compulsory education

- Maximum eight-hour day and 40-hour week. Some temporary exceptions allowed.[a]
- No night work between 22.00 hours and 06.00 hours or 23.00 hours and 07.00 hours. Certain exceptions permitted but never between midnight and 04.00 hours.[a]
- Daily rest period of 12 consecutive hours in every 24 hours.
- Weekly rest period of two days every seven days, consecutively if possible. Certain flexibility, provided compensatory rest is given.
- Thirty minutes rest break after 4.5 hours at work.

In full-time education

Maximum 12 hour week (over-13s).[a]

Note

[a]Adopted June 1994 for implementation in 1996. These items will not apply to the UK four years after implementation.

A6.6 CHILDREN AND YOUNG PEOPLE

Children and Young Persons Act 1933 (amended 1963, 1972 and 1989).

- No child[a] under age 13 should be employed.
- No employment before close of normal school hours.

- No employment before 07.00 hours or after 19.00 hours on any day; no employment for more than two hours on any school day and Sunday.
- No employment in a bar or betting shop.
- No lifting, carrying or moving anything heavy that may lead to injury.
- Restrictions on the various types of machinery that children may use.
- Local bye-laws may prohibit the employment of children in any specified occupation and may prescribe, subject to the restrictions imposed by the main legislation, daily or weekly hours, intervals for meals and rest, holidays or half-holidays and other conditions to be observed in relation to employment.
- Restrictions removed on hours of work (including night work) of young people (those between school-leaving age and age 18) (Employment Act 1989).
- Young persons may not clean dangerous machinery without adequate supervision, and must be instructed about the use of 'dangerous' machines (including food mixers, mincers, slicers and chippers) and their dangers (Offices, Shops and Railway Premises Act 1963).
- Change pending through Directive on the protection of young people at work.

Note

[a] A child is someone who is of compulsory school age. A child who is 16 between 1 September and 31 January inclusive is of compulsory school age until the end of the Spring term. A child who is 16 between 1 February and 31 August inclusive is of compulsory school age until the Friday before the last Monday in May in that year.

A6.7 SHOPS ACT 1950

- Where retail trade of business carried on, e.g. restaurants and public houses.
- Applies to 'shop assistants', which includes waiters and bar staff, and any person involved in serving customers, receipt of orders or the supply of goods, such as a kitchen assistant.
- Two sets of rules – all shops and catering trade premises (where refreshments are sold for consumption on the premises). Employers must choose one set of rules or the other.

Shop rules

- At least one half-day off (after 13.30 hours).
- Meal break of one hour for lunch between 11.30 hours and 14.30 hours (45 minutes if meal is taken on the premises) and half an hour for tea between 16.00 hours and 19.00 hours.
- Where employed on the sale of refreshments and/or intoxicating liquor, the break need not be between 11.30 hours and 14.30 hours if he or she is allowed a break immediately before or after that period.
- Continuous employment of more than six hours is prohibited without a 20-minute break.

- For every four hours worked on a Sunday, the employee must receive a whole day's holiday during either the previous or the following week.
- For every period less than four hours worked on Sundays (excluding bar staff) a half day's holiday must be allowed.
- Only three Sundays a month may be worked.

Catering trade rules

- Apply to all workers, not just shop assistants.
- Designed to allow greater flexibility.
- Maximum working week of 65 hours excluding meal breaks.
- Thirty-two weekday holidays per annum, at least two per month and one period of six consecutive days a year.
- Twenty-six Sundays off a year. No more than two consecutive Sundays may be worked.
- Meal breaks to be not less than 45 minutes during half-day working and not less than two hours in total during the whole day.
- Continuous employment of more than six hours is prohibited without a half-hour break.

A6.8 SUNDAY TRADING ACT 1994

- Right not to be dismissed for refusing to work on Sundays.
- Right not to be selected for redundancy for refusing to work on Sundays.
- Right not to suffer any other detriment[a] for refusing to work on Sundays.
- Workers can only give up their right not to work on Sundays by giving their employer a signed and dated written 'opting in notice' stating that they wish to work on Sunday or that they do not object to working on Sundays.
- Workers may opt out of Sunday working by giving the employer written notice and serving a three-month notice period.
- After the three-month period has ended a worker may complain to an industrial tribunal that any of the above rights has been infringed.

Note

[a]Not defined in the Act but could include the denial of overtime, promotion or training opportunities.

A6.9 STATUTORY SICK PAY (SSP)

Social Security Contributions and Benefits Act (SSCBA) 1992 amended by the Statutory Sick Pay Act 1994.

- SSP is payable for up to 28 weeks' absence owing to sickness or injury in any single 'period of entitlement'.

- Those paying Class 1 National Insurance contributions and married women/widows paying reduced contributions are eligible.
- Part-timers earning more than the current earnings limit (£57 per week in the 1994–5 tax year) are treated as full-timers – there is no minimum service qualification.
- Employees may be entitled to SSP under more than one contract or with more than one employer if relevant conditions are satisfied.
- Those ineligible for SSP can claim state sickness benefit.
- Those who have exhausted their entitlement to SSP can claim invalidity benefit (now called 'incapacity benefit').
- SSP is payable from the fourth day of incapacity for work, provided employees have complied with provisions requiring them to notify the employer of incapacity for work.
- Four or more consecutive days (including non-working days) of incapacity for work form 'a period of incapacity for work'.
- Periods of incapacity for work which are separated by eight weeks or less are treated as one so that employees do not have to wait another three days to qualify.
- A single incapacity cannot extend beyond three years.

There is no entitlement to SSP if:

- employees are over state pension age;
- the employee is engaged on a fixed-tern contract of three months or less and that period is not exceeded;
- the employee's gross weekly earnings are less than the current earnings limit for social security purposes;
- no work has been performed under a contract of service;
- in certain other circumstances including those related to maternity leave, a trade dispute and imprisonment.

Payment varies according to normal gross weekly earnings. From April 1994 to April 1995:

- £47.80 per week for earnings between £56 and £195 per week;
- £52.50 if normal weekly earnings exceed £195;
- daily rate is the appropriate weekly rate divided by the number of 'qualifying days' in the week.

REFERENCES

Health and Safety Executive (1992a) *Display Screen Equipment at Work: Health and Safety (Display Screen Equipment) Regulations 1992, Guidance on Regulations L 26*. London: HMSO.

Health and Safety Executive (1992b) *Management of Health and Safety at Work: Management of Health and Safety at Work Regulations 1992, Approved Code of Practice L 21*. London: HMSO.

Health and Safety Executive (1992c) *Manual Handling: Manual Handling Operations Regulations 1992, Guidance on Regulations L 23*. London: HMSO.

Health and Safety Executive (1992d) *Personal Protective Equipment at Work: Personal Protective Equipment at Work Regulations 1992, Guidance on Regulations L 25*. London: HMSO.

Health and Safety Executive (1992e) *Work Equipment at Work: Provision and Use of Work Equipment Regulations 1992, Guidance on Regulations L 22*. London: HMSO.

Health and Safety Executive (1992f) *Workplace Health, Safety and Welfare: Workplace (Health, Safety and Welfare) Regulations 1992, Guidance on Regulations L 24*. London: HMSO.

Lists of Legal Cases, Statutes and Regulations

CASES

The vast majority of cases cited throughout the book can be found in Industrial Relations Law Reports (IRLR). Notations used in the abbreviations are as follows:

CA	Court of Appeal
EAT	Employment Appeal Tribunal
ECJ	European Court of Justice
HL	House of Lords
IT	Industrial Tribunal
NIHC	Northern Ireland High Court

A few cases not reported in IRLR include those reported in Common Market Law Reports (CMLR), European Court Reports (ECR), Industrial Relations Law Bulletin (IRLB), Industrial Tribunal Reports (ITR) and IDS Brief. Numbers given in **bold** indicate page references for cases.

Aberdeen Steak Houses Group plc v *Ibrahim* [1988] IRLR 420 EAT, **277**.
Babar Indian Restaurant v *Rawat* [1985] IRLR 57 EAT, **277**.
Barber v *Guardian Royal Exchange Assurance Group* 262/88 [1990] IRLR 240 ECJ, **108**, **222**.
Bestuur van het Algemeen Burgerlijk Pensioenfonds v *Beune* C-7/93 [1994] *IDS Brief 527 ECJ*, **223**.
Birds Eye Walls Ltd v *Roberts*, C-132/91 [1994] IRLR 29 ECJ, **223**.
Bouchaala v *Trust House Forte Hotels Ltd* [1980] IRLR 382 EAT, **277**.
BSC Sports and Social Club v *Morgan* [1987] IRLR 391 EAT, **277**.
Burgess v *Bass Taverns Ltd* [1994] IRLB 499 EAT, **278**.
Capper Pass Ltd v *Lawton* [1976] IRLR 366 EAT, **221**.
Coloroll Pension Trustees Ltd v *Russell and others* C-200/91 [1994] IRLR 586 ECJ, **223**.
Commission for Racial Equality v *Dutton* [1989] IRLR 8 CA, **218**.

Commission of the European Communities v *United Kingdom of Great Britain and Northern Ireland* 61/81 [1982] IRLR 333 ECJ, **220**.
Commission of the European Communities v *United Kingdom of Great Britain and Northern Ireland* C-382/92 [1994] IRLR 392 ECJ, **177**, **191**, **199**, **203**, **241**, **267**, **274**.
Commission of the European Communities v *United Kingdom of Great Britain and Northern Ireland* C-383/92 [1994] IRLR 412 ECJ, **177**, **191**, **199**, **203**, **241**, **267**, **274**.
Defrenne v *Sabena (No. 2)* C-43/75 [1976] ECR 455 ECJ, **220**.
Dekker v *Stichting Vormingscentrum voor Jong Volwassenen (VJV–Centrum) Plus* 179/88 [1991] IRLR 27 ECJ, **225**, **286**.
Delaney v *Staples* [1992] IRLR 191 HL, [1991] 112 CA and [1990] IRLR 86 EAT, **194**, **197**, **200**.
Dhatt v *McDonald's Hamburgers* [1991] IRLR 130 CA, **218**.
Enderby v *Frenchay Health Authority and Secretary of State for Health* 127/92 [1993] IRLR 591 ECJ, [1992] 15 CA and [1991] IRLR 44 EAT, **221**.
Figael v *Fox (Inspector of Taxes)* [1990] (cited in Peters, 1992, p. 198), **119**.
Fisscher v *Voorhuis Hengelo BV and another* C-128/93 [1994] *IDS Brief 527* ECJ, **223**.
Ford v *Stakis Hotels and Inns Ltd* [1988] IRLR 46 EAT, **227**.
Frankovich v *Italian Republic* 6/90 [1992] IRLR 84 ECJ, **191**, **226**, **267**.
Garland v *British Rail Engineering Ltd* [1982] IRLR 111 ECJ and [1982] IRLR 257 HL, **222**.
Gateway Hotels Ltd v *Stewart and others* [1988] IRLR 287 EAT, **277**.
Gill and Coote v *El Vino Company Ltd* [1983] IRLR 206 CA, **216**.
Hall v *Honeybay Caterers Ltd* (1967) 2 ITR 538, **119**.
Hayes v *Malleable Working Men's Club and Institute* [1985] IRLR 367 EAT, **225**.
Hayward v *Cammell Laird Shipbuilders Ltd* [1988] IRLR 257 HL, [1987] IRLR 186 CA, [1986] 287 EAT and [1984] 463 IT, **220**.
Hendry and another v *Scottish Liberal Club* [1977] IRLR 5 IT, **219**.
Hertz v *Aldi Marked K/S (sub nom Handels- og Kontorfunktionaerernes Forbund i Danmark* v *Dansk Arbejdsgiverforening*) 199/88 [1991] IRLR 31 ECJ, **225**, **286**.
Hewcastle Catering Ltd v *Ahmed and another* [1991] IRLR 473 CA, **277**.
Hilton International Hotels (UK) Ltd v *Faraji* [1994] IRLR 267 EAT, **278**.
Hilton International Hotels (UK) Ltd v *Kaissi* [1994] IRLR 270 EAT, **226**.
Hilton International Hotels (UK) Ltd v *Protopapa* [1990] IRLR 316 EAT, **277**.
Hotson v *Wisbech Conservative Club* [1984] IRLR 422 EAT, **277**.
Keywest Club Ltd v *Choudhury and others* [1988] IRLR 51 EAT, **119**.
Les Ambassadeurs Club v *Bainda* [1982] IRLR 5 EAT, **277**.
Marleasing SA v *La Commercial Internacional de Alimentacion SA* [1992] 1 CMLR 305 ECJ, **268**.
Marshall v *Southampton and South-West Hampshire Area Health Authority (Teaching)* 152/84 [1986] IRLR 140 ECJ and [1983] IRLR 237 EAT, **222**.
Marshall v *Southampton and South-West Hampshire Area Health Authority (Teaching) (No. 2)* 271/91 [1993] IRLR 455 ECJ, **194**, **217**, **221**.
McConomy v *Croft Inns Ltd* [1992] IRLR 561 NIHC, **216**.
McGrath v *Rank Leisure* [1985] IRLR 323 EAT, **199**, **277**.
Mediguard Services Ltd v *Thame* [1994] IRLR 504 EAT, **220**, **224**.
Ministry of Defence v *Cannock and others* [1994] IRLR 509 EAT, **217**.
Moroni v *Firma Collo GmbH* C-110/91 [1994] IRLR 130 ECJ, **222**.
Moyes v *Hylton Castle Working Men's Social Club and Institute Ltd* [1986] IRLR 482 EAT, **277**.

Neath v *Hugh Steeper Ltd* C-152/91 [1994] IRLR 91 ECJ, **222**.
O'Kelly and others v *Trusthouse Forte plc* [1983] IRLR 369 CA, **66**, **277**.
Palmanor Ltd (t/a Chaplins Night Club) v *Cedron* [1978] IRLR 303 EAT, **119**.
Paola Faccini Dori v *Recreb Sri*, C-91/92 [1994] *The Times, 4 August, ECJ*, **191**.
R v *Secretary of State for Employment, ex parte EOC* [1994] IRLR 176 HL reversing [1993] IRLR 10 CA, **62**, **218**.
Rask and another v *ISS Kantineservice A/S* 209/91 [1993] IRLR 133 ECJ, **199**.
Roadchef Ltd v *Hastings* [1988] IRLR 142 EAT, **277**.
Showboat Entertainment Centre v *Owens* [1984] IRLR 7 EAT, **218**.
Smith and others v *Avdel Systems Ltd*, C-408/92 [1994] IRLR 602 ECJ, **223**.
Society of Licensed Victuallers v *Chamberlain* [1989] IRLR 421 EAT, **277**.
Soros and Soros v *Davison and Davison* [1994] IRLR 264 EAT, **278**.
Taylorplan Catering (Scotland) Ltd v *McInally* [1980] IRLR 53 EAT, **277**.
Ten Oever v *Stichting Bedrijfspensioenfonds voor het Glasenwassers en Schoonmaakbedrijf* 109/91 [1993] IRLR 601 ECJ, **222**.
Trust Houses Forte Leisure Ltd v *Aquilar* [1976] IRLR 251 EAT, **157**, **277**.
Trust Houses Forte Hotels Ltd v *Murphy* [1977] IRLR 186 EAT, **277**.
Trusthouse Forte (Catering) Ltd v *Adonis* [1984] IRLR 382 EAT, **157**, **277**.
Tsoukka v *Potomac Restaurants Ltd* (1968) 3 ITR 259, **119**.
Van den Akker and others v *Stichting Shell Pensioenfonds* C-28/93 [1994] IRLR 616 ECJ, **223**.
Vroege v *NCIV Instituut voor Volkshuisvesting BV and another C-57/93 [1994] IDS Brief 527* ECJ, **223**.
Webb v *EMO Air Cargo (UK) Ltd*, C-32/93 [1994] IRLR 482 ECJ, **225**, **267**, **286**.
Young v *Daniel Thwaites and Co* [1977] ICR 877, **199**.

References

Peters, R. (1992) *Essential Law for Hotel and Catering Students*. London: Hodder and Stoughton.

UK STATUTES AND REGULATIONS

Children and Young Persons Act 1933, **243**, **296**.
Companies Act 1985, **236**.
Contracts of Employment Act 1963, **192**, **269**.
Contracts of Employment Act 1972, **192**, **269**.
Control of Substances Hazardous to Health Regulations 1988, **240**, **292**.
Disabled Persons Employment Act 1944, **285**.
Disabled Persons Employment Act 1958, **285**.
Employment Act 1980, **269**, **288**.
Employment Act 1982, **179**, **269**.
Employment Act 1988, **269**
Employment Act 1989, **269**, **270**, **272**, **278**, **297**.
Employment Act 1990, **285**.
Employment Protection Act 1975, **192**, **208**, **238**, **269**, **270**, **271**, **272**, **273**, **274**, **285**.
Employment Protection (Consolidation) Act 1978, **193**, **195**, **198**, **200**, **269**, **270**, **271**, **272**, **273**, **274**, **275**, **276**, **278**.

OTHER STATUTES

Bibliography

Ahlstrand, B. and Purcell, J. (1988) Employee relations strategy in the multi-divisional company. *Personnel Review*, **17** (3), 3–11.

Analoui, F. (1987) An investigation into unconventional behaviours within the workplace. PhD thesis, Cranfield School of Management.

Armstrong, M. and Murlis, H. (1994) *Reward Management: a Handbook of Remuneration Strategy and Practice*, 3rd edition. London: Kogan Page Ltd (in association with the Institute of Personnel Management).

Bacon, N. and Storey, J. (1993) Individualization of the employment relationship and the implications for trade unions. *Employee Relations*, **15** (1), 5–17.

Baethge, M. and Oberbeck, H. (1988) Service society and trade unions. *The Service Industries Journal*, **8** (3), 389–400.

Baglioni, G. and Crouch, C. (eds), (1991) *European Industrial Relations: the Challenge of Flexibility*. London: Sage.

Baker, S. (1988) Hotel management – a disadvantaged profession for women. Paper presented to the International Association of Hotel Management Schools Autumn Symposium, Leeds.

Baum, T. (1993) Human resource concerns in European tourism: strategic response and the EC. *International Journal of Hospitality Management*, **12** (1), 77–88.

Baum, T. (ed.) (1993) *Human Resource Issues in International Tourism*. Oxford: Butterworth Heinemann.

Baum, T. and Mahesh, V.S. (1991) Sensitising front-line staff to the needs of customers in the hospitality industry. Paper presented to the Third International Journal of Contemporary Management Conference, October, Bournemouth.

Beardwell, I. (1991–2) The 'new industrial relations'? A review of the debate. *Human Resource Management Journal*, **2** (2), 1–7.

Beardwell, I. and Holden, L. (eds) (1994) *Human Resource Management: a Contemporary Perspective*. London: Pitman.

Boxall, P. (1992) Strategic human resource management: beginnings of a new theoretical sophistication? *Human Resource Management Journal*, **2** (3), 60–79.

Brownell, J. (1994) Women in hospitality management: general managers' perceptions of factors related to career development. *International Journal of Hospitality Management*, **13** (2), 101–17.

Brymer, R.A. (1991) Employee empowerment: a guest-driven leadership strategy. *Cornell Hotel and Restaurant Administration Quarterly*, **32**, 58–68.

Burgess, P. (ed.) (1990) *Recruitment*. European Management Guides. London: Incomes Data Services/Institute of Personnel Management.

Burgess, P. (ed.) (1991) *Industrial Relations*. European Management Guides. London: Incomes Data Services/Institute of Personnel Management.

Burgess, P. (ed.) (1991) *Terms and Conditions of Employment*. European Management Guides. London: Incomes Data Services/Institute of Personnel Management.

Burgess, P. (ed.) (1992) *Pay and Benefits*. European Management Guides. London: Incomes Data Services/Institute of Personnel Management.

Cannell, S. and Wood, S. (1992) *Incentive Pay – Impact and Revolution*. London: Institute of Personnel Management and NEDO.

Chapman, R. (1990) Personnel management in the 1990s. *Personnel Management*, January, 28–32.

Chivers, T.S. (1971) Chefs and cooks. PhD thesis, University of London.

Clark, J. (1993) Procedures and consistency versus flexibility and commitment in employee relations: a comment on Storey. *Human Resource Management Journal*, **4,** (1), 79–81.

Cockburn, C. (1989) The long and short agenda. *Industrial Relations Journal*, **20** (3), 213–25.

Daniel, W. W. (1987) *Workplace Industrial Relations and Technical Change*. London: Frances Pinter/Policy Studies Institute.

Dastmalchian, A. and Blyton, P. (1992) Organizational structure, human resource practices and industrial relations. *Personnel Review*, **21** (1), 58–67.

Dickenson, F. (1988) *Drink and Drugs at Work: the Consuming Problem*. London: Institute of Personnel Management.

Dickinson, A. and Ineson, E.M. (1993) The selection of quality operative staff in the hotel sector. *International Journal of Contemporary Hospitality Management*, **5,** (1) 16–21.

Dienhart, J.R., Gregoire, M.B., Downey, R.G. and Knight, P.K. (1992) Service orientation of restaurant employees. *International Journal of Hospitality Management* **11** (4), 331–46.

Drummond, K.E. (1990) *Human Resource Management for the Hospitality Industry*. New York: Van Nostrand Reinhold.

Dunn, S. (1990) Root metaphor in the old and new industrial relations. *British Journal of Industrial Relations*, **28** (1), 1–31.

Dunn, S. (1991) Root metaphor in industrial relations – a reply to Tom Keenoy. *British Journal of Industrial Relations*, **29** (2), 329–36.

Eaton, J. (1990/1) Human resource management and business policy. *Human Resource Management Journal*, **1** (2), 60–8.

Ferner, A. and Hyman, R. (eds) (1992) *Industrial Relations in the New Europe*. Oxford: Blackwell.

Gabriel, Y. (1985) *Pressure Cooking: an Investigation of Work in the Catering Industry*. Manchester: Manchester Polytechnic, Food Policy Unit, Hollings Faculty.

Goffee, R. and Scase, R. (1982) 'Fraternalism' and 'paternalism' as employer strategies in small firms. In G. Day *et al.* (eds), *Diversity and Decomposition in the Labour Market*, pp. 107–204. Aldershot: Gower.

Guest, D. (1989) Human resource management: its implications for industrial relations and trade unions. In J. Storey (ed.), *New Perspectives on Human Resource Management*, pp. 41–56. London: Routledge.

Guest, D. (1991) Personnel management: the end of an orthodoxy? *British Journal of Industrial Relations*, **29** (2), 149–75.

Guest, D. (1992) Employee commitment and control. In J.F. Hartley and G.M. Stephenson (eds), *Employment Relations: the Psychology of Influence and Control at Work*, pp. 111–35. Oxford: Blackwell.

Harris, V. and Johnson, K. (1988) Labour turnover in National Health Service domestic services departments pre- and post-competitive tendering. Paper presented to the International Association of Hotel Management Schools Autumn Symposium, Leeds.

Hayter, R. (1993) *Careers and Training in Hotels, Catering and Tourism*. Oxford: Butterworth Heinemann.

Hendry, C. and Pettigrew, A. (1986) The practice of strategic human resource management. *Personnel Review*, **15** (5), 3–8.

Hendry, C. and Pettigrew, A. (1990) Human resource management: an agenda for the 1990s. *The International Journal of Human Resource Management*, **1** (1), 17–43.

Hunt, J.W. (1985) Predicting differences in employee relations practice from values and beliefs of managers. *Personnel Review*, **16** (3), 9–15.

Hunter, L. and Thorn, G. (1991) External advisory services in labour management: a worm's eye view. *Human Resource Management Journal*, **2** (1), 22–41.

Incomes Data Services (1991) *Industrial Relations. European Management Guides*. London: Institute of Personnel Management (see Burgess).

Ineson, E.M. and Brown, S.H.P. (1992) The use of bio-data for hotel employee selection. *International Journal of Contemporary Hospitality Management*, **4** (2), 8–12.

Ingamells, W., Rouse, S. and Worsfold, P. (1991) Employment of the disabled in the hotel and catering industry: assumptions and realities. *International Journal of Hospitality Management*, **10** (3), 279–88.

Ingram, A. and Maclean, H. (1993) Human resources management: review of current texts and student guide. *International Journal of Hospitality Management*, **12** (3), 235–42.

Jagger, L. and Maxwell, G. (1988) Women in top jobs. Paper presented to the International Association of Hotel Management Schools Autumn Symposium, Leeds.

Jones, P. and Davies, A. (1991) Empowerment: a study of general managers of four star hotel properties in the UK. *International Journal of Hospitality Management*, **10** (3), 211–17.

Jones, P. and Pizam, A. (eds) (1993) *The International Hospitality Industry: Organizational and Operational Issues*. London: Pitman.

Jones, T., McEvoy, D. and Barrett, G. (1994) Labour intensive practices in the ethnic minority firm. In J. Atkinson and D. Storey (eds), *Employment, the Small Firm and the Labour Market*, pp. 172–205. London: Routledge.

Keegan, B.M. (1983) Leadership in the hospitality industry. In E. Cassee and R. Reuland (eds), *The Management of Hospitality*, pp. 69–93. Oxford: Pergamon Press.

Keenoy, T. (1991) The roots of metaphor in the old and new industrial relations. *British Journal of Industrial Relations*, **29** (2), 313–28.

Kelliher, C. (1993) Competitive tendering, management strategy and industrial relations in the public sector: the case of NHS catering. PhD thesis, London Business School, University of London.

Kirkbride, P.S. (ed.) (1994) *Human Resource Management in Europe: Perspectives for the 1990s*. London: Routledge.

Kochan, T.A., Katz, H. and McKersie, R.B. (1986) *The Transformation of American Industrial Relations*. New York: Basic Books.

Kohl, P.P. (1983) The impact of the Age Discrimination in Employment Act on employee benefit plans in the United States. *International Journal of Hospitality Management*, **2** (4), 203–6.

Legge, K. (1988) Personnel management in recession and recovery. *Personnel Review*, **17** (2), 1–72.

Lennon, J.J. and Wood, R.C. (1992) The teaching of industrial and other sociologies in higher education: the case of hotel and catering management studies. *International Journal of Hospitality Management*, **11** (3), 239–54.

Lucas, R.E. (1994) Part-time youth pay in catering: issues for the 1990s. Paper presented to the Third CHME Research Conference, Napier University, April.

McColgan, A. (1994) *Pay Equity – Just Wages for Women?* London: Institute of Employment Rights.

McDonald, M. (1988) Written disciplinary agreements and the Unfair Dismissals Act 1977. Paper presented to the International Association of Hotel Management Schools Autumn Symposium, Leeds.

McLoughlin, I. and Gourlay, S. (1991/2) Transformed industrial relations? Employee attitudes in non-union firms. *Human Resource Management Journal*, **2** (2), 8–28.

Marginson, P., Edwards, P., Martin, R., Purcell, J. and Sisson, K. (1988) *Beyond the Workplace: Managing Industrial Relations in a Multi-establishment Enterprise*. Oxford: Basil Blackwell.

Marsden, D. (ed.) (1992) *Pay and Employment in the New Europe*. Aldershot: Edward Elgar.

Marsh, A. (1983) *Employee Relations Policy and Decision Making*. Aldershot: Gower.

Mintzberg, H. (1983) *Structure in Fives: Designing Effective Organizations*. Englewood Cliffs, New Jersey: Prentice Hall.

Mok, C. and Finley, D. (1986) Job satisfaction and its relationship to demographics and turnover of hotel food-service workers in Hong Kong. *International Journal of Hospitality Management*, **5** (2), 71–8.

Mullins, L. (1988) Managerial behaviour and development in the hospitality industry. Paper presented to the International Association of Hotel Management Schools Autumn Symposium, Leeds.

Mullins, L. and Davies, I. (1991) What makes for an effective hotel manager? *International Journal of Contemporary Hospitality Management*, **3** (1), 22–5.

Mullins, L., Meudell, K. and Scott, H. (1993) Developing culture in short-life organizations. *International Journal of Contemporary Hospitality Management*, **5** (4), 15–19.

National Economic Development Council (1989) *Recruitment Challenges: Tackling the Labour Squeeze in Tourism and Leisure*. London: NEDC Tourism and Leisure Industries Sector Group.

Purcell, J. and Gray, A. (1986) Corporate personnel departments and the management of industrial relations: two case studies in ambiguity. *Journal of Management Studies*, **23** (2), 205–23.

Ramage, P. (1988) Labour turnover of graduates in the hotel and catering industry. Paper presented to the International Association of Hotel Management Schools Autumn Symposium, Leeds.

Reichel, A. and Pizam, A. (1984) Job satisfaction, lifestyle and demographics of US hospitality workers – versus others. *International Journal of Hospitality Management*, **3** (3), 123–34.

Riley, M. (1990) The labour retention strategies of UK hotel managers. *The Service Industries Journal*, **10** (3), 614–18.

Riley, M. (1992) Labour utilization and collective agreements: an international comparison. *International Journal of Contemporary Hospitality Management*, **4** (4), 21–3.

Riley, M. and Turam, K. (1988) The career paths of hotel managers: a developmental approach. Paper presented to the International Association of Hotel Management Schools Autumn Symposium, Leeds.

Rose, E. (1994) The 'disorganized paradigm' British industrial relations in the 1990s. *Employee Relations*, **18** (1), 27–40.

Ross, L.E. and Boles, J.S. (1994) Exploring the influence of workplace relationships on work-related attitudes and behaviors in the hospitality work environment. *International Journal of Hospitality Management*, **13** (2), 155–71.

Saunders, K.C. (1980) Head hall porters. *Employee Relations*, **2** (2), 12–16.

Saunders, K.C. (1981a) Measuring job stableness in employment: an experimental analysis based on life/work histories of workers with special reference to the hotel and catering industry. *Service Industries Review*, **1** (3), 25–43.

Saunders, K.C. (1981b) *Social Stigma of Occupations*. Farnborough: Gower.

Saunders, K.C. (1985) *Who Is Your Kitchen Porter?* London: Middlesex Polytechnic Research Monographs.

Seymour, D. (1985) An occupational profile of hoteliers in France. *International Journal of Hospitality Management*, **4** (1), 3–8.

Sisson, K. (1992) Industrial relations: challenges and opportunities. *Employee Relations*, **13** (6), 3–10.

Sisson, K. (ed) (1994) *Personnel Management: a Comprehensive Guide to Theory and Practice in Britain*, 2nd edition. Oxford: Blackwell.

Slattery, P. (1983) Social scientific methodology and hospitality management. *International Journal of Hospitality Management*, **2** (1), 9–14.

Stone, G. (1988) Personality and effective hospitality management. Paper presented to the International Association of Hotel Management Schools Autumn Symposium, Leeds.

Storey, J. (1992) *Developments in the Management of Human Resources*. Oxford: Blackwell.

Storey, J. and Bacon, N., with Edmonds, J. and Wyatt, P. (1993) The 'new agenda' and human resource management: a roundtable discussion with John Edmonds. *Human Resource Management Journal*, **4** (1), 63–70.

Teare, R. and Boer, A. (eds) (1991) *Strategic Hospitality Management: Theory and Practice for the 1990s*. London: Cassell.

Thurley, K. and Wood, S. (eds) (1983) *Industrial Relations and Management Strategy*. Cambridge: Cambridge University Press.

Torrington, D. (1994) *International Human Resource Management: Think Globally, Act Locally*. London: Prentice Hall.

Towers, B. (ed.) (1992) *A Handbook of Industrial Relations Practice*, 3rd edition. London: Kogan Page.

Towers, B. (ed) (1992) *The Handbook of Human Resource Management*. Oxford: Blackwell.

Venison, P. (1983) *Managing Hotels*. London: Heinemann.

Vest, J.M. and Murmann, S.K. (1992) Job seeker gender and its relationship to work-related attributes. *International Journal of Hospitality Management*, **11** (3), 269–78.

Walsh, T. (1991) 'Flexible' employment in the retail and hotel trades. In A. Pollert (ed.) *Farewell to Flexibility*, pp. 104–15. Oxford: Blackwell.

Wilkinson, A., Allen, P. and Snape, E. (1991) TQM and the management of labour. *Employee Relations*, **13** (1), 24–31.

Williams, P.W. and Hunter, M. (1992) Supervisory hotel employee perceptions of management careers and professional development requirements. *International Journal of Hospitality Management*, **11** (4), 347–58.

Witz, A. and Wilson, F. (1982) Women workers in service industries. *Service Industries Review*, **2** (2), 40–55.

Wood, R.C. (1992) Deviants and misfits: hotel and catering labour and the marginal worker thesis. *International Journal of Hospitality Management*, **11** (3), 179–82.

Wood, R.C. (1992) *Food for Thought: a Study of Hotel and Catering Workers' Enquiries to the Scottish Low Pay Unit*. Glasgow: Scottish Low Pay Unit Publications.

Wood, R.C. (1993) Status and hotel and catering work. *Hospitality Research Journal*, **16** (3), 3–15.

Wood, R.C. (1994) Hotel culture and social control. *Annals of Tourism Research*, **21**, 65–80.

Woods, R.H. (1991) Surfacing culture: the 'Northeast Restaurants' case. *International Journal of Hospitality Management*, **10** (4), 339–56.

Worsfold, P. (1989) Management selection in the hospitality industry. *Contemporary Hospitality Management*, **1** (1), 17–21.

Worsfold, P. (1989) A personality profile of the hotel manager. *International Journal of Hospitality Management*, **8** (1), 51–62.

Yamaguchi, Y.T. and Garey, J.G. (1993) The relationship between central life interest of restaurant managers and their level of job satisfaction. *International Journal of Hospitality Management*, **12** (4), 385–93.

INDEX